# Wide Bandgap Semiconductors

As Moore's law approaches its physical limit, research results beyond Moore's law based on novel materials, structures and devices is providing a new development direction for the semiconductor industry. GaN, SiC and other wide bandgap semiconductor materials are the third generation of semiconductor materials, following silicon, germanium, GaAs and InP. They are the preferred materials for manufacturing solid-state light sources, power electronic devices, and microwave RF devices due to their superior frequency, efficiency, voltage resistance, temperature resistance, and radiation resistance. The study of new materials can meet national strategic requirements for energy conservation and emission reduction, intelligent manufacturing and information security among others. It has become the forefront of global semiconductor technology research and a new industrial focus, and has a significant impact on industrial growth.

The book series "Wide Bandgap Semiconductors" is dedicated to clarifying the fundamental frontiers, key scientific and technological issues of wide bandgap semiconductors. It provides a comprehensive and prospective discussion of the characteristics of new generation semiconductor materials, fundamental physical properties, device fabrication, international R&D trends, key scientific and technological problems encountered, major R&D results achieved, device applications, and development prospects. This series represents the latest findings in fundamental semiconductor theory research and has significant application value.

The book series "Wide Bandgap Semiconductors" brings together the research experience and achievements of more than twenty of China's leading experts in this field. It is a creative exploration of the "synergy between fundamental and applied disciplines". This series fills a gap in the publication of monographs in this field both domestically and internationally, and can serve as a resource for scientific and technical workers engaged in the research and development of wide bandgap semiconductor electronic materials and devices, as well as for enterprise engineers and technicians seeking to strengthen their fundamental knowledge and investigate the problems encountered in practical scientific research.

Yuming Zhang · Hui Guo · Jinfeng Zhang ·
Chiwen Qian · Yapeng Liu

# Wide Band-Gap Semiconductor Nuclear Radiation Detector

Springer

Yuming Zhang
School of Electronic Engineering
Xidian University
Xi'an, Shaanxi, China

Hui Guo
School of Electronic Engineering
Xidian University
Xi'an, Shaanxi, China

Jinfeng Zhang
School of Electronic Engineering
Xidian University
Xi'an, Shaanxi, China

Chiwen Qian
School of Electronic Engineering
Xidian University
Xi'an, Shaanxi, China

Yapeng Liu
Beijing Microelectronics Technology Institute
Xi'an, Shaanxi, China

ISSN 2948-2607 ISSN 2948-2615 (electronic)
Wide Bandgap Semiconductors
ISBN 978-981-96-0339-8 ISBN 978-981-96-0340-4 (eBook)
https://doi.org/10.1007/978-981-96-0340-4

Jointly published with Xidian University Press
The print edition is not for sale in China (Mainland). Customers from China (Mainland) please order the print book from: Xidian University Press.
ISBN of the Co-Publisher's edition: 978-7-5606-6284-8

Translation from the Chinese Simplified language edition: "宽禁带半导体核辐射探测器" by Yuming Zhang et al., © Xidian University Press 2022. Published by Xidian University Press. All Rights Reserved.



This Springer imprint is published by the registered company Springer Nature Singapore Pte Ltd.
The registered company address is: 152 Beach Road, #21-01/04 Gateway East, Singapore 189721, Singapore

# Acknowledgments

The creation of *Wide Band-Gap Semiconductor Nuclear Radiation Detector* has been a collaborative endeavor, and I extend my deepest gratitude to all who contributed to this work.

First and foremost, I am profoundly indebted to Prof. Zhang Yuming, whose expertise in semiconductor physics and radiation detection technologies provided critical conceptual frameworks and methodological approaches for this book. Professor Zhang's rigorous academic guidance and visionary insights enabled me to navigate the complexities of wide-bandgap nuclear radiation detectors, laying a robust foundation for this research. His mentorship has been a beacon throughout my scholarly journey, inspiring both intellectual growth and perseverance in overcoming technical challenges.

My heartfelt thanks also go to Prof. Zhang Jinfeng for her indispensable contributions to the chapters on diamond-based detectors. Her profound knowledge in materials science and device engineering refined the theoretical frameworks, optimized experimental designs, and enriched the data interpretation. Her interdisciplinary collaboration expanded my perspective on the integration of advanced materials like diamond into radiation detection systems.

I would like to acknowledge Dr. Qian Chiwen for his contributions to the simulation and optimization of silicon carbide (SiC) nuclear radiation detector design and performance analysis. His expertise and advanced simulation skills enabled in-depth design optimizations and detailed performance analyses. His work provided critical theoretical support for this book and valuable insights for the development and application of related devices.

Special recognition is reserved for Dr. Liu Yapeng, who undertook the demanding task of full-text translation, ensuring the scientific integrity and linguistic precision of the English edition. His dedication to preserving technical nuances while enhancing readability has been instrumental in making this work accessible to a global audience.

Finally, I owe an immeasurable debt to my family for their unwavering spiritual support during this multi-year endeavor. Their patience and encouragement sustained my focus through countless challenges, making the completion of this book possible.

This work is dedicated to the pioneers advancing radiation detection technologies. May our collective efforts propel the industrialization of third-generation semiconductor detectors and illuminate new frontiers in nuclear science.

# Contents

# Chapter 1
# Introduction

## 1.1 The Function of Nuclear Radiation Detection in Extreme Environments and Strong Radiation Fields

The primary fields of development include nuclear power devices, space equipment, research and medical radiation devices, and nuclear radiation detection is widely used in these fields. Space exploration requires all-weather monitoring of cosmic rays and instant warning of strong pulses, new synchrotron light sources and other national scientific devices require irradiation detectors that are capable of withstanding high-dose irradiation and high-temperature environments for an extended period of time, the development of a new generation of nuclear energy and nuclear medicine applications also imposes new requirements for the charge collection efficiency, anti-irradiation characteristics, charge and positions capacity. These applications include the evolution of the universe's origin, the fundamental structure of matter, new materials, the sustainable development of the energy industry and nuclear medicine technology, as well as national security of new equipment technology, etc., which has become a major strategic direction that nations are competing to pursue.

The exploration of deep space has become a symbol of the nation's scientific and technological strength. Due to the lengthy duration and high intensity of irradiation, nuclear detectors with enhanced anti-irradiation capabilities are urgently required for the exploration of the radiation environment in deep space. In addition, the Chinese moon landing plan, space station plan, various satellites in orbit, and space high-energy physics detection devices have higher requirements for the in-orbit life span of the core components of detectors, providing a strong guarantee that China will improve its exploration of the cosmos.

The next generation of fission and fusion reactors, synchrotron light sources, large hadron colliders, neutron spallation source, and ultrafast pulsed field monitoring all place greater demands on the performance of detectors as nuclear science and technology continue to advance. Future nuclear detectors must prioritize dark current, saturation charge collection efficiency, energy resolution, impulse response rise time,



Y. Zhang et al., *Wide Band-Gap Semiconductor Nuclear Radiation Detector*,
Wide Bandgap Semiconductors, https://doi.org/10.1007/978-981-96-0340-4_1

and irradiation resistance. The transformation of the Large Hadron Collider (LHC) at European centre for nuclear research (CERN) into the Super Large Hadron Collider (SLHC) will result in a particle beam with a maximum energy of 13 TeV, which is crucial for the investigation of the origin of the universe and the advancement of high-energy physics, which places high demands on the anti-irradiation performance and the time response of nuclear detectors.

Chinese Inertial confinement nuclear fusion device the Eastern Super Ring (EAST) nuclear fusion device, the Eastern Super Ring (EAST) magnetic confinement fusion experimental device, China's hash neutron source and other large scientific devices with extremely high nuclear radiation intensity, reliable and accurate neutron detection in a wide dynamic range is one of the core key technical issues for the above applied research, which also places greater demands on the intensity, energy distribution, and time characteristics of nuclear detectors, for example: In plasma fusion research, neutron emissivity increases from $10^8$ n/s in deuterium–tritium operation to $10^{20}$ n/s in deuterium–tritium reaction, which is approximately $10^{10}$ times higher than the intensity of existing neutron calibration sources. The measurement accuracy provided by neutron diagnostics should be greater than 10%, and the detectors should not demonstrate significant performance degradation. A new generation of synchrotron radiation sources can focus X-rays to fractions of a few microns, necessitating accurate measurement of the position of the beam relative to the sample. Achieving in-situ measurement of beamlines poses new challenges for X-ray position monitoring detectors, and new synchrotron radiation sources can have luminous fluxes of $10^{12}$/pulse or more.

Therefore, the accurate detection of rays in extreme environments and strong radiation fields is not only about the future development of science and technology, but also about the strategic development of the nation, which necessitates nuclear detectors that can adapt to the detection of targets in the aforementioned environments and provide a solid foundation for the future development of nuclear technology.

## 1.2 Introduction to the Development of Nuclear Radiation Detectors

There have been three major phases in the development of nuclear detectors: gas detectors, scintillator detectors, and semiconductor nuclear detectors; both scintillator and semiconductor detectors are classified as solid-state detectors.

In numerous radiation detection applications, solid state detectors offer significant benefits. For measurements of high energy electrons or gamma rays, the detector size is significantly smaller than that of an equivalent gas detector because the density of solids is approximately hundreds of times that of gases; another disadvantage of gas detectors is that their ionization energy is significantly greater than that of semiconductor detectors. The average ionization energy of an electron–hole pair generated in an air ionization chamber can reach 33.8 eV, and other gas detectors such as Ar, He, $H_2$, $N_2$, and $O_2$, etc. can also be distinguished to 26.4, 41.3, 36.5, 34.8, 30.8 eV, indicating that the signal strength obtained is weaker for the same radiation energy; other aspects, because the gas detectors are affected by temperature, pressure, device sealing characteristics, the device itself varies greatly under different radiation environments, indicating the energy resolution is not high, response speed is slow, Growingly unsuitable for the advancement of nuclear detection technology.

Scintillator detectors perform indirect detection in which the radiation is absorbed by the scintillator, the scintillator emits photons, and after a series of circuit system processing, the photons are converted to electrical signals. A too-long system link prevents the scintillator detector from being used by itself. For scintillators, there is also the problem of a large average ionization energy, which is typically on the order of 100 eV. As a result, the statistical fluctuations caused by the device's volatility have a significant impact on the energy resolution, which is only 6% energy resolution for the detection of 0.662 MeV gamma rays.

As the lower average ionization energy can generate a larger number of non-equilibrium carriers, which can improve energy resolution performance, there are significant advantages to preparing nuclear detectors with semiconductor materials in order to develop more effective nuclear detectors. These characteristics include a small dimension, relatively quick time characteristics, and the ability to adjust the effective sensitivity thickness based on application requirements. Nuclear detectors made of silicon and germanium continue to be the most common varieties, and the average ionization energy of silicon-based materials is only 3.62 eV, which is significantly less than that of gas and scintillator detectors. However, the disadvantages of silicon and germanium-based nuclear detectors are severe performance degradation due to radiation damage and relatively high sensitivity to environmental changes, which are unfavorable for applications in extreme environments, strong radiation fields, and higher performance detection fields. Wide band gap semiconductors such as silicon carbide and diamond have attracted widespread interest from national research institutions in recent years due to their superior anti-irradiation properties, higher carrier mobility and saturation rate, and enhanced resistance to extreme environments. The rapid development of this type of technology has made it possible to create nuclear detectors utilizing broad-band semiconductors. Researchers and significant institutions from various nations have focused their research on this topic, which is the primary focus of this book. As the lower average ionization energy can generate a larger number of non-equilibrium carriers, which can improve energy resolution performance, there are significant advantages to preparing nuclear detectors with semiconductor materials in order to develop more effective nuclear detectors. These characteristics include a small dimension, relatively quick time characteristics, and the ability to adjust the effective sensitivity thickness based on application requirements. Nuclear detectors made of silicon and germanium continue to be the most common varieties, and the average ionization energy of silicon-based materials is only 3.62 eV, which is significantly less than that of gas and scintillator detectors. However, the disadvantages of silicon and germanium-based nuclear detectors are severe performance degradation due to radiation damage and relatively high sensitivity to environmental changes, which are unfavorable for applications in harsh environments, strong radiation fields, and higher performance detection fields. Broadband semiconductors such as silicon carbide and diamond have attracted widespread interest from national research institutions in recent years due to their superior anti-irradiation properties, higher carrier mobility and saturation rate, and enhanced resistance to extreme environments. The rapid development of this type of technology has made it possible to create nuclear detectors utilizing broad-band semiconductors. Researchers and significant institutions from various nations have focused their research on this topic, which is the primary focus of this book.

## 1.3 Overview of General Progress in Broadband Semiconductor Nuclear Radiation Detectors (Current Status of Research)

Compared to conventional gas detectors and scintillator detectors, semiconductor nuclear radiation detectors have the advantages of high detection efficiency, good linear range, simple structure, small size, fast response, and good energy resolution, etc., and are increasingly used for the detection of $\alpha$, $\beta$, $\gamma$ rays and neutrons. However, the intrinsic material properties of gas, scintillator, and conventional semiconductor detectors dictate that their detection performance will decrease sharply with increasing temperature and radiation intensity, and as a result, they are unsuitable for applications involving extreme environments, such as reactor core monitoring, high-energy physics, and deep space research. To overcome the shortcomings of conventional semiconductor nuclear radiation detectors, such as low irradiation resistance and temperature-sensitive performance, it is necessary to replace conventional Si/germanium materials with semiconductors that have higher atomic ionization thresholds and wider forbidden band widths. The third-generation wide band semiconductor materials represented by silicon carbide (SiC) and diamond have excellent characteristics such as wide band width, high critical breakdown electric field, fast carrier saturation drift, and strong irradiation resistance; thus, in recent years, numerous types of SiC- and diamond-based detectors have been the subject of extensive study.

### *1.3.1 The Development and Application of Diamond Nuclear Radiation Detectors*

1. The advantages of manufacturing nuclear detectors from diamond

Nuclear technology has been used extensively in cosmic navigation, space physics, controlled nuclear reactions, national defense, border defense, medicine, biological protection, disease prevention, and food testing. Nuclear detectors are the primary components of nuclear measurement devices, which use the measurement signals generated by the interaction of radiation and matter to reflect information such as radiation dose, intensity, and time-varying properties. Semiconductor nuclear detectors comprised of semiconductor materials and microelectronics technology have the advantages of small size, high energy resolution, high linearity, and fast impulse response, making them the most important objects for academic research and practical applications of nuclear detection, and playing a crucial role in science and engineering fields.

As a result of the ongoing advancement of science and technology, nuclear power devices, space equipment, scientific research, and medicinal radiation devices are the primary areas of development concentration at the present time. However, nuclear power facilities, space probes, and radiation detectors have stringent performance requirements for nuclear detectors. The upgrade of the Large Hadron Collider (LHC) at the European center for Nuclear Research (CERN) to the High Brightness Large Hadron Collider (HL-LHC), in which the particle beam will be stronger and the maximum energy will reach 14 TeV, places high demands on the anti-irradiation performance and time response of nuclear detectors. China's laser confinement inertial fusion device, the Eastern Super Ring (EAST) all-superconducting non-circular cross-section tokamak nuclear fusion experimental device, China's hash neutron source, etc. also put forward higher requirements on intensity, energy distribution and time characteristics; Space exploration requires all-weather monitoring of cosmic rays and instant warning of strong pulses; new synchrotron light sources and other national large scientific devices require nuclear detectors that can withstand high dose irradiation and high temperature environment for an extended period of time; and the development of a new generation of nuclear energy and nuclear medicine applications also require nuclear detectors The new generation of nuclear energy development and nuclear medicine applications also put forward new requirements on the charge collection efficiency, irradiation resistance, energy and position resolution, time response and other performance of nuclear detectors. These applications include the evolution of the universe's origin, the fundamental structure of matter, and new materials at the level of basic science, the sustainable development of energy industry and nuclear medicine technology for national livelihood, and new equipment technology for national security, which have become a major strategic direction for which nations compete globally today. Traditional semiconductor nuclear detectors comprised of silicon, germanium, and other semiconductor materials cannot meet the current demands of scientific advancement, and there is an imperative need to conduct research and development on new nuclear detectors.

Due to its superior physical properties, diamond is an ideal material for the next generation of nuclear detectors with powerful radiation fields. Table 1.1 shows the results of comparing the properties of diamond with other typical semiconductor materials [1]. Diamond has an atomic number (Z) of 6. The lower atomic number reduces the irradiation damage caused by high energy cascades and multiple scattering during high-energy physics experiments, furthermore, because the atomic number of diamond is close to human muscle and tissue ($Z \approx 7.5$), it can more accurately reflect the radiation damage characteristics of the human body and has a higher human equivalence, making it suitable for medical dosimetry measurements.

The diamond band gap (Eg) is up to 5.47 eV, indicating that diamond possesses exceptional anti-irradiation properties and very low intrinsic carrier concentration (intrinsic diamond resistivity is up to $10^{16}$ Ω cm at normal temperature). As a result, diamond can obtain very low dark current and exhibit high sensitivity and signal-to-noise ratio for low intensity radiation. The diamond electron mobility(μ) is 4500 $cm^2/V$ s ($\mu_n$ electrons) and 3800 $cm^2/V$ s ($\mu_p$ carriers), while the diamond relative permittivity (r) is only 5.7, and the parasitic capacitance at the same dimension is half that of silicon and 0.6 times that of silicon carbide, resulting in ultra-fast time response in picoseconds and reduced noise during the signal readout process. Diamond's thermal conductivity ($K$) reaches 22 W/cm K, which is 4 times that of normal heat dissipation materials copper/silver and can therefore be used as a detector material without thermal effects. The ionization energy ($w$) of diamond is greater than other semiconductors, as is the energy required to generate an electron–hole pair, however, this issue can be solved by amplification of the readout electronics system. In addition to the above material properties, diamond has a melting point up to 3550–4000 °C, slightly higher than tungsten, the highest Mohs' scale of hardness (10), the coefficient of thermal expansion is only $0.8 \times 10^{-6}$/K, and it is chemically inert, resistant to acid and alkali corrosion, meaning it can withstand the harsh environmental impact. The small size of diamond has a negligible impact on the radiation field, and its non-toxicity poses no threat to the human body. Through the above analysis, it can be seen that diamond has obvious advantages over silicon (Si), gallium arsenide (GaAs), silicon carbide (SiC) and gallium nitride (GaN) materials, making it an ideal material for the future development of nuclear detectors.

The rapid development of microelectronics process levels and advancements in CVD diamond material growth technology have facilitated the development of diamond nuclear detectors. In order to accommodate the future development of high-energy physics, the next-generation LHC equipped by CERN has set stringent requirements for the performance of nuclear detectors, including the following: the nuclear detectors should be able to withstand heavy ion bombardment with a total injection volume of up to $10^{15}$ $cm^{-2}$; be able to withstand an injection rate of up to $10^7$ $cm^{-2}$ $s^{-1}$; be able to provide a position resolution of less than 10 μm; and have a time response of less than 25 ns [2]. The most significant issue for silicon-based devices is the degradation or destruction of device performance due to radiation damage. In order to overcome the issues of increased leakage current, signal degradation, and poor time response caused by irradiation damage in silicon-based devices under strong radiation fields, the RD42 research group developed nuclear detectors based on CVD polycrystalline diamond, with the majority of the organization's personnel from the United States and Europe. Other research institutions that have made significant contributions to the development of diamond nuclear detectors include those in the United Kingdom (King's College, Imperial College), the United States (Los Alamos National Laboratory, Lawrence Livermore Laboratory, Ohio State University), the Italian Institute of Materials Physics (INFM), the European Atomic Energy Community (EURATOM), CEA-LETI (French Atomic Energy Commission), etc. China's research began later than that of other nations, and it has focused primarily on preliminary exploration using polycrystalline diamond; there is a significant disparity

**Table 1.1** Material parameters of the main semiconductor materials

| Material | Si | Ge | GaAs | 4H-SiC | 6H-SiC | GaN | Diamond |
|---|---|---|---|---|---|---|---|
| Atomic number $Z$ | 14 | 32 | 31,33 | 14,6 | 14,6 | 31,7 | 6 |
| Band gap $E_g$ (eV) | 1.12 | 0.66 | 1.4 | 3.26 | 3.0 | 3.39 | 5.47 |
| Relative permittivity $\varepsilon_s$ | 11.8 | 16 | 12.8 | 9.66 | 10 | 9 | 5.7 |
| Electron mobility $\mu_n$ (cm$^2$ V$^{-1}$ s)$^{-1}$ | 1350 | 3900 | 8500 | 1000 | 500 | 900 | 4500 |
| Hole mobility $\mu_p$ (cm V$^{-1}$ s)$^{-1}$ | 480 | 1900 | 400 | 120 | 40 | 30 | 3800 |
| Critical breakdown field strength $E_{BR}$ (MV/cm) | 0.3 | 0.1 | 0.4 | 3.0 | 2.4 | 3.3 | > 10 |
| Saturation drift velocity $v_{sat}$ (cm/s) | $1 \times 10^7$ | $5 \times 10^6$ | $2 \times 10^7$ | $2 \times 10^7$ | $2 \times 10^7$ | $2.5 \times 10^7$ | $3 \times 10^7$ |
| Thermal conductivity $K$ (W cm$^{-1}$ K)$^{-1}$ | 1.5 | 0.6 | 0.5 | 4.5 | 4.5 | 1.3 | 22 |
| Mass density $\rho$(g/cm)$^3$ | 2.33 | 5.35 | 5.32 | 3.2 | 3.2 | 6.1 | 3.5 |
| Atomic excursion threshold $E_d$ (eV) | 13–20 | 16–20 | 8–20 | 22–35 | 22–35 | 17–23 | 40–50 |

between the research level and device performance. Numerous domestic and international studies have been conducted on the energy spectrum characteristics (charge collection efficiency and energy resolution), current–voltage (I–V) characteristics, and time response characteristics of nuclear detectors, which are the most crucial performance parameters.

2. Development status

(a) Response of diamond core detectors to different rays

Response study of charged particles: typically, charged particles are used as radiation sources to study the charge collection efficiency (CCE) and energy resolution of diamond nuclear detectors. Charged particles include α particles, β particles, protons, heavy ions, etc. Among them, α particles are the most common charged particles. The energy of α particles is large, up to several MeV, but due to their large mass and weak penetration, it is safe to conduct experiments, and it is simple to obtain irradiation sources through isotope sources; therefore, energy spectrum studies of diamond nuclear detectors employing particles are the most widely utilised internationally.

In 1996, the RD42 group used α and β particles investigated diamond nuclear detectors with a thickness of 300 μm and a charge collection distance (CCD) of approximately 60 μm, which equates to a CCE of approximately 20% [2]. In 1997, the U.S. Brookhaven National Laboratory produced three nuclear detectors using

MPCVD diamond, DCCVD diamond, and natural diamond, respectively. The efficacy of three nuclear detectors fabricated with MPCVD diamond, DCCVD diamond, and natural diamond was evaluated using α particles and found that the MPCVD diamond made the highest performance nuclear detector, which had a thickness of 30 μm, a CCD of 10 μm, a CCE of about 33%, and a mobility lifetime product of $3 \times 10^{-8}$ cm$^2$/V [3]. In 2002, the French CEA-LETI used MPCVD to epitaxy diamond with thicknesses of 20 m and 150 m, respectively, using particles and 2 MeV protons as particle sources. The measured detector CCEs were 50% and 30%, respectively, with energy spectrum half-height width exceeding 50%, and the energy resolution parameter was meaningless and lacked energy spectrum measurement capability [4]. All of the aforementioned studies used CVD polycrystalline diamond to fabricate nuclear detectors. The grain boundaries, inconsistencies, and material defects of polycrystalline diamond limit its use for energy spectroscopy, and the polarisation effect in the diamond body also causes the energy spectral peak position to shift with time, which degrades the performance of diamond nuclear detectors. However, the large area of polycrystalline diamond nuclear detectors, which can be more than 2 inches or even 4 inches, can be used for part of the detection process. In 2004, the RD42 group investigated the efficacy of CVD single-crystal diamond nuclear detectors with metal-intrinsic semiconductor–metal (MiM) structure for the first time using β particles and CVD single-crystal diamond supplied by Element Six. The report does not provide the complete results of the analysis of the energy spectrum, and the energy resolution is undetermined. In 2012, Hokkaido University and the Japan Advanced Institute of Industrial Technology (AIST) epitaxialized four CVD single-crystal diamonds with epitaxial layer thicknesses 37 μm, 44 μm, 67 μm and 178 μm on HPHT Type IIa substrates. The highest cavity CCE was 98% (sample thickness 44 μm) and the lowest was 5.4% (sample thickness 178 μm) for electron hole using α-particle testing; The highest cavity CCE was 92% (sample thickness 44 μm) and the lowest was 5.2% (sample thickness 178 μm) for electron using α-particle testing; For the energy resolution, the energy resolution of the 44 μm sample reached 3.8% [5]. 2014, the University of Huelva in Spain evaluated the performance of a 50 μm-thick commercial diamond nuclear detector using α particles and obtained an energy resolution of 19–21 keV, whereas the energy resolution of the silicon-based detector used in the same experiment was 34–38 keV [6]. In 2016, CERN created a diamond nuclear detector with an all-carbon structure utilising graphite electrodes and tested the CCE of this detector with a super proton gas accelerator producing a beam current of 120 GeV to achieve $42 \pm 4\%$ [7]. In 2017, Japanese researchers Sato et al. tested their 50 μm-thick diamond nuclear detector with 26 MeV $O^{5+}$ particles and 45 MeV $Si^{7+}$ particles, achieving a CCE of 86.2% for electron holes and a carrier mobility lifetime product of $(1.8 \pm 0.2) \times 10^{-6}$ cm$^2$/V; For electrons, the CCE attained 60.4%, and the carrier lifetime mobility product was $(6.4 \pm 0.8) \times 10^{-7}$ cm$^2$/V [8]. In the same year, Tallaie et al. in France studied the performance of CVD diamond core detectors grown on various substrates and discovered that the CVD diamond performance attained using HPHT Type IIa substrate was approximately 10% [9] higher than the CCE of HPHT Type Ia substrate. In 2018, the National Institute for Materials Research (NIMS), Japan, conducted a study on diamond nuclear detectors of the p–n junction type that were incapable of achieving full absorption of α-particle energy with an energy resolution of 30% [10]. In the same year, Italian researchers Verona et al. reported p-i-M Schottky-type diamond nuclear detectors for miniature dose-rate metres that can detect charged particles such as $^{1}H$, $^{4}He$, $^{7}Li$, $^{12}C$, $^{16}O$. The nuclear detectors have an intrinsic layer thickness of about 0.8 m and a CCE of approximately 7% [11]. In 2019, the Helmholtz Centre for Heavy Ion Research in Germany, along with several Italian and French scientists, studied CVD heteroepitaxial single-crystal diamond nuclear detectors, utilising a 4-inch Ir/YSZ/Si(100) substrate to grow CVD single-crystal diamond and fabricate dozens of nuclear detectors, with optimal results of 95% CCE for electron holes and about 0.3% energy resolution, but the CCE for electrons was reduced to 40%.

X-rays response studies: The use of X-rays to study the performance of diamond nuclear detectors is typically divided into three aspects, the first of which is as an X-rays monitoring detector due to the excellent position sensitivity and high stability of diamond nuclear detectors, which have a low X-rays absorption coefficient and can achieve in-situ measurement of the position and flux of X-rays. The second is that diamond can be used as a medical dose-rate meter in radiotherapy because of its superior human equivalence. In such applications, the main concern is the I–V characteristics of the device, including dark current, signal-to-noise ratio, dose rate dependence and linearity, responsivity, sensitivity and specific sensitivity, photoconductive gain, repeatability, etc. [12]. The third is that due to diamond's rapid response speed, it can be used for X-ray pulse measurement, with the primary parameters being pulse response rise time, pulse breadth, fall time, etc.

In 2012, Marinelli et al. examined the properties of a p-i-M type diamond X-ray position monitoring detector (XBPM) at the Harwell synchrotron light source in the United Kingdom. They discovered that the detector possessed exceptional position sensitivity and high stability [13]. In 2014, Desjardins et al. investigated diamond ultra-thin XBPM based on the diamond stripping process at the SOLEIL synchrotron light source in France, with a transmission rate of greater than 50% for 1.3 keV X-rays, a signal-to-noise ratio of more than 105, a position resolution of less than 1 m, and a sampling rate of greater than 1 kHz [14]. In recent years, diamond X-ray position and flux detectors with nitrogen-coupled super-nanocrystal contacts [15], X-ray double transverse position-sensitive detectors [16], and rapid X-ray beam detectors with discrimination times of 100 ps have been developed.

In 2009, the Brookhaven National Laboratory studied the use of pulsed bias technique to mitigate the effect of charge trap effect on the detection performance of diamond nuclear detectors in X-ray dosimetry. Under the bias conditions of electric field strength of $4 \times 10^5$ V/m, frequency of 100 Hz, and duty cycle of 98–99%, the responsiveness to 0.25–7 keV soft X-rays reached 0.064 A/W [17]. In 2010, the group also studied the electron multiplication phenomenon and collection characteristics of diamond nuclear detectors under soft X-ray (50 eV–6 keV) irradiation, and discovered that trap and composite centres and electrode to charge transport have significant effects on electron multiplication, and the responsiveness to holes can reach 0.065 A/W, while the responsiveness to electrons has approached 10 A/W, which is significantly higher than 0.075 A/W per unit gain [18]. In 2011, the University of Surrey,

UK, investigated the X-ray response characteristics of different electrode metals at dose rates of 5.45 cGy/min and 26 cGy/min using a 50 kV X-ray tube and found that the signal-to-noise ratios of nuclear detectors with Pt electrode and Al/Pt electrode structures exceeded 1000, with linearities of 0.975 and 0.968, sensitivity of 735 and 600 nC/Gy, and repeatability of 0.268% and 0.345%, respectively [12]. In 2017, the University of the Free State of South Africa examined the performance of diamond nuclear detectors for mammography (26–30 kVp) and the effect of material defects on their performance, with a sensitivity of 0.4–6.7 nC $Gy^{-1}$ $mm^{-3}$ [19]. In 2019, Italian researchers initiated the development of 3D diamond nuclear detectors for clinical dosimetry, paving the way for future diamond nuclear detectors in X-ray imaging [20]. In X-ray pulse measurements, diamond nuclear detectors exhibit an exceptionally rapid time response, with an impulse response rise time of only 90 ps for pulsed X-rays of 8.5 keV [21].

Gamma ($\gamma$) radiation response studies: the use of $\gamma$-rays provides the study of anti-irradiation properties as well as the response properties of pulsed and steady-state gamma rays. In 2001, Brambilla of CEA-LETI, France, conducted experiments on the dose rate response of diamond detectors to gamma rays in a strong radiation environment, with dose rates ranging from 1 mGy/h to 1 kGy/h, and total doses up to $5 \times 10^5$ Gy, and the operating temperature increased from 25 to 250 °C, demonstrating the stable operation of diamond detectors under high dose rates [22]. In 2007, Gastélum et al. investigated the dose rate effect of diamond films and discovered that diamond films have exceptional dose rate linearity properties [23]. In 2016, Baranova et al. investigated the response of diamond detectors to gamma radiation at pulse mode dose rates between $10^{-3}$ and 0.4 Gy/h and current mode dose rates between $10^{-1}$ and $2 \times 10^4$ Gy/h, with nonlinear characteristics maintained within 10%. The results obtained suggest that diamond detectors can be used to prevent emergency accidents in nuclear power plants as well as for daily control monitoring of nuclear power plants. Duke University has also investigated diamond's response to 2–7 MeV 16.8 ns ultrafast Gamma ray pulses in 2017 [24]. All of the aforementioned investigations have demonstrated that diamond is an effective detector of gamma rays.

Neutron response studies: The detection of neutrons by diamond can be categorised primarily as rapid neutron detection and thermal neutron detection. Fast neutrons have a small reaction cross section, and when they are detected, only a small quantity of energy is deposited, while the majority of energy is scattered by thermonuclear reactions. Using MiM structures, diamond neutron detectors have effectively measured fast neutrons with energies of 14.8 MeV [25] and between 17 and 34 MeV [26]. To detect sluggish neutrons, a conversion layer must have a large absorption reaction cross section and convert the neutron reaction into charged particles. Commonly used materials for diamond neutron detector conversion layers are $^{6}$LiF, $^{10}$B, $^{235}$U isotopes. Today, detectors with $^{10}$B conversion layer [27] and detectors with $^{6}$LiF conversion layer [28] have been manufactured successfully. Currently, the best results for diamond neutron detectors are 0.4% energy resolution for $\alpha$ particles and 2.9% energy resolution for neutrons, with nearly 100% charge collection efficiencies. However, the percentage of reported high-performance diamond neutron detectors is extremely low, and further research is required to develop critical technologies for high-performance diamond nuclear detectors.

(b) Resistance to irradiation characteristics study

When very high total injection/absorbed dose rates of energetic particles, X/ rays, neutrons, etc. are accumulated with high injection/absorbed dose rates, the performance of diamond nuclear detectors maintains minimum operating conditions. CERN research has shown that diamond nuclear detectors can operate reliably for at least ten years under high dose irradiation conditions. Due to diamond's extremely high displacement threshold energy, it can withstand irradiation with $10^{15}$ protons/cm$^2$, 250 Mrad photons, and $3 \times 10^{15}$ neutrons/cm$^2$ with only slight variations in detection performance [29]. In recent years, a comparative evaluation of the performance of diamond nuclear detectors for various particles at varying irradiation doses produced the results shown in Table 1.2. It can be seen that diamond has a much higher resistance to irradiation than silicon-based devices, making it ideal for the detection of ultra-intense radiation fields.

(c) Detector structure design

**Table 1.2** Effect of irradiation dose on the performance of diamond nuclear detectors

| Particle type | Energy range | Total dose | Performance changes | Literature |
|---|---|---|---|---|
| Gamma rays ($^{60}$Co) | 1.25 MeV | 100 kGy | None | [30] |
| Electron | 2.2 MeV | 1 MGy | None | [30, 31] |
| Proton | 500 MeV | $10^{14}$ cm$^{-2}$ | None | [32] |
| Proton | 24 GeV | $10^{15}$ cm$^{-2}$ | None | [33] |
| Proton | 24 GeV | $3 \times 10^{15}$ cm$^{-2}$ | 10% attenuation | [33] |
| Proton | 24 GeV | $5 \times 10^{15}$ cm$^{-2}$ | 40% attenuation | [34] |
| Neutron | 1 MeV | $3.2 \times 10^{14}$ cm$^{-2}$ | None | [35] |
| Neutron | 1 MeV | $8 \times 10^{14}$ cm$^{-2}$ | 15% attenuation | [36] |
| Neutron | 1 MeV | $10^{15}$ cm$^{-2}$ | 20% attenuation | [37] |
| Neutron | 1 MeV | $1.35 \times 10^{15}$ cm$^{-2}$ | 40% attenuation | [34] |
| Neutron | Superheated neutrons | $10^{16}$ cm$^{-2}$ | 90% attenuation | [38] |
| α particles | 5 MeV | $10^{13}$ cm$^{-2}$ | None | [30] |
| α particles | 5 MeV | $10^{15}$ cm$^{-2}$ | 80% attenuation | [30] |

The structural design of diamond nuclear detectors is also a significant area of international research. In terms of vertical structure, they can be divided into MiM structure, p-i-n structure, p–n junction structure, and p-i-M structure; in addition to these, there are also diamond nuclear detectors with planar structure and micro-strip detectors that can detect particle positions [39]. In recent years, as the demand for irradiation imaging has increased, the development of diamond pixel array nuclear detectors [40] has been initiated, which has led to new requirements on the size, pixel density, uniformity, and back-end sampling system of diamond detectors. Currently, only the detector unit of diamond pixel array detectors is made of diamond [41], while the storage and amplification readout of back-end signals are realized by external device, which cannot take advantage of diamond irradiation resistance and fast response time. It requires continued research in diamond MOSFETs and other aspects to develop a fully integrated diamond monolithic pixel array nuclear detector.

(d) Temperature characteristics study

The wide band gap of diamond and the extremely low concentration of thermally excited intrinsic carriers can produce excellent dark current characteristics; the material's high melting point and chemical inertness make it suitable for use in extreme environments. High-temperature environments are exceedingly damaging to silicon-based detectors, and environments with strong radiation fields are typically accompanied by high-temperature environments; however, the superior properties of diamond materials can compensate for silicon's shortcomings. Japan's Hokkaido University studied the CCE properties of single-crystal diamond at high temperatures in 2015, and at 573 K, the detector was still able to obtain a CCE of 93.8% and an energy resolution of 2.6% [42]. In 2017, the Indian Centre for Atomic Energy Research (ICAR) met the ITER experiment requirement of detecting neutrons at 300–500 °C. They examined the energy resolution of the Al electrode, Cr/Au electrode, diamond nuclear detector at 300 °C, which was approximately 2% for $\alpha$ particles and 3.5% [43] for the 8.5 MeV peak of 14 MeV fast neutrons. Angelone examined the current response of diamond nuclear detectors to gamma rays emitted from 60 Co isotopes at temperatures between 300 and 625 K in Italy, 2019. It was discovered that the detectors' response current was affected by the carrier saturation rate, and that the response current began to saturate once the temperature reached 600 K. Additionally, the I–V characteristics of the detectors had good linearity.

3. Current status of domestic development

With the rapid improvement of China's comprehensive national power and its rapid economic development, there are increased demands in numerous disciplines pertaining to the national economy and the advancement of science and technology. Nuclear power devices, space exploration, scientific research, and medical radiation devices have been elevated to the forefront of national development. Our planetary exploration programme, moon landing programme, space station programme, and orbiting satellites have imposed stringent irradiation resistance requirements on various semiconductor materials and devices. Space exploration necessitates continuous monitoring of cosmic radiation and immediate warning of powerful pulses.

Several large scientific devices in China, such as the new synchrotron light source, the Eastern Super Ring (EAST), Shen Guang III, the strong current deuterium–tritium fusion neutron generator (HINEG), and the China Hash Neutron Source, require nuclear detectors that can withstand high doses of irradiation and high temperature environments for extended periods. In terms of charge collection efficiency, irradiation resistance characteristics, energy and position resolution, and time response, the new generation of nuclear energy development and nuclear medicine applications put new demands on the performance of nuclear detectors. These requirements have stimulated China's research and development of diamond nuclear detectors.

In 1997, Lanzhou Institute of Physics utilised HFCVD diamond to manufacture detectors and studied the response characteristics of detectors to α particles; however, no energy spectrum characteristics [44] were reported. In 1998, China Academy of Engineering Physics measured the soft X-ray radiation power using natural Type IIa diamond [45]. Subsequently, from 2010 to 2018, the Chinese Academy of Engineering Physics conducted research on CVD diamond nuclear detectors for X-ray energy spectra [46], time response [47–50], and sensitivity [51]; the average sensitivity of diamond detectors to X-rays was $1.196 \times 10^{-5}$ C/J with a response time of less than 1 ns. From 2005 to 2015, Shanghai University conducted diamond nuclear detectors utilising CVD diamond, including energy spectrum characterization of α particles [52, 53], I–V response characterization of X-ray [53], neutron detection [54], and ohmic contact study of nuclear detectors [55], yielding a CCE of 45.1%, energy resolution of 3.3%, and response currents of 16.8 nA for X-rays and 16.8 nA for dark currents. From 2006 to 2017, the Northwest Institute of Nuclear Technology conducted CVD diamond nuclear detector response experiments to $^{60}$Co gamma rays with an output response current of 3.31 nA and a response rise time of 560.4 ps to femtosecond pulsed UV lasers [56]; carried out studies on the energy spectrum characteristics of α particles, with a CCE up to 33.5% [57]; carried out studies on the energy spectrum characteristics of γ rays and neutron detection, with a CCE of 55.47% [58]; also studied The response characteristics of diamond nuclear detectors to fast pulsed X-rays and pulsed electrons have been studied, and the response rise time of the developed polycrystalline diamond nuclear detector is 5 ns, which is twice faster than that of Si-PIN detector, and the response rise time of the developed single crystal diamond nuclear detector is 1.0 ns. In 2017, Harbin Institute of Technology developed a CVD single crystal diamond nuclear detector with a graphite electrode structure, and the energy resolution reached 3.5% [59]; in 2019, Wuhan University developed a single-crystal diamond nuclear detector with a CCE of 97% and the energy resolution of 3.7% [60]; and in 2020, Beijing University of Science and Technology reported a diamond nuclear detector with an optimal energy resolution of 2.25% [61].

China has been conducted research in the field of diamond nuclear detectors based on the energy spectrum characteristics, I–V characteristics, and time response characteristics of α particles, γ rays, X-rays, and neutrons, but there is still a large disparity between the domestic level and the international level. Currently, Chinese research on diamond nuclear detectors focuses primarily on measurement research and less on the devices' response mechanisms. Regarding diamond materials, the major domestic research institutions in the field of diamond are primarily focused on the study of diamond's large size growth and doping technology, and no "detector" level diamond material growth process has been developed. Moreover, there is no material screening method established for diamond core detectors. There are few reports on the development of high-performance diamond nuclear detectors, and there is still a need for in-depth research on the device structure, process, material growth, material characterization methods, and device physical mechanism.

### *1.3.2 Development and Application of SiC Nuclear Radiation Detectors*

On the basis of the material properties of third-generation semiconductors, SiC detectors are anticipated to solve the problem of nuclear radiation detection under conditions of high temperature and intense radiation. As Si semiconductors have gradually approached their theoretical material limits in the field of power devices since the turn of the twenty-first century, third-generation semiconductor materials have gained increasing interest in all academic fields. Chinese and international manufacturers and research institutions have devoted themselves to the research and production of SiC materials and devices, which has promoted the commercialization of high-quality, large-size SiC crystal materials and the development of epitaxial layer growth technology and device preparation process, thereby accelerating the growth of SiC nuclear radiation detectors.

1. Material basis of SiC nuclear radiation detectors

To meet the requirements of high-performance detection in high-temperature, high-radiation, and other harsh environments, the main body of nuclear radiation detection semiconductor materials must possess the following properties [62]:

(1) Wide bandgap: Even at high fields, the material can achieve a very low reverse leakage current. Low leakage current is a prerequisite for low noise and the key to producing detectors with high energy resolution at low radiation energies.
(2) sufficiently small electron–hole pair generation energy $\varepsilon$: ensure that the number of electron–hole pairs generated at a given ionizing radiation is large enough to obtain a high signal-to-noise ratio. It should be emphasized that $\varepsilon$ is proportional to the band gap, so conditions (1) and (2) need to be considered as a compromise.
(3) Low permittivity: enables the detector to have a low operating capacitance, thereby reducing the noise component of the front-end electronics to achieve a higher energy resolution.
(4) High purity, homogeneous, defect-free single-crystal materials: The homogeneity of the material and low defect density ensure charge collection efficiency, low leakage current, and better energy resolution, while single-crystal materials are designed to avoid the harmful effects of grain boundaries and other extended defects.

(5) High intrinsic carrier mobility lifetime product: The product of carrier mobility and carrier lifetime is ($\mu \cdot \tau$), where $\mu$ represents carrier mobility and $\tau$ represents carrier lifetime. The detector generally operates under higher electric field conditions, when carrier velocity saturation occurs, so ($\mu - \tau$) is more reasonably expressed by $L_d = (v_d - \tau)$, where $v_d$ is the carrier drift velocity; $L_d$ is the average carrier drift length (capture length), which represents the average of the distance the carrier is swept by the electric field before it is captured. The high $L_d$ value ensures high charge collection efficiency with good energy resolution.

(6) High atomic delocalization threshold: Some radiation-induced lattice damage can function as a carrier capture centre, resulting in incomplete charge collection and diminished energy resolution. β particles or X and γ rays cause relatively minor damage, whereas neutrons and highly charged particles can cause severe damage. The main defects produced by high-energy particle bombardment are vacancies (V), interstitial (I), Frenkel pairs (V + I), and antipodal sites. The generation of defects must involve the displacement of atoms from lattice junctions; consequently, the atomic dislocation threshold $E_d$ is an important parameter to describe the radiation damage of the material, as it represents the minimum amount of energy that must be delivered to the lattice atoms by colliding particles for the formation of any of the defects listed above. Therefore materials with a high delocalization threshold, such as diamond ($E_d \approx 40$–50 eV) and SiC ($E_d \approx 22$–35 eV), should also have a high radiation resistance.

(7) High coefficient of thermal conductivity: Cooling the detector itself is one method to increase the radiation resistance of a semiconductor detectors, which necessitates the use of materials with excellent cooling properties. In addition, the high coefficient of thermal conductivity aids in the regulation of the system's operating temperature when the front-end electronics are adjacent to or in direct contact with the detector.

Table 1.1 illustrates the main parameters of the widely researched semiconductor materials [1]. Considering the requirements listed above, the intrinsic material properties of traditional Si, Ge, and Gallium Arsenide (GaAs) materials determine that their detection performance will decrease sharply with an increase in temperature and radiation intensity, making them unsuitable for use in extreme environments such as reactor cores, high-energy physics research, and deep space research, whereas the third generation of SiC as the representative material is suitable for such environments. SiC, which represents the third generation of wide bandgap semiconductor materials, is entirely compatible with the requirements for radiation detection in harsh environments. Comparing 4H-SiC and Si, 4H-SiC has a substantially larger band gap, which increases its operating temperature and reduces its leakage current. Without considering the junction temperature increase, the maximum allowable operating temperature for Si will not exceed 300 °C, while the theoretical maximum allowable operating temperature for SiC is 1240 °C [63]. Simultaneously, the larger critical breakdown field and saturation rate indicate that SiC devices are better suited for high frequency operation and can measure narrower pulses, and the higher coefficient of thermal conductivity means better self-cooling capabilities and high temperature operating characteristics, which reduces size, cost, and environmental constraints. Smaller permittivity equates to smaller capacitance, which further reduces the additional capacitance during detection. The greater atomic displaced threshold of $E_d$ increases the irradiation resistance of SiC, thereby extending the detector's lifetime. In the case of fission fragment monitoring, for example, SiC devices have a lifetime approximately four times [64] longer than Si devices. In addition, when compared to other third-generation semiconductors such as GaN and next-generation semiconductor materials such as diamond, SiC's electrical properties do not stand out in terms of material properties alone. In contrast to GaN materials, which have low thermal conductivity, small substrate sizes and costly, and immature diamond materials, SiC materials with high quality and large size have reached industrialization level and are more conducive to the preparation of large size devices.

In conclusion, SiC is regarded as one of the most promising materials for nuclear radiation detectors, and various varieties of SiC-based detectors have received considerable attention and research in recent years. The efficacy of traditional semiconductor materials in power device applications is close to the theoretical limit due to the limitations of Si, Ge, and Gallium Arsenide (GaAs). Optimising the design, process, and integration of Si-based devices and their corresponding modules to further exploit their performance potential is no longer sufficient to meet the high efficiency and low power consumption requirements of today's power electronics technology. As a typical example of third-generation semiconductor materials, the significant development of SiC materials and devices over the past few years has also paved the way for the development of high-performance SiC nuclear radiation detection.

2. Status of SiC nuclear radiation detector research

Babcock et al. developed the first SiC radiation detector in 1957 and analysed its performance for α particles, discovering that the detector could yield a response signal to α radiation at zero bias and continued to detect alpha particles up to 700 °C. In 1972, Tikhomirova et al. [65] fabricated a Be doped 6H-SiC PN junction radiation detector, obtained an energy resolution of 9% for α particles (4.8 MeV) at 1–10 V bias with a leakage current of several tens of nano amps, and measurements showed that it was resistant to neutron irradiation with a flux of $5 \times 10^{13}/cm^2$ and fission fragment bombardment with $7 \times 10^8/cm^2$, and it was discovered that the SiC detector performed better than the Si-based surface barrier detector under the same radiation conditions. These early results were limited by the level of SiC materials and device processes at the time and did not fully reflect the excellent properties of SiC materials. Furthermore, SiC nuclear radiation detectors were not further developed in subsequent years due to the limitations of SiC single crystal growth, chemical processing, and mechanical processing. With the commercialization of high-purity SiC materials and the maturation of device processes (e.g., oxidation, metal–semiconductor contacts, doping, and encapsulation), research on SiC-based particle radiation detectors progressively became active again in the 1990s. The preparation of SiC radiation detectors, as well as their detection capabilities in a variety of scenarios, as well as the damage and performance degradation of this detector, have been intensively studied by scientific teams from a number of nations, with abundant experimental results.

(a) Status of foreign research

U.S. researchers F.H. Ruddy, A.R. Dulloo, and their affiliated teams have produced a significant number of international studies on SiC nuclear irradiation detectors. This team's research is more comprehensive and consistent [1]. In 1998, the research team was the first to use the Chemical Vapour Deposition (CVD) epitaxial growth method to fabricate two types of SiC nuclear radiation detectors, the SBD type and the PN junction type [66], with an N-type epitaxial layer doping concentration of $1 \times 10^{15}$ cm$^{-3}$, as illustrated in Fig. 1.1. The test results show that the energy resolution of the SBD-type detector is 5.8% for α-particles (294 keV) and the energy resolution of the PN junction-type detector is 6.6% for α-particles (260 keV). The operating temperature range of 22 to 89 °C has no effect on the performance of the radiation detector, and it has a strong gamma ray screening ability, making it suitable for high temperature and high radiation field application environments. In 2002, the team published the application of the same structured device incorporating the LiF neutron conversion layer to simulate nuclear reaction power. After strong irradiation (fast neutron dose of $1.7 \times 10^{17}$ cm$^{-2}$ and tritium ($^3$H) particle irradiation dose of $1.1 \times 10^{15}$ cm$^{-2}$), the device still works. As shown in Fig. 1.2, such irradiation doses exceed the upper limits of conventional Si or Ge detectors, and the detectors can maintain consistent detection efficacy over a wide range of neutron fluxes. From 2006 to 2009, the team used the prepared N-type low-doped ($< 2.5 \times 10^{14}$cm$^3$) thick epitaxial layers (100 μm) to carry out a series of studies as follows:

(1) The response to 14 MeV neutrons was tested with a 4H-SiC PiN diode detector, and good linear agreement was obtained, confirming its suitability for fission/fusion nuclear reactions, nuclear fuel monitoring.
(2) The research results of a high-resolution α particle detector were published, and an SBD-type detector with an epitaxial layer thickness of 100 m was developed; the energy resolution of the detector for alpha particles with an energy of 5.499 MeV was 0.37% [68].
(3) Measurement of very low neutron injection rates to simulate detection of hidden nuclear materials with a 6 mm diameter, 100 μm depletion layer thickness SiC diode, yielding a background resolution of 2000 n/γ, ideal for fission/fusion nuclear reactions, nuclear fuel monitoring [69].

Other teams from around the world have also carried out more extensive research on the performance of SiC nuclear radiation detectors and related applications.

In 2009, A.M. Ivanov, N.B. Strokan, A.A. Lebedev, et al. of Russia [70] developed an epitaxial SiC semiconductor detector with normal detection of alpha particles at 375 °C and studied the energy spectral response of 5.8 MeV alpha particles in the field of charged particle detection. The results showed an energy resolution of 1.35%. In 2012, Chaudhuri S. K's research team at the University of South Carolina, USA [71] prepared a 4H-SiC SBD detector by depositing a 10 nm thick Ni electrode (12 mm$^2$) on the surface of a 4H-SiC epitaxial layer with a thickness of 50 μm and a doping

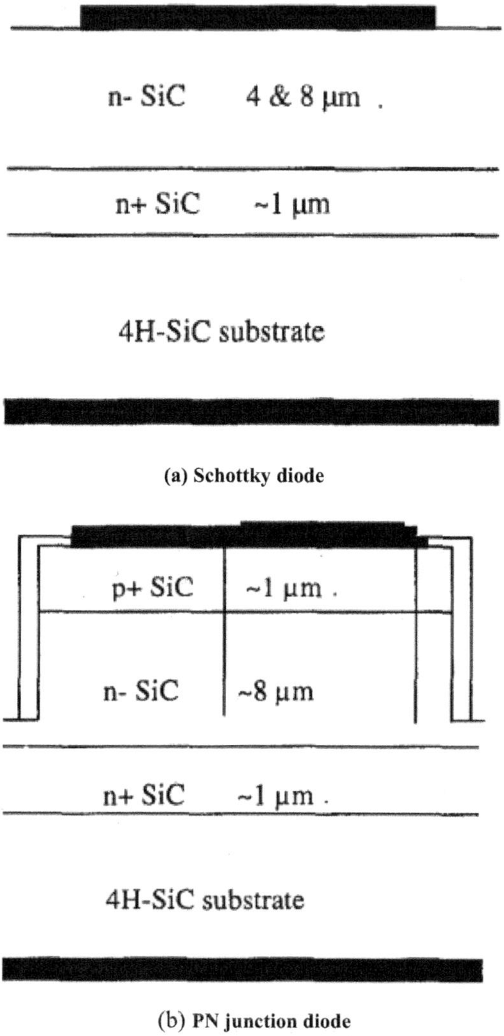


**Fig. 1.1** Detector structure diagram [66]

concentration of $1.1 \times 10^{15}$ cm$^{-3}$. At a reverse voltage of 100 V, the leakage current of the detector was 0.2 nA, and the energy resolution for α particles of 5.486 MeV reached 2.7%. In 2013, the team [72] reduced the thickness of the epitaxial layer of the 4H-SiC SBD detector to 20 um and the doping concentration to $2.4 \times 10^{14}$ cm$^{-3}$; at a reverse voltage of 110 V, the leakage current of the detector was 14pA, and the energy resolution for α particles of 5.486 MeV reached 0.29%, the low microtubule density (< 1 cm$^{-2}$) and the reduction of the effective doping concentration of the epitaxial layer substantially improved the performance of the detector. In 2014, the research team used PECVD to deposit SiO2 and SI films with a 400 nm thickness on the surface of SiC, respectively, and compared with the surface unpassivated

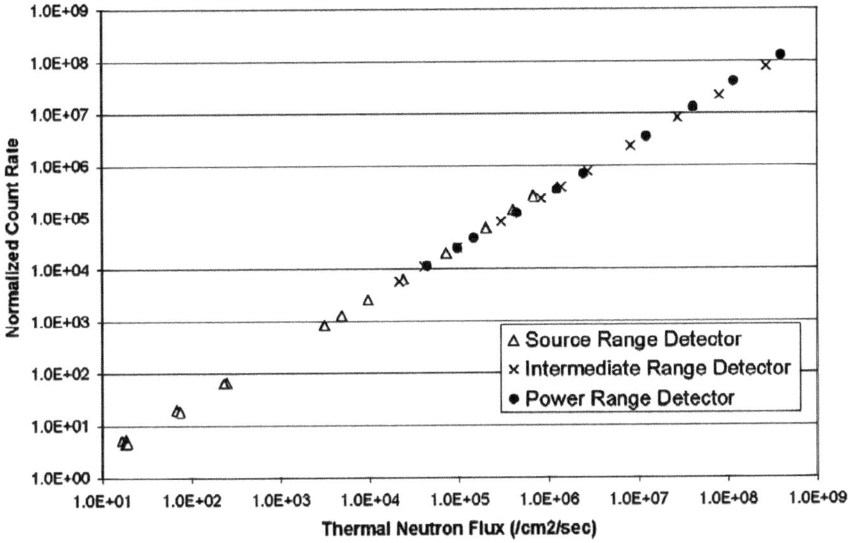


**Fig. 1.2** Variation of thermal neutron count rate with flux for three different thermal neutron fluxes [67]

detector, it will reduced surface leakage current by surface passivation. The surface-passivated detectors' leakage currents were lowered from 5 nA to 86 pA ($SiO_2$) and 71 pA ($Si_3N_4$) at a reverse voltage of 200 V, with an energy resolution of 0.38% and 0.39%, respectively. The epitaxial layer doping concentration was lowered to $1.0 \times 10^{14}$ cm$^{-3}$ in 2015 by the Zat'ko B research team of the Slovak Academy [73]. They also used 15 nm Ni/Au (0.3 mm$^2$) as Schottky electrode, and the detector's leakage current was only 2 pA at the reverse voltage of 700 V. The energy resolution of 5.486 MeV alpha particles reached 0.25%, ranking as the most advanced in the world. In 2017, the University of Messina, Italy's Torrisi L research team [63] prepared 4H-SiC SBD-type detectors by sinking 20 nm thick $Ni_2Si$ as Schottky electrodes on 4H-SiC epitaxial layers with a thickness of 25 nm and a doping concentration of $1.0 \times 10^{14}$ cm$^{-3}$. The energy resolution of 0.85% for α particles of 5.15 MeV. With similar energy resolution to Si detectors, SiC nuclear radiation detectors can monitor low-energy alpha particles produced by single or multi-layer metal targets with range of 300–600 keV. In tough environmental circumstances including high temperature, intense light, and large doses of ionising radiation, SiC nuclear radiation detectors can replace traditional Si detectors and can accommodate the needs of long-term monitoring.

Large area (7 and 20 mm$^2$) superthermal neutron detectors that cab work under 250 V bias conditions were created in the field of neutron detection by Manfredotti et al. [74, 75] in Italy between 2005 and 2007. This detector reaches a detection efficiency of 3‰ and an energy resolution of 1.2% when the 50 μm epitaxial layer is fully depleted in the operating state. In 2009, Korea, Jang Ho Ha et al. developed a metal/n/N + self-biased fast neutron detector operating with an internal strong electric field. As shown in Fig. 1.3, It has many properties, including low power consumption, irradiation resistance, charge collection efficiency (CCE) exceeding 80% at 0 V bias, and CCE up to 100% at bias voltages greater than 20 V. In the meantime, the Am-Be neutron source was used to measure the neutron response spectrum at 5.0 MeV, 1.1 × $10^6$n/cm$^2$·s. It was found that the detection efficiency with the conversion layer is 4 times higher than that without it. Another PiN 4H-SiC neutron detector with four layers was created in 2011 [77], and it had a 5.1 × $10^{-4}$ absolute detection efficiency for neutrons with a muon energy of 5.0 MeV at ambient temperature, zero bias, and self-sufficient energy. In 2009, Ko et al. [78] developed a diode detection array consisting of dozens of Schottky contacts integrated on a SiC substrate. Its Schottky barrier height, leakage current, and ideal factor remain unchanged after 10 MeV, 2.75 × $10^{11}$ n/cm$^2$ neutron irradiation. This study provides an effective method for enhancing the detector's radiation resistance, detection sensitivity, and detection efficacy. In 2013, D. Szalkai, F. Issa et al. proposed an alternative plan which is to use the SiC neutron detector as a neutron detector for the International Thermonuclear Experimental Reactor (ITER) and reported the results of its use. Comparing the simulation results of 14 MeV fast neutrons shown in Fig. 1.4 with the measured results of the device, the characteristic peaks of certain nuclear reactions of the measured energy spectrum are not as obvious as those obtained from the simulation, but the overall match is better, particularly for the $^{12}C(n,\alpha)^9Be$ reaction.

In X-ray and γ-ray detection, in 2009, Bertuccio et al. [80] grew unintentionally doped high-purity epitaxial layers as thick as 115 μm on a 2-inch 4H-SiC substrate with an exceptionally low N-type residual doping concentration of 3.7 × $10^{13}$ cm$^{-3}$. With a reverse bias leakage voltage of 200 V, the Schottky junction detectors (illustrated in Fig. 1.5) fabricated at 22 °C have a current density as low as 1 pA/cm$^2$. It is more than two orders of magnitude below the most advanced Si detectors of

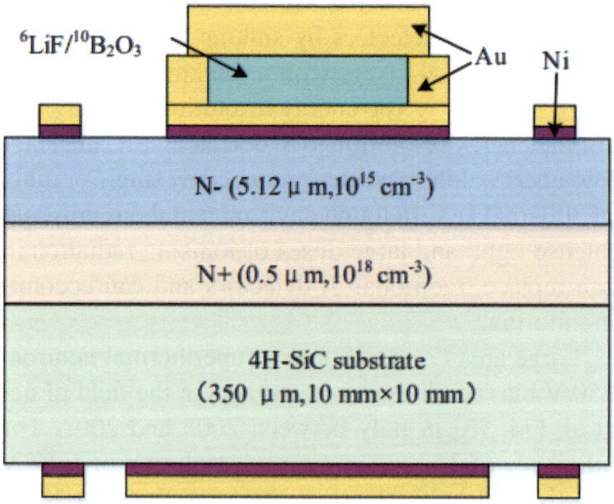


**Fig. 1.3** Schematic diagram of the self-biased SiC-based SBD neutron detector [76]

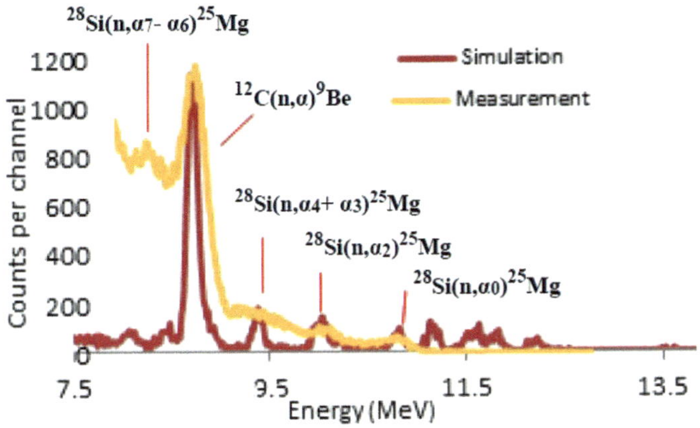


**Fig. 1.4** Comparison of the simulated and measured results at several characteristic peak positions [79]

the same period. The team also prepared a 4 × 4 array detector with a pixel size of 400 × 400 μm$^2$, as shown in Fig. 1.6a, b shows the energy spectrum of the array detector obtained under X-ray irradiation at energies ranging from 3 to 30 keV. The measured results indicate that the equivalent noise energy of this array detector is 195 eV FWHM, which corresponds to an electronic noise of 10.6 electrons r.m.s., which is the greatest performance of compound semiconductor detectors operating at room temperature reported during the same time period. In 2013, Krishna et al. [81] fabricated SBD radiation detectors on 4H-SiC epitaxial layers with substantially better soft X-ray detection performance than commercially SiC UV diodes of the same time period, and the detection results for γ-rays are shown in Fig. 1.7, with an energy resolution of 2.1% (1.2 keV FWHM) for 59.5 keV γ-rays, which is comparable to a high-quality CdZnTe detector.

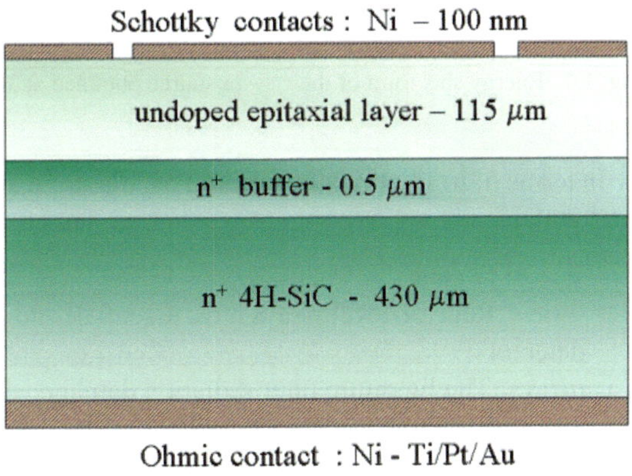


**Fig. 1.5** SBD detector structure diagram, the radiation incidence window is located on the side of the Schottky electrode [80]

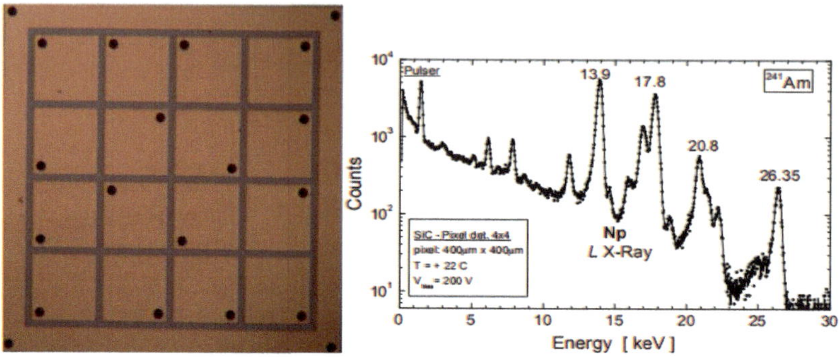


(a) SiC array detector real image, acquired by size (b) array SiC nuclear radiation detector at roomtemperatureA 4x4 matrix of 400x400 μm$^2$ pixels $^{241}$Am source X-ray spectra

**Fig. 1.6** Array detector and its X-ray energy spectrum [80]

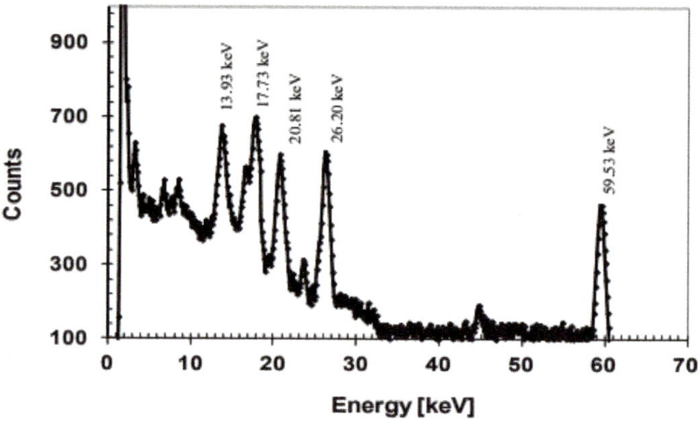


**Fig. 1.7** Energy spectrum of the $^{241}$Am source obtained at 300 K, 250 V bias [81]

In terms of irradiation damage and irradiation resistance characteristics, SiC irradiation detectors are exposed to irradiation damage under experimental conditions from several sources:

(1) γ-rays: Reactor-level γ-rays have essentially no effect on SiC nuclear irradiation detectors;
(2) α-rays: The literature on α-radiation damage generally agrees that α irradiation has limited effect on the device;
(3) β-rays (electron beam); Nava et al. [62] have reported successively on β-rays of different energies. Only at energies up to 8 MeV and injection doses up to 9.5

$\times$ $10^{14}$/$cm^2$ are significant carrier compensation phenomena observed, which originate from profound energy level defects introduced by electron irradiation in the energy range 2 to 8 MeV;

(4) Protons: For proton-only irradiation, the studied energy range is between 6.5 MeV and 24 GeV, and the studied irradiation dose is between $10^{11}$ and 1.4 $\times$ $10^{16}$ $cm^{-2}$. Numerous investigations and analyses have led to the following conclusions: ① The numerous parameters of irradiated defects, including ionisation energy, capture cross section, and defect structure, are comparable to those of high-energy β-ray irradiation; ② The proton energy level has a significant impact on the generation rate of deep defect energy levels; as the proton energy level increases, the capture cross section of Si and C atoms decreases, and the defect energy level generation rate falls. ③ As long as the irradiation dose is sufficient at any proton energy level (6.5 MeV–24 GeV), the concentration of deep energy level centres increases while the oligon diffusion length Lp and charge collection efficiency (CCE) decrease considerably.

(5) Neutrons: Neutron irradiation has a more significant effect on damage than proton irradiation. Figure 1.8 shows the CCE of the SiC SBD β-particle detector after irradiation with 24 GeV/1.4 $\times$ $10^{16}$ $cm^{-2}$ protons or the equivalent 1 MeV/7 $\times$ $10^{15}$ $cm^{-2}$ neutrons.

At high doses of α/β/γ radiation, or at 8 MeV proton doses less than 3 $\times$ $10^{14}$ $cm^{-2}$ or at equivalent 1 MeV neutron doses less than 5 $\times$ $10^{13}$ $cm^{-2}$, the charge collection efficiency of the device does not degrade significantly. The aforementioned dosages of irradiation correspond to defect concentrations of 2 to 3 $\times$ $10^{12}$ $cm^{-3}$. As long as the dose of irradiation does not exceed the above threshold, the material's conductive

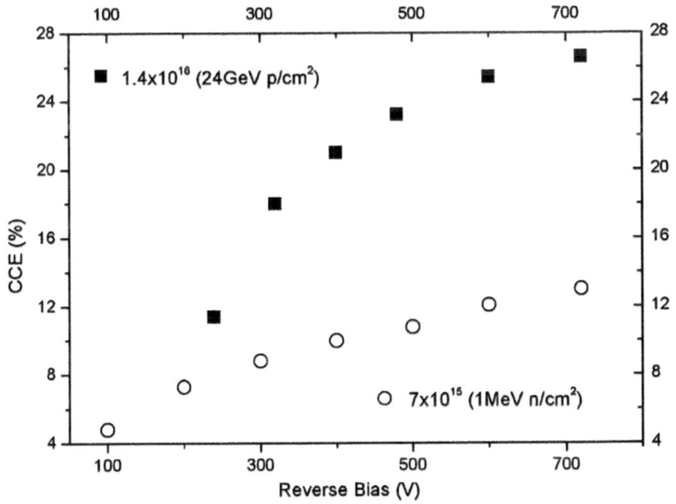


**Fig. 1.8** Comparison of CCE of devices irradiated with the same type and dose of work [62]

properties will not compensate and the CCE can be maintained at 95% after irradiation. Since the defect energy centres produced by high doses of irradiation also appear during material growth, the next generation of higher-quality 4H-SiC should have a higher threshold to ensure nearly 100% CCE.

(b) Status of domestic research

China's research on SiC materials and devices was delayed due to limitations in material and device processing. Until the mid-1990s, a small number of universities in China were engaged in the study of SiC materials and devices, including Xidian University, Institute of Semiconductors, Chinese Academy of Sciences and China Electronics Technology Group Corporation 55th Institute and so on. In recent years, the country's support for the development of wide bandgap semiconductor technology, particularly SiC power electronics, has progressively increased. China has established a comprehensive development of SiC single crystal, epitaxial growth, device design, device production, and application promotion. Numerous research institutions have investigated and developed key technologies, such as 4H-SiC low-doped epitaxial growth, Schottky/Ohmic contact preparation, oxidation annealing, and high-voltage protection, laying the groundwork for the development of SiC semiconductor detectors.

In 2013, Chen Yu et al. from Sichuan University [82] developed a sandwich structure neutron detector based on SiC Schottky diode and neutron conversion material $^{6}$LiF, in which the Schottky diode has the best energy resolution of up to 3.4% for 5.486 MeV $\alpha$ particles, and the sandwich neutron detector shown in Fig. 1.9 responds well to thermal neutrons generated by the critical device, and the count rate is linearly related to power of the critical device. It is demonstrated that, for SiC diodes, decreasing the thickness of the Schottky contact metal can reduce the energy loss before the incident particles enters the sensitive region, which is advantageous for the detection of low-energy charged particles and leads to the enhancement of energy resolution, detection sensitivity, and detection efficiency; Increasing the thickness of the epitaxial layer raises the loss of incident particle energy in the sensitive region, even to the point of total energy loss; Improving the quality of epitaxial layers and substrates can decrease the probability of carrier capture and compounding in the sensitive region, thereby enhancing the energy resolution; Thus, SiC diodes can be used to measure the energy of charged particles. Combined with the irradiation resistance of SiC diodes, SiC nuclear radiation detectors coated with6 LiF film can be developed as a new type of neutron detector for reactor neutron monitoring. Figure 1.10 illustrates that in 2015, the energy resolution of SiC nuclear radiation detectors prepared by Wu Jian and Jiang Yong of the China Academy of Engineering Physics [83] for $\alpha$ particles in the energy range of 4.8–7.7 MeV is 0.61–0.90%, which is comparable to the energy resolution of the internationally reported high-resolution SiC -particle detectors. In 2016, Du et al. [84] of The Institute of High Energy Physics of the Chinese Academy of Sciences used the 4H-SiC SBD detector shown in Fig. 1.11 to measure 59.5 keV $\gamma$-rays produced by americium-241

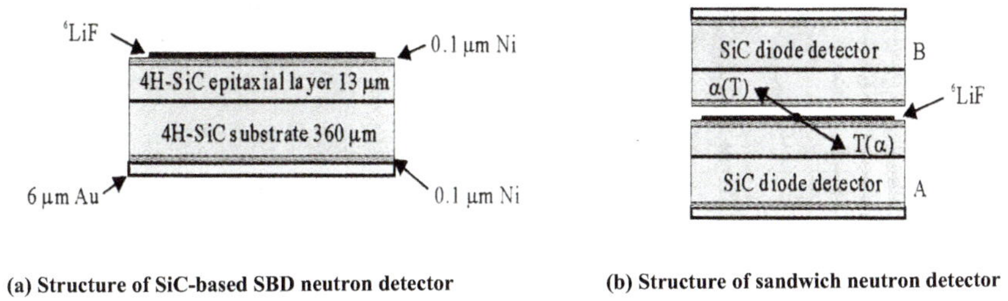


**Fig. 1.9** Structure and principle of SiC-based neutron detector [82]

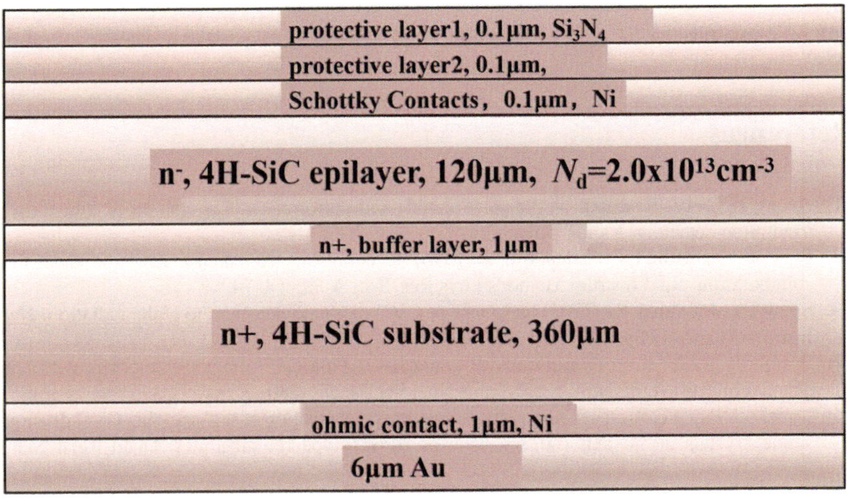


**Fig. 1.10** Schematic diagram of 4H-SiC Schottky diode cross section [83]

($^{241}$Am), and the results demonstrated that the detector has good operating characteristics, with an energy resolution of less than 10% at a reverse bias voltage of 300 V.

In general, nations around the world have dedicated an enormous number of time and money to research on SiC nuclear radiation detectors. Due to the limitations of SiC material and device technology, the structure of SiC nuclear radiation detectors is still relatively simple compared to that of Si radiation detectors, in spite of the fact that the performance of SiC nuclear radiation detectors has been continuously enhanced by the advancement of SiC material and device technology. With the maturation of SiC technology, it is reasonable to assume that SiC nuclear radiation detectors have substantial development potential. In recent years, China has produced an abundance of research results on SiC nuclear radiation detectors, which have progressively begun to catch up to the international advanced level.

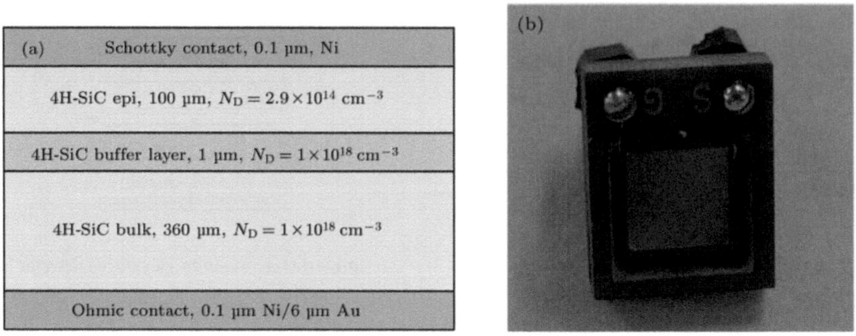


**(a) schematic diagram of cross section** **(b) photo of encapsulated detector[98]**

**Fig. 1.11** Schematic diagram of SiC Schottky diode structure and packaged SiC nuclear radiation detector

# References

1. Huang H (2019) Research on the performance and performance degradation of SiC particle irradiation detectors. Xidian University
2. Bauer C, Baumann I, Colledani C et al (1996) Recent results from the RD42 diamond detector collaboration. Nucl Instrum Methods Phys Res, Sect A 383:64–74
3. Souw EK, Meilunas RJ (1997) Response of CVD diamond detectors to alpha radiation. Nucl Instrum Methods Phys Res, Sect A 400(1):69–86
4. Brambilla A, Tromson D, Bergonzo P et al (2002) Thin CVD diamond detectors with high charge collection efficiency. IEEE Trans Nucl Sci 49:277–280
5. Kaneko J, Fujita F, Konno Y et al (2012) Growth and evaluation of self-standing CVD diamond single crystals on off-axis (001) surface of HP/HT type IIa substrates. Diam Relat Mater 26:45–49
6. Duenas JA, De La Torre PJ, Martin Sanchez A et al (2014) Diamond detector for alpha-particle spectrometry. Appl Radiat Isot 90:177–180
7. De Feudis M, Caricato AP, Chiodini G et al (2016) Characterization of surface graphitic electrodes made by excimer laser on CVD diamond. Diam Relat Mater 65:137–143
8. Sato S-I, Makino T, Ohshima T et al (2017) Transient current induced in thin film diamonds by swift heavy ions. Diam Relat Mater 75:161–168
9. Tallaire A, Mille V, Brinza O et al (2017) Thick CVD diamond films grown on high-quality type IIa HPHT diamond substrates from new diamond technology. Diam Relat Mater 77:146–152
10. Shimaoka T, Koizumi S, Tanaka MM (2018) Diamond photovoltaic radiation sensor using pn junction. Appl Phys Lett 113(9):093504
11. Verona C, Magrin G, Solevi P et al (2018) Toward the use of single crystal diamond based detector for ion-beam therapy microdosimetry. Radiat Meas 110:25–31
12. Abdel-Rahman MAE, Lohstroh A, Sellin PJ (2011) The effect of annealing on the X-ray induced photocurrent characteristics of CVD diamond radiation detectors with different electrical contacts. Physica Status Solidi (a) 208(9):2079–2086
13. Marinelli M, Milani E, Prestopino G et al (2012) X-ray beam monitor made by thin-film CVD single-crystal diamond. J Synchrotron Radiat 19(Pt 6):1015–1020
14. Desjardins K, Pomorski M, Morse J (2014) Ultra-thin optical grade scCVD diamond as X-ray beam position monitor. J Synchrotron Radiat 21(Pt 6):1217–1223
15. Zou M, Gaowei M, Zhou T et al (2018) An all-diamond X-ray position and flux monitor using nitrogen-incorporated ultra-nanocrystalline diamond contacts. J Synchrotron Radiat 25(Pt 4):1060–1067

16. Desjardins K, Bordessoule M, Pomorski M (2018) X-ray position-sensitive duo-lateral diamond detectors at SOLEIL. J Synchrotron Radiat 25(Pt 2):399–406
17. Keister JW, Smedley J (2009) Single crystal diamond photodiode for soft X-ray radiometry. Nucl Instrum Methods Phys Res, Sect A 606(3):774–779
18. Keister JW, Smedley J, Dimitrov D et al (2010) Charge collection and propagation in diamond x-ray detectors. IEEE Trans Nucl Sci 57(4):2400–2404
19. Ade N (2017) An investigation of the role of defect levels on the radiation response of synthetic diamond crystals when used as sensors for the detection of mammography X-rays. Appl Radiat Isot 127:237–244
20. Kanxheri K, Aisa D, Alunni Solestizi L et al (2020) Intercalibration of a polycrystalline 3D diamond detector for small field dosimetry. Nucl Instrum Methods Phys Res, Sect A 958:162730
21. Lyoussi A, Gallin-Martel ML, Abbassi L et al (2018) A large area diamond-based beam tagging hodoscope for ion therapy monitoring. EPJ Web Conf 170:09005
22. Brambilla A, Tromson D, Aboud N et al (2001) CVD diamond gamma dose rate monitor for harsh environment. Nucl Instrum Meth Phys Res Sect A 458(1–2):220–226
23. Gastelum S, Cruz-Zaragoza E, Favallf A et al (2007) Dose rate effects on the thermoluminescence kinetics properties of MWCVD diamond films. Phys Status Solidi A-Appl Mat 204(9):3053–3058
24. Williams T, Martens A, Cassou K et al (2017) Novel applications and future perspectives of a fast diamond gamma ray detector. Nucl Instrum Methods Phys Res, Sect A 845:199–202
25. Balducci A, Marinelli M, Milani E et al (2005) Synthesis and characterization of a single-crystal chemical-vapor-deposition diamond particle detector. Appl Phys Lett 86(21):213507
26. Majerle M, Angelone M, Krása A et al (2020) The response of single crystal diamond detectors to 17–34 MeV neutrons. Nucl Instrum Methods Phys Res, Sect A 951:163014
27. Jagannadham K, Verghese K, Butler JE (2001) Neutron transmutation of 10B isotope-doped diamond. Appl Phys Lett 78(4):446–447
28. Marinelli M, Milani E, Prestopino G et al (2006) High performance Li6F-diamond thermal neutron detectors. Appl Phys Lett 89(14):143509
29. Pernegger H (2006) High mobility diamonds and particle detectors. Physica Status Solidi (a) 203(13):3299–3314
30. Bauer C, Baumann I, Colledani C et al (1995) Radiation hardness studies of CVD diamond detectors. Nucl Instrum Meth Phys Res Sect A 367(1–3):207–211
31. Han S (1997) New developments in photoconductive detectors (invited). Rev Sci Instrum 68(1):647–652
32. Meier D, Adam W, Bauer C et al (1999) Proton irradiation of CVD diamond detectors for high-luminosity experiments at the LHC. Nucl Instrum Meth Phys Res Sect A 426(1):173–180
33. Grilj V, Skukan N, Jaksic M et al (2013) Irradiation of thin diamond detectors and radiation hardness tests using MeV protons. Nucl Instrum Methods Phys Res Sect B 306:191–194
34. Adam W, Bauer C, Berdermann E et al (1999) Review of the development of diamond radiation sensors. Nucl Instrum Meth Phys Res Sect A 434(1):131–145
35. Husson D, Bauer C, Baumann I et al (1997) Neutron irradiation of CVD diamond samples for tracking detectors. Nucl Instrum Meth Phys Res Sect A 388(3):421–426
36. Weilhammer P, Adam W, Bauer C et al (1998) Recent results on CVD diamond radiation sensors. Nucl Instrum Meth Phys Res Sect A 409(1–3):264–270
37. Friedl M, Adam W, Bauer C et al (1999) CVD diamond detectors for ionizing radiation. Nucl Instrum Meth Phys Res Sect A 435(1–2):194–201
38. Spielman RB, Ruggles LE, Pepping RE et al (1997) Fielding and calibration issues for diamond photoconducting detectors. Rev Sci Instrum 68(1):782–785
39. Bauer C, Baumann I, Colledani C et al (1995) Recent results from diamond microstrip detectors—diamond detectors collaboration (RD42). Nucl Instrum Meth Phys Res Sect A 367(1–3):202–206
40. Muraro A, Giacomelli L, Nocente M et al (2016) First neutron spectroscopy measurements with a pixelated diamond detector at JET. Rev Sci Instrum 87(11):11D833

41. Claps G, Murtas F, Foggetta L et al (2018) Diamondpix: a CVD diamond detector with Timepix3 chip interface. IEEE Trans Nucl Sci 65(10):2743–2753
42. Tsubota M, Kaneko JH, Miyazaki D et al (2015) High-temperature characteristics of charge collection efficiency using single CVD diamond detectors. Nucl Instrum Meth Phys Res Sect A 789:50–56
43. Kumar A, Kumar A, Topkar A et al (2017) Prototyping and performance study of a single crystal diamond detector for operation at high temperatures. Nucl Instrum Methods Phys Res, Sect A 858:12–17
44. Yining S, Jingqi L, Wantu G et al (1997) Diamond film high-energy particle detector. Vacuum Sci Technol 05:336–339
45. Hongbin W, Kexu S, Daoyuan T (1998) Measurement of soft x-ray radiation power using diamond detector. J At Mol Phys 01:3–5
46. Lifei H, Fang L, Yongteng Y et al (2010) Measurement of soft x-ray spectrum by chemical vapor deposition diamond detector. Acta Physica Sinica 59(02):1137–1142
47. Li F, Hou L, Su C et al (2010) Development of fast-response chemical vapor deposition diamond soft x-ray detector. High Power Laser and Particle Beams 22(06):1404–1406
48. Lifei H, Fang L, Shenye L et al (2011) Chemical vapor deposition diamond x-ray detector. High Power Laser and Particle Beams 23(07):1873–1876
49. Bo Y, Bolun C, Lifei H et al (2013) Measurement of hard x-rays from radiation-driven implosion by chemical vapor deposition diamond detector. Acta Physica Sinica 62(05):470–475
50. Hou L, Du H, Che X et al (2018) Preliminary study on temporal characteristics of a novel CVD diamond x-ray detector. Acta Optica Sinica 38(07):39–43
51. Lifei H, Zhichao L, Yongteng Y et al (2012) Relative calibration of chemical vapor deposition diamond x-ray detector. High Power Laser and Particle Beams 24(08):1871–1873
52. Wang L, Lou Y, Su Q et al (2005) Study on device structure of diamond film detector. Laser Optoelectron Progress (12):32+30
53. Jianmin L, Yiben X, Linjun W et al (2006) Development and performance study of CVD diamond film x-ray detector. Nucl Electron Detect Technol 05:669–672
54. Hujiang S, Linjun W, Jian H et al (2009) Research on radiation detector based on (100)-oriented diamond film. J Inorg Mater 24(06):1254–1258
55. Pan, Qin, Zhou et al (2015) The diamond radiation detector with an Ohmic contact using diamond-like carbon interlayer. Mater Res Innov 19(sup5)
56. Xiaoping O, Lan W, Ruyu F et al (2006) Development of diamond film detector. Acta Physica Sinica 05:2170–2174
57. Lan L, Xiaoping O, Na C et al (2009) Charge collection efficiency of chemical vapor deposition diamond film detector for alpha particles. High Power Laser and Particle Beams 21(07):1083–1087
58. Liu LY, Ouyang XP, Zhang ZB et al (2012) Polycrystalline chemical-vapor-deposited diamond for fast and ultra-fast neutron detection. Sci China Technol Sc 55(9):2640–2645
59. Bolshakov AP, Zyablyuk KN, Kolyubin VA et al (2017) Thin CVD diamond film detector for slow neutrons with buried graphitic electrode. Nucl Instrum Methods Phys Res, Sect A 871:142–147
60. Liu JW, Chang JF, Zhang JZ et al (2019) Design, fabrication and testing of CVD diamond detectors with high performance. AIP Adv 9(4):045205
61. Guo Y-Z, Liu J-L, Liu J-W et al (2020) Comparison of α particle detectors based on single-crystal diamond films grown in two types of gas atmospheres by microwave plasma-assisted chemical vapor deposition. Int J Miner Metall Mater 27(5):703–712
62. Nava F, Bertuccio G, Cavallini A et al (2008) Silicon carbide and its use as a radiation detector material. Meas Sci Technol 19(10):102001
63. Torrisi L, Foti G, Giuffrida L et al (2009) Single crystal silicon carbide detector of emitted ions and soft x rays from power laser-generated plasmas. J Appl Phys 105(12):463
64. Tikhomirova VA, Fedoseeva OP, Bol'shakov VV (1973) Silicon carbide detectors as fission-fragment counters in reactors. Meas Tech 16(6):900–901

65. Tikhomirova VA, Fedoseeva OP, Kholuyanov GF (1972) Properties of silicon-carbide counters of highly ionizing radiations produced by beryllium diffusion. Fiz Tekh Poluprovn 6(5):957–959
66. Ruddy FH, Dulloo AR, Seidel JG et al (1998) Development of a silicon carbide radiation detector. IEEE Trans Nucl Sci 45(3):536–541
67. Ruddy FH, Dulloo AR, Seidel JG et al (2002) Nuclear reactor power monitoring using silicon carbide semiconductor radiation detectors. Nucl Technol 140(2):198–208
68. Ruddy FH, Seidel JG, Sellin P (2009) High-resolution alpha spectrometry with a thin-window silicon carbide semiconductor detector. In: Nuclear science symposium conference record. IEEE, pp 2201–2206
69. Ruddy FH, Flammang RW, Seidel JG (2009) Low-background detection of fission neutrons produced by pulsed neutron interrogation. Nucl Instrum Methods Phys Res 598(2):518–525
70. Ivanov AM, Kalinina EV, Strokan NB et al (2009) 4H-SiC nuclear radiation p-n detectors for operation up to temperature 375 °C. Mater Sci Forum 615–617:849–852
71. Chaudhuri SK, Krishna RM, Zavalla KJ et al (2013) Schottky barrier detectors on 4H-SiC n-type epitaxial layer for alpha particles. Nuclear Inst Methods Phys Res A 701(Feb 11):214–220
72. Chaudhuri SK, Zavalla KJ, Mandal KC (2013) High resolution alpha particle detection using 4H–SiC epitaxial layers: fabrication, characterization, and noise analysis. Nuclear Inst Methods Phys Res A 728(Nov 11):97–101
73. Zat'Ko B, Dubecky F, Šagátová A et al (2015) High resolution alpha particle detectors based on 4H-SiC epitaxial layer. J Instrument 10(04):C04009–C04009
74. Manfredotti C, Lo Giudice A, Fasolo F et al (2005) SiC detectors for neutron monitoring. Nucl Instrum Methods Phys Res, Sect A 552(1–2):131–137
75. Giudice AL, Fasolo F, Durisi E et al (2007) Performances of 4H-SiC Schottky diodes as neutron detectors. Nucl Instrum Methods Phys Res 583(1):177–180
76. Ha JH, Kang SM, Park SH et al (2009) A self-biased neutron detector based on an SiC semiconductor for a harsh environment. Appl Radiat Isot 67(7–8):1204–1207
77. Jang Ho HA, Kang SM, Kim HS et al (2011) 4H-SiC PiN-type semiconductor detector for fast neutron detection. Progr Nuclear Technol 1:237–239
78. Ko G, Kim HY, Bang J et al (2009) Electrical characterizations of Neutron-irradiated SiC Schottky diodes. Korean J Chem Eng 26(1):285–287
79. Szalkai D, Issa F, Klix A et al (2013) First tests of silicon-carbide semiconductors as candidate neutron detector for the ITER Test Blanket Modules. In: International conference on advancements in nuclear instrumentation measurement methods & their applications. IEEE, pp 1–4
80. Bertuccio G, Caccia S, Nava F et al (2009) Ultra low noise epitaxial 4H-SiC x-ray detectors. Mater Sci Forum 615–617:845–848
81. Mandal KC, Muzykov PG, Chaudhuri SK et al (2013) Low energy x-ray and γ-ray detectors fabricated on n-type 4H-SiC epitaxial layer. IEEE Trans Nucl Sci 60(4):2888–2893
82. Yu C, Yong J, Jian W et al (2013) Response of SiC-based neutron detectors to thermal neutrons. High Power Laser Particle Beams (10):2711–2716 (2013)
83. Jian W, Yong J, Lei G et al (2015) High energy resolution alpha particle detector based on 4H-SiC. High Power Laser Particle Beams 1:151–154
84. Yuanyuan D, Chunlei Z, Xuelei C (2016) Gamma ray detector based on 4H-SiC Schottky barrier diode. Acta Physica Sinica 65(20):216–223

# Chapter 2
# Detection Principles of Wide Bandgap Semiconductor Nuclear Radiation Detectors

Semiconductor materials have a rapid pulse rise time, which is vastly superior to the radiation resisting ability of gas media and vastly lower than the average ionization energy of gas media. As a result, semiconductor detectors have excellent energy resolution, a wide energy linear range, a rapid time response, and a small size, among other benefits. Since the 1960s, when semiconductor materials began to be used as radiation detection media, semiconductor detectors have been widely used in a variety of radiation detection fields. They are superior to gas detectors and scintillation counters in terms of precision energy spectrum measurements, but have the disadvantages of low resistance to radiation damage, temperature-sensitive performance, and the requirement that most be used at low temperatures. As mentioned in Sect. 1.3, for semiconductor detectors, the choice of the original material is one of the most important factors, and the use of wide-bandgap semiconductor materials as radiation detection media is anticipated to improve the high temperature and radiation resistance of conventional Si semiconductor detectors while preserving their excellent characteristics. SiC and diamond detectors as semiconductor detectors, and commonly used Si-based radiation detectors have the same basic principle, that is, in the particle irradiation detection, the radiation particles through the detector generate reaction, excitation, ionization, and other processes, and finally in the form of electrical signals output and analysis of the detection system to measure these signals contain a variety of characteristics of the radiation particles, including the number of radiation particles, intensity, energy, direction. There are numerous varieties of semiconductor detectors, with the most common device structures being junction type, junctionless type, and Metal–Insulator–Semiconductor (MIS) type. Different structures can be chosen based on the characteristics of the semiconductor material and the application needs. For wide-bandgap materials, junction and junctionless types are very common. Considering that the epitaxial material used for junctionless devices is typically superior to the semi-insulator material used for junctionless devices in terms of quality, defects, minority lifetime, and carrier collection. Furthermore the process of preparing electrodes on epitaxial materials is simpler and more



Y. Zhang et al., *Wide Band-Gap Semiconductor Nuclear Radiation Detector*,
Wide Bandgap Semiconductors, https://doi.org/10.1007/978-981-96-0340-4_2

mature than that of semi-insulator materials, SiC materials are considered first for junction structure preparation in detectors. As diamond has a high resistivity and is difficult to dope, junctionless structures are typically utilized.

## 2.1 Basic Properties of Nuclear Radiation

Nuclear radiation, or radiation as it is commonly known, is a stream of subatomic particles emitted by the nucleus of an atom during its transition from one structure or energy state to another. Nuclear radiation is also known as ionizing radiation, and its interaction with matter to cause ionization or excitation of matter is the foundation of radiation detection. Ionizing radiation is also classified into direct and indirect subtypes. Direct ionizing radiation consists of charged particles, including protons. Indirect ionizing radiation consists of photons and neutrons, which are uncharged particles. It is customary to classify nuclear radiation into four main categories as follows [1]:

- charged particles radiation
  - heavy charged particles： α particles, fission fragment etc.
  - light charged particles: fast electrons, β-particle(electrons) etc.
- uncharged particle radiation
  - X-ray, γ-ray etc.
  - neutrons

(1) The alpha particle is a heavy particle that is emitted during nuclear decay. It is composed of two protons and two neutrons and contains no electrons outside of its nucleus, so it has two positive charges, which is equivalent to the nucleus of He, $He^4$. The α particle is a positively charged, high-energy particle that swiftly loses energy as it passes through a medium and cannot penetrate very far. The most commonly used source of alpha particles is $^{241}$Am.

(2) Fast electrons include β particles emitted in decay, positive or negative electrons and electrons with considerable energy produced by other processes, such as photoelectrons. β particles are the electrons that are released during β decay. β rays have continuous energy, often less than 2 MeV, and while their energy is lower than that of alpha particles, their speed is substantially higher than that of alpha particles due to the smaller mass of beta particles.

(3) X radiation and γ radiation are electromagnetic radiation; γ radiation is the ray that emitted from atomic nucleus de-excitation or particle annihilation process; its wavelength is less than 0.001 nm. X radiation is the ray from electronic transition or bremsstrahlung, which is also electromagnetic wave with a wavelength between 0.001 and 10 nm. The radiation emitted when electrons in specific

orbits of an atom's nucleus move from higher to lower energy levels is referred to as characteristic X-rays.

(4) Neutron is one of the elements that comprise an atom's nucleus. It is unstable in its free state and decays into protons, electrons, and antineutrinos. Neutrons' half-life is 10 to 25 min, so free neutrons do not exist in nature. Neutrons are produced by nuclear reactors, gas pedals, and radioisotopes in laboratories.

## 2.2 Interaction of Particles and Substances

1. Neutron

Neutrons are neutral, unaffected by Coulomb repulsion, and do not directly induce ionization when passing through matter. Therefore, neutrons are detected through intense interactions between neutrons and atomic nuclei, i.e., nuclear reactions, which generate charged particles or photons, which are then detected. The interactions that can be used include: (i) the emission of charged particles as a result of a nuclear reaction between neutrons and nuclei; (ii) neutrons collide with the nucleus resulting in elastic scattering, causing the nucleus to recoil; (iii) the neutron's capture initiates the activation of the nucleus. (iv) initiation of nuclear fission.

Different detection techniques are required for neutrons of different energies due to their varying probabilities of various types of interactions, such as thermal or slow neutrons entering the nucleus and being easily captured and fast neutrons that primarily undergo elastic collisions with the nucleus and interact with heavier nuclei to cause inelastic scattering and produce $\gamma$-rays. For example, Typically, the slow neutron flux is measured through nuclear reactions; nuclear recoil is a technique for detecting fast neutrons with energies larger than 0.3 MeV. In addition, there are additional measurements, such as the nuclear fission for measuring neutron flux in strong $\gamma$ environments and the nuclear activation method for measuring neutron injection rates.

2. Charged particles

Charged particles with a much larger mass than electrons, also known as heavy charged particles, interact with matter and lose energy primarily through Coulombic interactions with the shell electrons of matter atoms, expending their own energy to ionize and excite the shell electrons. The energy loss curve of the heavy charged particle along its track is known as the Bragg curve at this time, as shown in Fig. 2.1 for the $\alpha$ particle Bragg curve. There is typically a minor difference between the Bragg curves of a single particle and a particle beam of the same particle. The ionization energy loss rate of a heavy charged particle—$\mathrm{d}E/\mathrm{d}x$ (energy lost per unit distance) is given by the Bethe-Bloch equation [2]

Fig. 2.1 Variation of energy loss rate along the α-particle path [1]

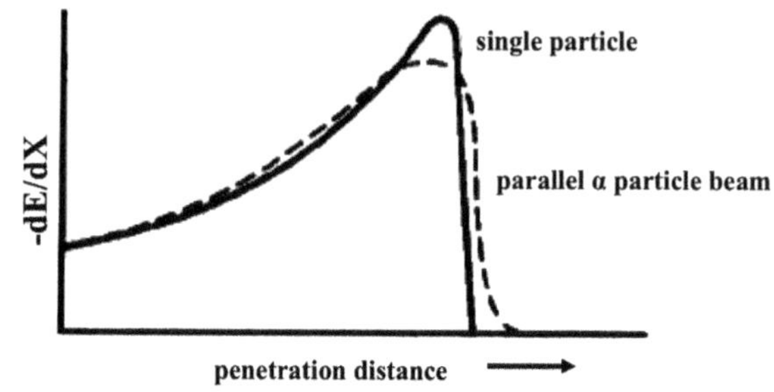


$$-\left(\frac{dE}{dx}\right) = Kz^2\frac{Z}{A}\frac{1}{\beta^2}\left[\frac{1}{2}\ln\frac{2m_ec^2\beta^2\gamma^2T_{max}}{\varepsilon^2} - \beta^2 - \frac{\delta(\beta\gamma)}{2}\right]$$
$$K = 4\pi N_A r_e^2 m_e c^2\left(r_e = e^2/4\pi\varepsilon_0 m_e c^2\right)$$
$$T_{max} = \frac{2m_ec^2\beta^2\gamma^2}{1 + 2\gamma\frac{m_e}{m_0} + \left(\frac{m_e}{m_0}\right)^2} \tag{2.1}$$

$T_{max}$ is the maximum energy transfer in a single collision, if $m_e \ll m_0$, $2\gamma m_e/m_0 \ll 1$ at low energy, approximated by $T_{max} = 2m_ec^2\beta^2\gamma^2$; $N_A$ is Avogadro constant; $Z$ is the atomic number of the medium atom; $z$ is the charge of the incident particle; $m_e$ is the electron mass; $m_0$ is the incident particle mass; $A$ is the atomic number of the medium atom; $\varepsilon$ is the average ionization excitation energy of the medium atom; $\delta$ is the density correction effect; $\beta$ = particle *velocity/c*; $\gamma = (1 - \beta^2)^{-2}$.

On the other hand, the scenario involving particles with low charges, such as electrons and positrons, presents a significantly more intricate situation. In addition to the loss of ionization energy, there is also the loss of radiation energy. The phenomenon of ionization energy loss exhibits similarities to that of heavily charged particles. However, it is noteworthy that due to the nearly equivalent masses of lightly charged particles and the particles they collide with (typically shell electrons), a substantial amount of energy can be dissipated in a single interaction. Furthermore, the incident particles may experience significant alterations in their direction of motion. On the other hand, radiation energy loss pertains to the dissipation of energy resulting from the inelastic collision between lightly charged particles and atoms of matter. During the process of light charged particles approaching the nucleus, their velocity and trajectory undergo alterations as a result of the Coulomb force, while simultaneously emitting electromagnetic waves in an outward direction. It is important to note that this part of the energy loss has an inverse relationship with the square of the mass of the incident particle. Consequently, this energy loss is consistently insignificant for particles with a high charge density.

Furthermore, it should be noted that for Z > 2, the cations that have undergone electron loss are referred to as heavy ions. The aforementioned high-energy charged particles, such as protons and alpha particles, are classified as fast light ions. Their interaction with matter primarily involves the Coulomb interaction between the nucleus of the light ion and the electrons and nuclei inside the atoms of the target material. The Coulomb contact between the bound electrons of a heavy ion and the target nucleus is a factor that contributes to the blocking principle, although its effect is significant only when the heavy ion is moving at a very slow velocity. At elevated velocities, heavy ions continue to transfer energy to the target material mostly via ionization excitation. The energy dissipation mechanism of heavy ions in a medium is not fundamentally distinct from that of lighter ions, such as protons and α particles. One notable distinction between the energy dissipation mechanisms of heavy ions and light ions lies in the fact that heavy ions possess a stronger binding to their inner electrons. Consequently, heavy ions typically face limitations in their ability to completely shed all of their orbital electrons. Additionally, the charge state of heavy ions may undergo alterations during their traversal through the target material. Hence, it is evident that certain aspects that have been overlooked in the energy loss mechanism pertaining to the interaction between light ions and matter, namely charge exchange and nuclear blocking, are not only non-negligible but also have a significant influence in the context of heavy ion-matter interaction.

3. Photons (X/γ rays)

X/γ rays are electrically neutral and do not possess the ability to directly ionize or excite atoms within materials. In order to initiate detection, it is necessary for the photons to engage with the atoms of the material, thereby transferring a portion or the entirety of their energy to an electron within the absorbing substance. Subsequently, the electron's interaction leads to ionization and excitation, ultimately enabling the detecting process. The phenomenon of photons interacting with atoms of matter encompasses the photoelectric effect, the Compton effect, and the electron pair effect. The three effects exhibit competition and can coexist concurrently, with their respective significance being dependent on the photon energy and atomic number of the material, as depicted in Fig. 2.2.

**Fig. 2.2** The relationship between the mode of absorption of electromagnetic radiation and the photon energy hv and the atomic number of the substance Z

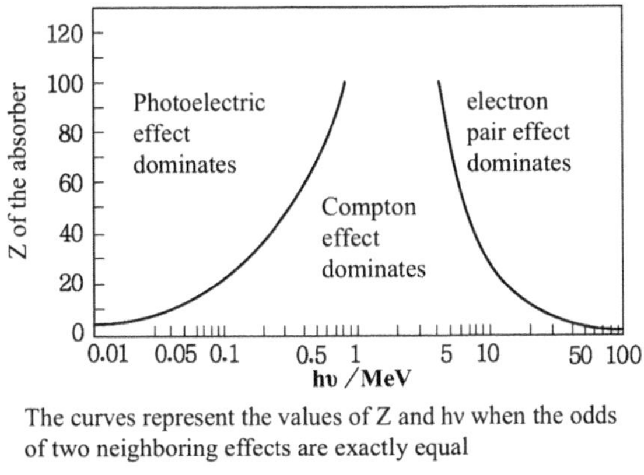


The curves represent the values of Z and hv when the odds of two neighboring effects are exactly equal

## 2.3 Basic Structure and Operating Principle of Wide Bandgap Semiconductor Nuclear Radiation Detectors

Semiconductor detectors predominantly involve junction and junctionless semiconductor material architectures. To achieve superior semiconductor materials for detector fabrication, the preparation of SiC and diamond crystals utilized in semiconductor detectors often involves chemical vapor deposition (CVD) epitaxy. In the context of SiC materials, the epitaxial growth process offers a convenient means to achieve high-quality doping of both P-type and N-type. Additionally, the process of fabricating electrodes on epitaxial materials is comparatively simpler and more well-established when compared to semi-insulating materials. Consequently, the formation of junction semiconductor structures is easier. Furthermore, the rectification properties of the space charge layer in junction semiconductor material structures, along with the existence of the built-in electric field, contribute to the accomplishment of superior performance characterized by low dark current, high collection efficiency, and excellent temporal features. Hence, SiC junction structures are frequently prioritized for the fabrication of detectors. In the case of diamond, it is worth noting that while P-type diamond can be produced through the introduction of boron (B) dopants and N-type diamond can be achieved through the incorporation of phosphorus (P) dopants, N-type diamond materials tend to exhibit a high degree of instability. Additionally, the acceptor level associated with P-type diamond poses challenges for its effective operation at room temperature. Consequently, the fabrication of junctional CVD diamond materials becomes a complex and challenging task. Furthermore, it is notable that the intrinsic diamond resistivity exceeds $10^{11}$ Ω·cm, which satisfies the necessary parameter requirements for homogeneous semiconductor detectors. Additionally, the breakdown field strength of diamond reaches $10^7$ V cm$^{-1}$, further meeting the desired criteria. In comparison to other semiconductor detectors like silicon (Si), detectors constructed with intrinsic diamond materials exhibit significantly reduced noise current. Moreover, the utilization of intrinsic diamond materials circumvents the issue of detector lifespan reduction caused by junction aging. Consequently, junctionless devices based on chemical vapor deposition (CVD) diamond are commonly employed for the fabrication of detectors.

### *2.3.1 Detector Structure and Working Principle*

1. Semiconductor junction type detector

Semiconductor structures of the junction type can often be categorized into two main types: Schottky junctions and PN junctions. In this section, we provide a concise overview of the operational mechanism of a semiconductor detector, using the PN junction type detector as a representative example, as depicted in Fig. 2.3.

When a semiconductor material undergoes a transition from N-type to P-type, the disparity in carrier concentrations between the two regions leads to a net diffusion

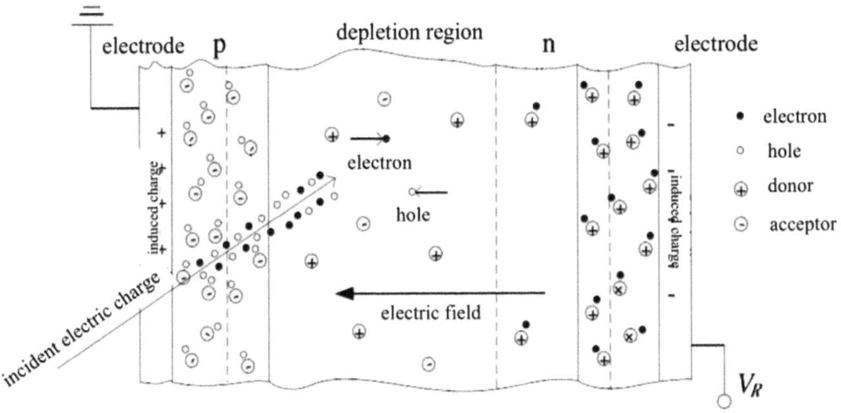


**Fig. 2.3** Schematic diagram of the working principle of PN type detector [3]

of free-migratable carriers. Specifically, the polytronic electrons in the N-region diffuse into the P-region, while the polytronic holes in the P-region diffuse into the N-region. Simultaneously, the donor ions with positive charges and the acceptor ions with negative charges remain immobile, resulting in the creation of an inherent electric field that is directed from the n-region to the p-region. The area characterized by a lack of mobile charges and an inherent electric field is commonly referred to as the space charge region or depletion region. Thermal equilibrium circumstances are characterized by the presence of a built-in electric field that effectively limits the net diffusion of carriers, hence generating a state of charge distribution that is dynamically balanced. In the scenario where the transition of the semiconductor from N-type to P-type is characterized by an abrupt and uniform doping profile, the width of the space charge area (denoted as W) at thermal equilibrium, specifically at zero bias, can be determined:

$$W = \sqrt{\frac{2\varepsilon_s\varepsilon_0}{e}\left(\frac{N_a + N_d}{N_a N_d}\right)\left(\varphi_{bi} - \frac{2kT}{e}\right)} \tag{2.2}$$

$$\varphi_{bi} \approx \frac{kT}{e}\ln\left(\frac{N_a N_d}{n_i^2}\right) \tag{2.3}$$

In the given context, the symbol "k" represents the Boltzmann constant, "T" denotes the temperature in Kelvin, "$\varepsilon_s$" signifies the relative dielectric constant of semiconductor material, "$\varphi_{bi}$" represents the built-in potential, "Nd" and "Na" refer to the doping concentrations in the n- and p-regions respectively, and "$n_i$" represents the intrinsic carrier concentration. At this point, applying a positive voltage $V_R$, i.e., reverse bias, to the n-region compared to the p-region, $W$ can be further written as:

$$W = \sqrt{\frac{2\varepsilon_s\varepsilon_0}{e}\left(\frac{N_a + N_d}{N_a N_d}\right)\left(\varphi_{bi} + V_R - \frac{2kT}{e}\right)} \tag{2.4}$$

PN-type detectors typically function in a state of reverse bias. When radiation particles interact with the semiconductor material and lose energy, if this energy loss happens in the depletion area, the extra carriers (electron–hole pairs) that are created by the excitation move towards the electrodes due to the presence of the built-in electric field. When energy loss takes place in an electrically neutral zone, excess carriers progressively diffuse in a radial manner. The displacement of the excess carriers results in a modification of the induced charge on the poles, so generating a pulse signal in the output loop. The operational mechanism of the aforementioned phenomenon is referred to as the drift–diffusion theory. The collection efficiency of carriers is higher in the depletion layer compared to the neutral zone due to the presence of a bigger electric field. This electric field facilitates the efficient separation of produced electrons and holes, resulting in reduced losses of composite. Typically, the PN structure of the detector is not directly employed. Instead, the extended structure of the PiN is utilized, wherein a layer of greater thickness and significantly lower doping concentration, known as the I layer, is sandwiched between the P region and the N region. The I-layer experiences quick depletion and the concentration of the electric field primarily occurs within the I-layer when reverse bias is applied at both ends, owing to its significantly low doping level. The I-layer experiences quick depletion and the concentration of the electric field primarily occurs within the I-layer when reverse bias is applied at both ends, owing to its significantly low doping level. The thickness of the I layer is greater in comparison to the depletion layer typically seen in a PN junction. Consequently, this characteristic enables the device to achieve a heightened level of charge collection efficiency.

2. Homogeneous Conductivity detector

Diamond homogeneous conductivity detectors, classified as solid-state detectors, possess the ability to detect various particles or rays that have the capacity to generate free carriers (electron–hole pairs) within diamond. These include ultraviolet rays, X-rays, γ-rays, charged particles, neutrons, mesons, and other high-energy particles. The fundamental principle of radiation detection is independent of the specific emitting particle, provided that its energy exceeds the band gap of the diamond material. The principle and structure of diamond detectors are shown in Fig. 2.4 [4]. The layout of the device is characterized by a straightforward configuration like that of a sandwich. It consists of a layer of high resistivity diamond positioned between two metal electrodes. The application of a voltage across the metal electrodes enables the creation of an electric field within the diamond material. When the detector is traversed by a beam of high-energy particles, a significant quantity of electron–hole pairs are produced along the trajectory of the incident particles. These unbound charges then migrate towards the poles due to the influence of an externally provided electric field. Consequently, an amplified electrical signal is generated and subsequently acquired.

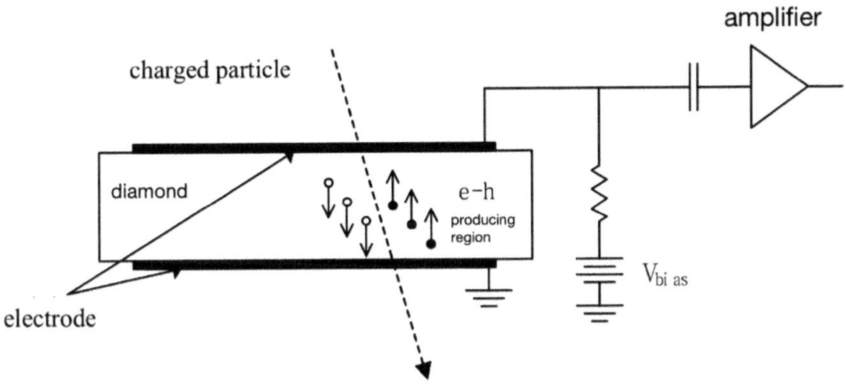


**Fig. 2.4** Schematic diagram of diamond detector principle [4]

The Charge Collection Efficiency (CCE) is quantitatively expressed as the proportion of the detected power to the generated power. The efficiency of the diamond film detector, known as the Charge Collection Efficiency (CCE), is less than 100% as a result of impurities and defects present in the diamond film. These impurities and flaws significantly decrease the average distance that carriers can drift and hence diminish the collection of charges. The parameter denoted as the charge collection distance, Ld, represents the mean displacement of a liberated charge carrier prior to its eventual capture. Carrier trapping is influenced by various factors, including impurities, defects, and material-related properties such as carrier momentum and mobility. Furthermore, the presence of grain boundaries in polycrystalline diamond films can lead to the formation of high-density defects. Consequently, the growth of diamond films characterized by larger grains can effectively mitigate the density of grain borders and subsequently decrease the concentration of impurities. The acquisition of diamond films of superior quality, characterized by significant grain size, can be achieved through iterative refinement of process parameters to attain substantial Ld values.

3. Charge transport

As previously stated, the semiconductor detector encompasses two essential components, namely carrier creation and carrier transport, which collectively influence the detector's efficacy in detecting nuclear radiation particles. As previously stated, the semiconductor detector encompasses two vital parts, namely carrier creation and carrier transport, which collectively influence the detector's efficacy in detecting nuclear radiation particles. Section 2.2 provides a detailed description of the carrier production method and the factors that influence the reaction cross-section. These factors play a crucial role in determining the energy deposition magnitude of nuclear radiation particles in the detector medium and the quantity of carriers generated. In contrast, the process of carrier transport plays a crucial role in determining the fraction of carriers generated by irradiation that are ultimately collected by the two electrodes. Additionally, it influences the time required for these carriers to traverse to the electrodes. This process significantly affects various aspects of the detector, such as sensitivity and temporal characteristics. The charge-collection efficiency (CCE) is a useful metric for characterizing this process. As previously stated, the charge collection efficiency (CCE) is quantitatively defined as the ratio between the charge accumulated by the external circuit of the detector ($Q_c$) and the charge associated with the carriers generated by radiation ($Q_g$).

$$\text{CCE} = \frac{Q_c}{Q_g} \tag{2.5}$$

The charge collecting efficiency is influenced by multiple factors, including carrier movement, trap density, and electric field strength.

(a) Carrier scattering and drift velocity

As carriers traverse a semiconductor, they undergo frequent collisions with thermally vibrating lattice atoms and impurity atoms, resulting in scattering phenomena. The amplitude and direction of carrier velocity exhibit continual fluctuations. The diagram presented in Fig. 2.5 illustrates the non-uniform thermal movement of electrons within a semiconductor material. When no electric field is applied, these charge carriers do not exhibit a collective macroscopic flow in any specific direction, resulting in the absence of a current. Consequently, they cannot be gathered by the detector electrodes to provide an output signal. When an electric field is present, whether it is an applied electric field or a built-in electric field, carriers experience a force in the direction of the electric field, causing them to move. However, at the same time, carriers are subjected to scattering, resulting in constant changes in the direction of their movement. This phenomenon is illustrated in Fig. 2.6. In this manner, the drift velocity acquired as a result of the electric field's influence is consistently dispersed in various directions, preventing it from accumulating indefinitely. The accelerated movement of the charge carriers, caused by the electric field force, is only present during the intervals between scattering events. Subsequently, the carriers lose the additional velocity they had acquired. Consequently, the collective impact of the electric field force and scattering phenomena induces the carriers to undergo a drift motion in alignment with the force, exhibiting a specific average velocity. Hence, the drift velocity of the carriers is contingent upon the magnitude of the diverse scattering phenomena experienced by the carriers within the semiconductor. The impact of scattering on the drift velocity of the carriers can be quantified by means of the carrier mobility, denoted as μ, which represents the average drift velocity of the electrons per unit field strength. Consequently, the drift velocity of the carriers can be derived as

$$v\text{d} = \mu(E)E \tag{2.6}$$

Scattering phenomena commonly observed in semiconductors including ionizing impurity scattering, lattice vibration-induced scattering, and scattering arising from

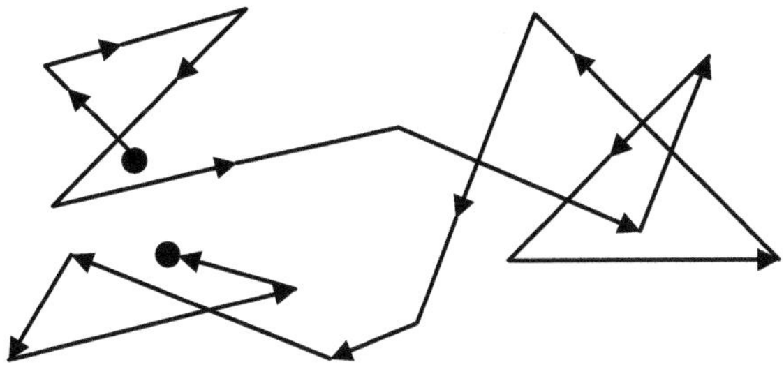

**Fig. 2.5** Schematic diagram of the thermal motion of carriers figure

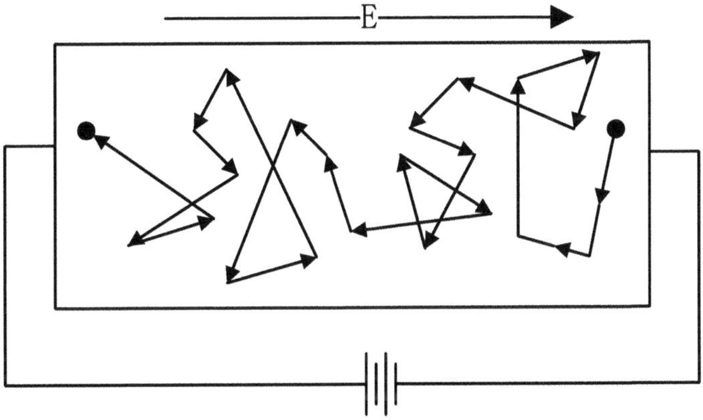


**Fig. 2.6** Drifting motion of electrons under the action of electric field

other sources. The individual action of carrier mobility in the semiconductor can give rise to a diverse range of scattering processes. The overall mobility of carriers in the semiconductor, denoted as μ, can be represented by the equation

$$\frac{1}{\mu} = \sum \frac{1}{\mu_i} \tag{2.7}$$

The symbol $\mu_i$ represents the mobility under various scattering modes. Furthermore, when the electric field is incrementally intensified, the carriers experience an increase in energy acquisition, resulting in an elevation of the average energy of the carriers beyond the lattice energy. Consequently, as the energy gained becomes greater, the mean free time of the carriers diminishes, leading to a reduction in mobility. When the electric field reaches a level of enhancement comparable to the optical phonon energy, it initiates the process of energy dissipation by the emission of optical phonons. Therefore, a significant portion of the energy acquired by the carriers is subsequently lost, resulting in the attainment of an average velocity saturation point. This implies that the carrier drift velocity does not continue to increase indefinitely under high electric fields, since it is limited by a saturation velocity.

(b) Carrier compounding

The process of irradiation in semiconductors results in the generation of carriers known as non-equilibrium carriers. The non-equilibrium carrier composite process can be categorized into three main types: direct recombination, indirect recombination, and auger recombination. Direct recombination refers to the process in which electrons and holes form recombine directly through the direct transition of electrons between the conduction and valence bands. Indirect recombination refers to the process of recombination facilitated by recombination centers, which are localized impurities or faults within the semiconductor material. A recombination center is typically defined as a defect energy level situated within the middle of the forbidden band, exhibiting nearly equivalent electron capture cross-section and hole capture cross-section. Electrons and holes are commonly combined through the utilization of recombination centers, with the SHR model frequently being used to explain this recombination process including recombination centers, as depicted in Fig. 2.7. The SRH model starts with the assumption that there is a deeply dominant energy level $E_t$ in the forbidden band and that there are two charging states (0/−), neutral or charged with one unit negative. Similarly, the deep donor energy level charge state is (+/0). In the state of thermal equilibrium, the carrier transitions occurring at the defect energy level can be described by the following four fundamental processes: (a) conduction band electrons are captured by the defect energy level; (b) electrons are thermally excited from the recombination center back to the conduction band; (c) electrons fall from the recombination center into the valence band to recombine with holes; (d) the empty recombination center is filled with valence band electrons. Non-equilibrium carriers, which are produced as a result of irradiation and survive after the recombination process, do not instantaneously vanish. Instead, they exhibit a finite survival duration inside the conduction and valence bands. The average duration for which non-equilibrium carriers survive is referred to as the non-equilibrium carrier lifetime, denoted as $\tau$.

(c) Carrier capture

The existence of capture centers exerts a direct influence on the process of carrier transport. When non-equilibrium carriers move inside the crystal, the carriers are captured by the capture center. Trap centers are a class of deeper defect energy levels found in semiconductors, characterized by their greater capture cross sections and longer de-capture periods. In the context of impurity and defect energy levels within the forbidden band, when the likelihood of thermal excitation of a trapped carrier to a free carrier at a specific energy level surpasses the probability of the trapped carrier recombining with the opposite carrier before excitation, the energy level predominantly functions as a trap. Carrier capture is a phenomenon whereby carriers are trapped by deep energy levels, resulting in a reduction in the pulse height observed at the detector output. If carrier de-capture is slow, the captured charge generates a built-in electric field, resulting in a polarization effect. In scenarios where the de-capture time (td) is shorter than the carrier crossing time (tr), the primary consequence of the capture process is observed as a reduction in the effective rate of carrier fracture.

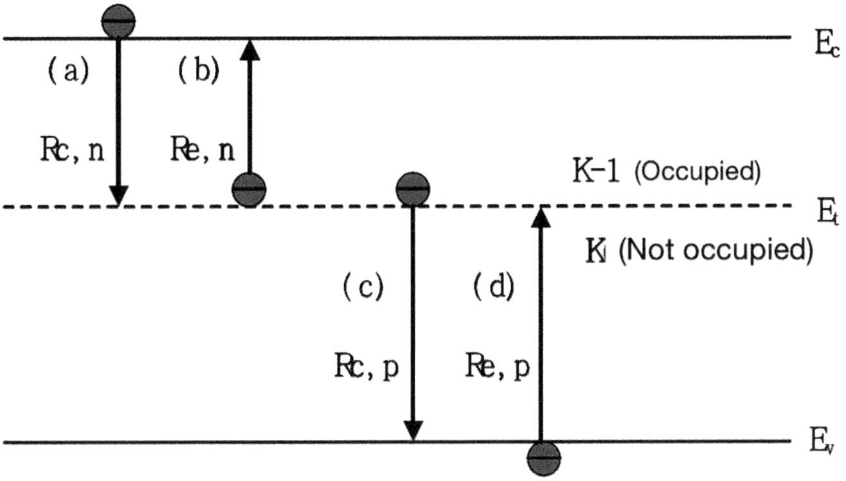


**Fig. 2.7** SRH recombine model

(d) Space charge effect

Due to the comparatively long de-capture durations of deep energy levels in semiconductors, the capture of carriers by these levels often leads to the generation of space charges within the crystal. The presence of space charge has a significant impact on the internal electric fields distribution. An example of this phenomenon can be observed in the presence of a space charge effect near the electrodes of a conductivity detector with homogeneous conductivity. The additional electric field generated by this effect can be attributed to its ability to counteract a portion of the external electric field. In extreme situations, this counteracting effect can hinder the electrodes' ability to efficiently collect the carriers produced by irradiation, ultimately rendering the detector non-functional.

(e) Charge collection

In the context of a homogeneous conductivity detector with a uniform electric field, the electrons and holes excited by radiation in the semiconductor material leads to their respective drift towards the positive and negative electrodes. This drift is facilitated by the presence of the electric field. Consequently, charge signals are generated as a result of the sensing process occurring at the two electrodes of the detector. The signal development process observed in semiconductor detectors bears similarity to the signal generating process exhibited by the output of gas ionization chambers. Assuming a distance L between the two electrodes of the detector, and considering the relative movement of each electron–hole pair by a distance $x(x < L)$, the resulting cumulative charge induced at the two electrode plates of the detector is:

$$Q = \frac{ex}{L} \tag{2.8}$$

The equation's derivative with respect to time yields the induced transient current that is created on the pole plate. In the event that the electron–hole pair undergoes drift until it reaches the two electrodes, specifically at position $x = L$, the total charge accumulated at the two plates of the detector is equivalent to the elementary charge denoted as $e$. When electrons and holes become trapped or captured by a trap while moving $x'$ relative to each other, and remain confined for an extended duration, their ability to contribute to the generation of the electrical signal of the detector is diminished. The concept of charge collection distance precisely characterizes the mean value of $x'$. The impact of carrier mobility, carrier recombination, and carrier capture, as previously discussed, on the charge collecting distance and, consequently, on the charge collection by the detector is obvious. The charge collection distance $L_d$ can be expressed as:

$$L_{\rm d} = (\mu_{\rm e}\tau_{\rm e} + \mu_{\rm h}\tau_{\rm h})E = (v_{\rm e} + v_{\rm h})E \tag{2.9}$$

In the broader context, Shockley [5] and Ramo [6] conducted a comprehensive investigation in 1938 on the alterations in the potential energy of a system resulting from the movement of charged particles within a spatial domain. As a result of their analysis, they put up the principle known as the Shockley-Ramo principle: In the context of a parallel plate capacitor, when one electrode is maintained at a unit potential, specifically when it is externally biased with a voltage of $U = 1$ V, and the other electrode is grounded, the resulting transient current that arises from the movement of space charge on the electrode at unit potential is [7]:

$$i = E_{\rm v}qv \tag{2.10}$$

In this equation, $i$ is the transient current collected at the electrode due to the charge motion, $q$ is the electric charge quantity, $v$ is the instantaneous velocity of a charge in motion, and $Ev$ is the quantity of the charge's instantaneous electric field in the direction of the charge's velocity of motion. For the junction detector structure shown in Fig. 2.8, where $W1 - W2$ are the depletion regions, the cumulative charge induced by the detector bipolar plates in d$t$ time $Q$ can be obtained as:

$$\mathrm{d}Q = i\mathrm{d}t = q\frac{\mathrm{d}x}{W_2 - W_1} \tag{2.11}$$

where d$x$ is the distance moved by the carriers in the depletion region in d$t$ time, and the instantaneous induced current at time $t$ is:

$$i(t) = \frac{q}{W_2 - W_1}v[x(t)] \tag{2.12}$$

where the drift velocity $v[x(t)] = \mathrm{d}x(t)/\mathrm{d}t$. For carriers generated in the body region outside the depletion region, an induced current is generated only when they enter the depletion region by diffusion.

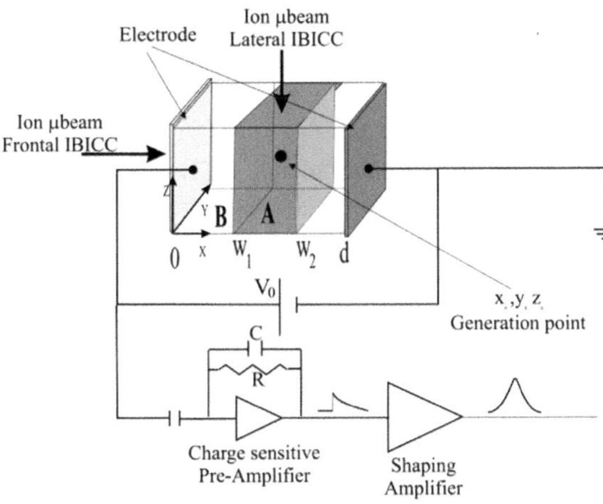


**Fig. 2.8** Junction detector charge collection model [8]

4. Key parameters of the particle irradiation detector

The parameters utilized for evaluating and calculating the efficacy of particle irradiation detectors with distinct configurations and functionalities exhibit variability. The parameters utilized for evaluating and calculating the efficacy of particle irradiation detectors with distinct configurations and functionalities exhibit variability. Some of the more common metrics are detection efficiency, CCE, leakage current, energy resolution, response time, and so on [3].

(a) Detection efficiency

Typically, two distinct categories of detection efficiencies can be determined, namely the absolute detection efficiency ($\varepsilon_{\text{abs}}$) and the intrinsic detection efficiency ($\varepsilon_{\text{int}}$). The term "$\varepsilon_{\text{abs}}$" denotes the ratio between the effective count value and the total number of particles released by the irradiation source:

$$\varepsilon_{\text{abs}} = \frac{\text{quantity of effective impulse}}{\text{total number of particles emitted by the radiation source}} \tag{2.13}$$

The term "$\varepsilon int$" denotes the ratio of the effective count value in relation to the overall number of particles incident on the detector:

$$\varepsilon_{\text{int}} = \frac{\text{quantity of effective impulse}}{\text{total number of particles incident on the detector}} \tag{2.14}$$

The detection effectiveness of a system is primarily influenced by many key elements. These factors include the characteristics of the detector material and its structural features, the type and energy of the particles being irradiated, and the positional relationship between the detector and the source of radiation. Moreover, it is worth noting that the quantity of effective impulse typically tends to be lower than the overall pulse count due to the constant presence of noise. This holds true for any detection system, particularly a spectral detection system, which necessitates the establishment of a threshold to filter out noise-induced pulses.

(b) CCE

The charge collection efficiency CCE is defined as the proportion of charge output from the detector relative to the overall charge produced by the particles that are stimulated within the detector:

$$\mathrm{CCE} = \frac{\text{the output charge of detector}}{\text{total amount of charge excited in the detector}} \tag{2.15}$$

The main factors influencing charge collection efficiency (CCE) include the carrier mobility of the detector material, the carrier lifetime, the defect density, and the operating state (bias) of the detector. In practical applications, the ratio between two voltages is commonly employed as a means to describe Charge Collection Efficiency (CCE), mostly due to the inherent challenges associated with directly measuring the output charge of a detector. These two voltage values are the pulse voltage amplitude $V_{\text{pulse}}$ of the pulse current output from the detector through the subsequent amplification and shaping circuit and the pulse voltage amplitude $V_{\text{ideal}}$ of the current pulse containing a known charge through the same amplification and shaping circuit under some ideal conditions.

$$\mathrm{CCE} = \frac{V_{\text{pulse}}}{V_{\text{ideal}}} \tag{2.16}$$

(c) Leakage current

For semiconductor devices, the existence of leakage currents is very common. it includes not only the reverse saturation current of junction devices, but also the leakage resulting from various non-ideal circumstances, such as tunneling leakage and surface leakage. An increase in leakage current not only affects the energy resolution of the detector, but also has the potential to affect the detector's operating state and operating safety. In the case of detectors that are exposed to prolonged periods of intense irradiation, the extent of damage inflicted upon the detector can often be assessed by monitoring alterations in the leakage current.

(d) Energy resolution

The term "energy resolution" is frequently used to describe the capacity to differentiate between spectrum lines that possess comparable energy levels. The definition of the ratio between the Full Width at Half Maximum (FWHM) and the peak value ($X_0$) is illustrated in Fig. 2.9.

$$\text{energy resolution} = \frac{\mathrm{FWHM}}{\mathrm{X_0}} \tag{2.17}$$

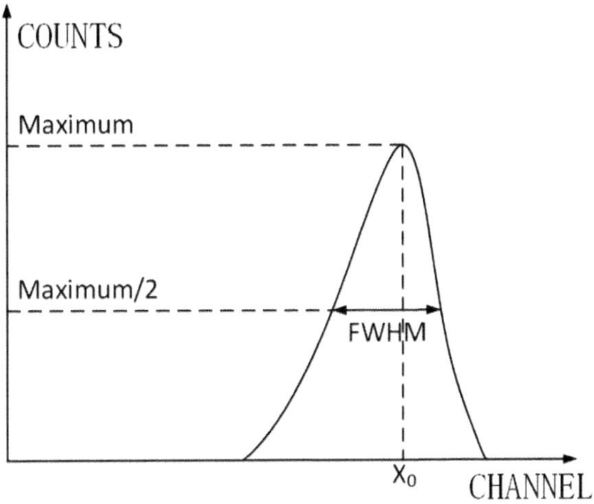


**Fig. 2.9** Schematic of the definition of energy resolution

The full width at half maximum (FWHM) refers to the spectral range between two channels where the count value is equal to half of the maximum count value. The expression of energy resolution is commonly presented as a percentage. A smaller resolution value indicates a higher capacity to differentiate between spectrum lines that possess similar energy levels. Semiconductor detector energy resolution is affected by three main aspects:

(1) The phenomenon of statistical fluctuation occurring when a detected particle undergoes energy loss, resulting in the creation of an electron–hole pair;
(2) Noise generated by phenomena such as leakage in electrical systems and detectors;
(3) Other factors, e.g., changes in particle deflection and energy loss due to the incident barrier layer.

(e) Response time

The response time of the detector equipment indicates the efficacy of its capacity to track the incoming signal. The response time of a detector can be partitioned into two distinct components, namely the rise time and the fall time. The rise time is commonly described as the duration required for the amplitude of an output pulse to increase from 10 to 90% of its maximum value. In contrast, the fall time is commonly described as the duration required for the amplitude of the output pulse to decrease from 90 to 10% of its maximum value.

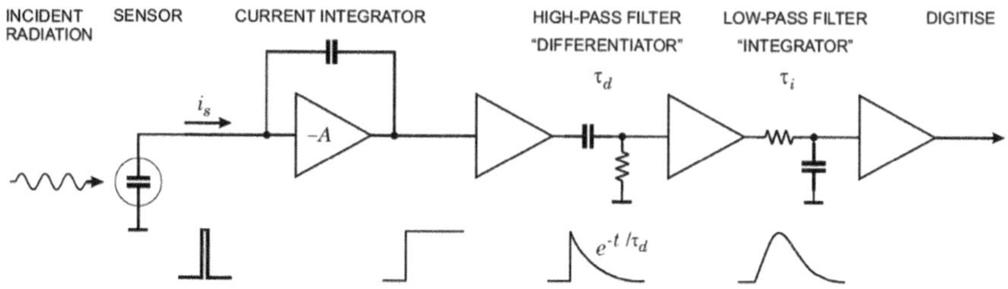


**Fig. 2.10** Schematic flow diagram of pulse forming

### *2.3.2 Basic Irradiation and Irradiation Detection Systems*

1. Energy spectrum measurement system

The measurement of energy spectrum lines is a widely employed technique for investigating the characteristics of radiation. By conducting an analysis of the energy distribution of incident particles and integrating it with theoretical derivations, the irradiation source can be effectively responded to. Given that the signals sent by the detector are often in the form of pulses and possess low intensity, it is necessary to amplify and shape them in order to facilitate their future preservation and analysis.

A typical device utilized for the measurement of irradiation energy spectra, comprising a detection device and a signal processing system. The detection devices is capable of absorbing irradiation and subsequently converting it into an electrical signal. The signal processing system performs the integration, shaping, and digitization of electrical signals in order to quantify the quantity of particles, verify the energy of the rays, and classify the particle species. The schematic diagram of this process is shown in Fig. 2.10.

Typically, the total charge output of the detector exhibits a direct relationship with the energy of the incident particles. Consequently, when these particles affect upon the detector, it becomes stimulated and generates a pulsed current. This current includes a quantity of charge that is directly proportional to the overall charge generated by the detector. The capacitor of the charge-sensitive amplifier, depicted as the current integrator in Fig. 2.10, receives the pulsed current and performs the integration process to generate a step voltage signal representing the integrated pulsed current signal from the detector. The step voltage's magnitude exhibits a direct relationship with the overall charge quantity present in the pulse current. Typically, to avoid capacitor damage, an integrating capacitor is connected in parallel with a high-resistance resistor to enable slow discharge. During this process, the voltage signal at the capacitor's terminals gradually diminishes over time. To reduce the accumulation of these signals, the aforementioned signals are subjected to a high-frequency filter. This filter accelerates the reduction of voltage signals while maintaining the signal amplitude at a relatively constant level, as depicted in Fig. 2.11. In order to enable the process of signal digitization, it is necessary for the output voltage pulse to have a generally smooth behavior in the region of its maximum amplitude. One

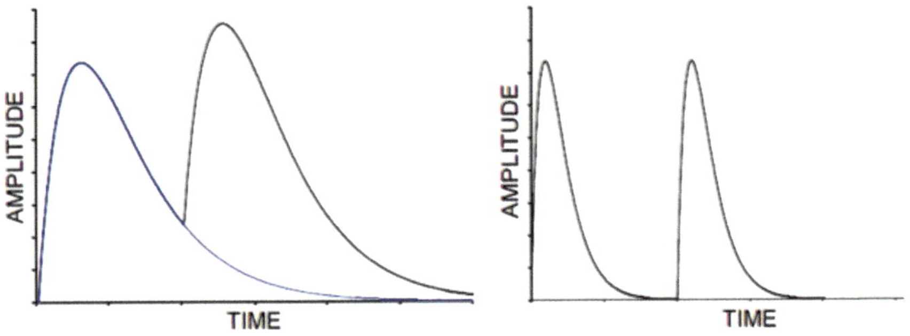


**Fig. 2.11** Signal buildup phenomenon and signal decay time optimization

common method involves inputting the signal into a low-pass filter, which results in an output signal that exhibits characteristics similar to a Gaussian distribution. This approach also maintains the signal amplification at a somewhat consistent level. The magnitude of the resulting voltage signal is fundamentally proportional to the energy of the incident particle.

In the final analysis, the aforementioned signals undergo digital capture and filtration within a multi-channel analysis system. This process aims to identify and quantify the voltage intervals within which the magnitude of the current pulse amplitude resides.

2. Transient pulse measurement system

The time width of pulsed radiation is often on the order of ps–μs, but the intensity often far exceeds that of steady state or quasi-steady state sources. For instance, several sources such as spallation neutron source, pulsed reactors, and pulsed radiation sources from accelerator have the capability to promptly generate irradiated particles at an intensity of up to $10^{21}$ $s^{-1}$. The nuclear explosion pulsed radiation source field, whose neutron and $\gamma$ radiation intensity is even higher than $10^{32}$ $s^{-1}$.

Given the presence of such violent transient processes, contemporary electronic systems lack the capability to discern and classify individual particles. Consequently, the only feasible approach for investigating these systems is to analyze the collective behavior of particle clusters. In contrast to energy spectroscopy, which focuses on providing information about the energy of individual particles while disregarding their temporal characteristics, transient pulse measurements capture the time course of irradiation in a comprehensive manner, though at the expense of averaging the energy information of the particles. For the special characteristics of pulsed radiation, pulsed measurement systems are needed:

(1) During the detection, transmission, and signal recording, the goal is to reduce or potentially eliminate temporal distortion of the signal, thereby accurately reproducing the timing characteristics of the pulse process;
(2) Pulsed radiation detectors with high irradiation resistance;
(3) The current output of a pulsed radiation detector exhibits a linear relationship with the intensity of radiation across a wide range of radiation field strengths;

(4) The pulsed radiation detector itself or its components have some particle-resolving capability.

## References

1. Ding H (2010) Nuclear radiation detectors. Harbin Engineering University Press
2. Groom DE, Klein SR (2014) Passage of particles through matter. Chin Phys C 38(9):398–412
3. Huang H (2019) Research on the performance and performance degradation of SiC particle irradiation detectors. Xidian University
4. Jianguo L, Shi L, Yiyi L et al (2005) Research progress on diamond film radiation detectors. Mater Rev 19(003):7–9
5. Shockley W (1938) Currents to conductors induced by a moving point charge. J Appl Phys 9(1):635–636
6. Ramo S (1939) Currents induced by electron motion. Proc IRE 27(9):584–585
7. He Z (2001) Review of the Shockley-Ramo theorem and its application in semiconductor gamma-ray detectors. Nucl Instrum Methods Phys Res 463(1–2):250–267
8. Vittone E, Fizzotti F, Giudice AL et al (2000) Theory of ion beam induced charge collection in detectors based on the extended Shockley-Ramo theorem. Nucl Inst Meth Phys Res B 161–163(1):446–451

# Chapter 3
# Diamond Nuclear Radiation Detectors

The charge collection efficiency, dark current, energy resolution, I–V characteristics, and time response characteristics are critical performance parameters for diamond nuclear detectors. Numerous research institutes are actively engaged in enhancing these parameters. However, it is important to note that there exists a significant disparity between the current achieved values and the theoretical benchmarks. The size and shape of polycrystalline diamond are bigger compared to monocrystalline diamond, hence facilitating a wider range in the detection area of the device. However, both domestic and international research endeavors have shifted towards investigating monocrystalline diamond nucleus detectors due to their poor performance. One of the primary concerns in the examination of diamond nucleus detectors pertains to the substantial variability in their performance, which remains poorly understood in terms of its underlying mechanism. The limited proportion of high-performance diamond nuclear detectors is particularly apparent in key metrics such as charge collecting efficiency, energy resolution, current–voltage, and time-response characteristics. This constraint significantly hinders the advancement of diamond nuclear detector technology. Furthermore, it is evident that there remains a significant gap in the comprehensive exploration and comprehension of the fundamental aspects pertaining to materials, devices, performance, and mechanisms of diamond nuclear detectors. Furthermore, it is clear that there remains a significant gap in the comprehensive exploration and comprehension of the fundamental aspects pertaining to materials, devices, performance, and mechanisms of diamond nuclear detectors. Thus, this chapter aims to present a comprehensive overview of diamond nuclear detectors, encompassing the development of diamond materials, characterization of material properties, device preparation procedure, performance evaluation, and mechanism analysis.



Y. Zhang et al., *Wide Band-Gap Semiconductor Nuclear Radiation Detector*,
Wide Bandgap Semiconductors, https://doi.org/10.1007/978-981-96-0340-4_3

## 3.1 Method of Growing Nuclear Radiation Detector Grade Diamond Material

### 3.1.1 Carrier Transport in Nuclear Radiation Detectors

After exposing diamond to nuclear radiation, the diamond undergoes the generation of non-equilibrium carriers. The subsequent transit of these non-equilibrium carriers is closely related to their mobility and lifetime. In general, semiconductor materials do not exhibit flawless crystalline structures. These materials are typically characterized by the presence of impurities, line defects, and other crystal defects. Furthermore, the surface characteristics of the material have significant effects on the preparation process and performance of diamond core detectors.

The phenomenon of nuclear radiation interacting with a nuclear detector is responsible for determining both the amount of energy absorbed by the detector and the quantity of non-equilibrium electron–hole pairs generated. The application of an electric field to the detector causes the movement of non-equilibrium carriers towards the electrodes, driven by the applied electric field, resulting in an electrical signal. The successful capture of the signal necessitates a nuclear detector with low leakage, commonly referred to as dark current. Otherwise, the weak electrical signals produced by the rays will be overshadowed by the leakage, necessitating a detector with strong resistance properties. In the context of silicon-based devices, the utilization of reverse bias in the p–n junction results in the formation of a space charge area. This region is an efficient solid ionization chamber. At this point, the p–n junction operating under reverse bias is a highly insulated region. When examining the properties of diamond, it is commonly acknowledged that intrinsic diamond possesses characteristics of a high-resistance insulating crystal, primarily attributed to its significant forbidden band gap. The Gudden model [1], as depicted in Fig. 3.1, provides an explanation for the charge acquired at the electrode when a ray induces non-equilibrium carriers in a high-resistance insulating crystal.

If the thickness of the detector is $d$, and if the ray produces a charge of $Q_{\mathrm{in}}$ inside the detector at a distance $x$ from the anode, and if electrons are able to be transported to the anode, the charge $Q_{\mathrm{c}}$ collected by the external circuit is:

$$Q_{\mathrm{c}} = Q_{\mathrm{in}} \frac{x}{d} \tag{3.1}$$

Actually, since the non-equilibrium carrier are scattered by various impurities and defects inside the crystal, the actual transport distance may be only $L$, when the charge collected by the external circuit is only:

$$Q_{\mathrm{c}} = Q_{\mathrm{in}} \frac{L}{d} = Q_{\mathrm{in}} \frac{\mu \tau E}{d} \tag{3.2}$$

Fig. 3.1 Transport of non-equilibrium carriers in nuclear detectors

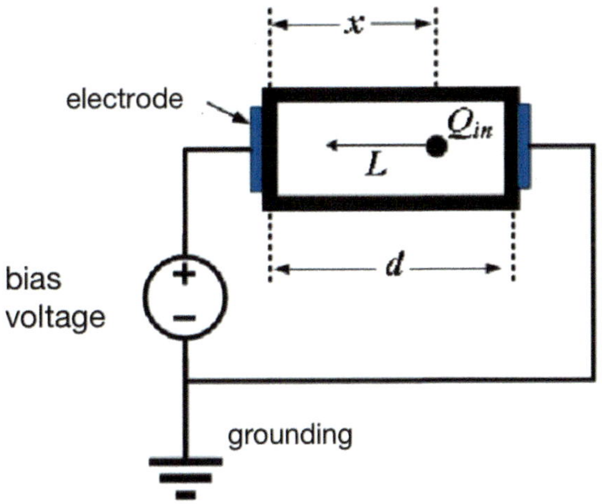


The symbol $\mu$ here is the carrier mobility, $\tau$ is the average lifetime of non-equilibrium carriers, and $E$ is the electric field strength applied to the detector. The influence of the electrode contact and the interface state of the semiconductor material are not taken into account in this analysis. However, the correlation between the carrier transport mechanism within the nuclear detector and the charge accumulated by the external circuit is given from the point of view of the body material characteristics.

### 3.1.2 *High Quality CVD Single Crystal Diamond Material Growth*

A diverse range of diamond materials exists, and the presence of imperfections in these materials can have a negative effect on carrier mobility and lifetime. Consequently, the performance of diamond-based nuclear detectors may be impaired.

Polycrystalline diamond exhibits a significant presence of grain boundaries, which serve as recombination and trap centers. These grain boundaries have a substantial impact on the lifetime product of carrier mobility. Additionally, the internal grain sizes of polycrystalline diamond exhibit variations, posing challenges in accurately measuring energy spectra through statistical distribution observations. Hence, it can be concluded that polycrystalline diamond is not a viable option for manufacturing nuclear detectors at the spectrum level. Nevertheless, the cost and size benefits associated with polycrystalline diamond can be used to perform measurements in the current manner. The primary challenge associated with HPHT diamond is to the uncontrollable impurities that are produced during the synthesis process. These impurities exhibit a complex types, and their introduction leads to the creation of deep energy levels, which significantly diminish the carrier mobility and lifetime. On the other hand, the size of HPHT diamond is limited, and it is difficult to synthesize large-size single-crystal diamonds exceeding 12 mm. Therefore, HPHT diamond is also unsuitable for making high-performance spectroscopic diamond core detectors.

At present, it can be claimed that CVD diamond has the highest level of suitability for the production of diamond detectors with exceptional performance, cost-effectiveness, and homogeneity. Nevertheless, there exists a diverse range of chemical vapor deposition (CVD) diamonds, and the efficacy of nuclear detectors fabricated from even the most superior Type IIa diamonds exhibits considerable variability. Previous studies have demonstrated that the presence of crystal dislocations and trace impurities in single-crystal diamond has a substantial role in influencing the performance of devices. Penetrating dislocations in chemical vapor deposition (CVD) diamond typically arise along the growth direction, primarily originating from defects located at the interface between the high-pressure, high-temperature (HPHT) substrate and the growth process, as shown in Fig. 3.2. Therefore, one strategy to reduce dislocations is to utilize a substrate with a low defect density, and another way is to etch the surface of the HPHT substrate to eliminate defects [2] at the growth interface. The primary impurities present in chemical vapour deposition (CVD) monocrystalline diamond are $NV^0$, $NV^-$, and $SiV^-$ centers. Nitrogen impurities typically arise from leakage in the growth chamber, impurities parasitized in the growth gas, or contamination during the manufacturing process. On the other hand, silicon impurities originate from the quartz observation window of the growth chamber due to plasma etching, leading to a minor introduction of silicon impurities [3]. Consequently, this requires focused growth of nuclear detector grade CVD single crystal diamonds.

1. CVD diamond growth equipment

The industrial expansion of chemical vapor deposition (CVD) diamond production is primarily facilitated by MPCVD (microwave plasma-assisted CVD) equipment. Notably, the Japanese company SEKI and France's PLASSYS company are the main

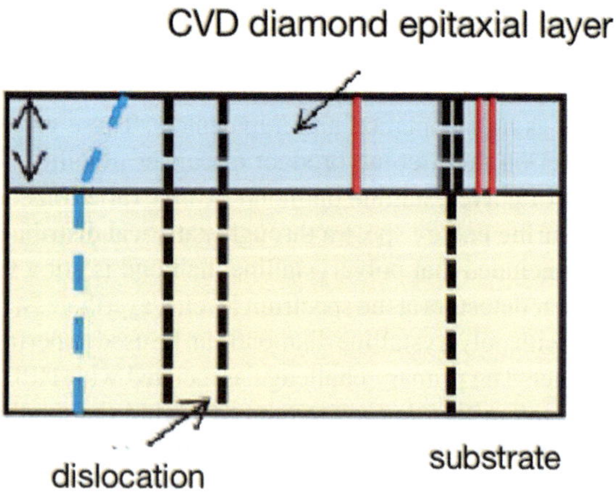


**Fig. 3.2** Main sources of dislocations in homogeneous epitaxy of CVD single crystal diamond

manufacturers in this field. Additionally, China has also made significant strides in developing MPCVD equipment, contributing to a favorable landscape for this technology. Figure 3.3 illustrates the MPCVD machine SSDR150, manufactured by PLASSYS in France. The aforementioned apparatus utilizes plasma with high power density to facilitate the growth of diamond epitaxial layers with exceptional purity and at accelerated rates. The SSDR150 is a dependable and sturdy unit that can be customized to suit the requirements of laboratory research and development as well as industrial expansion, owing to its well-optimized microwave and plasma design. The chamber of this apparatus is designed for easy cleaning and is equipped with a replaceable quartz viewing bell housing. The SSDR150 exhibits a peak power output of 6 kW, operates at a microwave frequency of 2.45 GHz, and utilizes a combination of five gases, namely hydrogen ($H_2$), methane ($CH_4$), nitrogen ($N_2$), oxygen ($O_2$), and argon (Ar). The sample stage incorporates a motorized lifting mechanism that enables automatic adjustment of the lifting mechanism's height based on temperature data obtained from the infrared thermometer during the growth process. This feature ensures the fulfillment of temperature control requirements. The sample stage is equipped with a rotating function to ensure uniformity in the growth process of all sides of the substrate.

2. Growth optimization

(a) Substrate selection and surface polishing

The utilization of a homogeneous epitaxial layer has been found to yield a higher quality of chemical vapor deposition (CVD) single crystal diamond. It is worth noting that the quality of the high-pressure, high-temperature (HPHT) substrate and

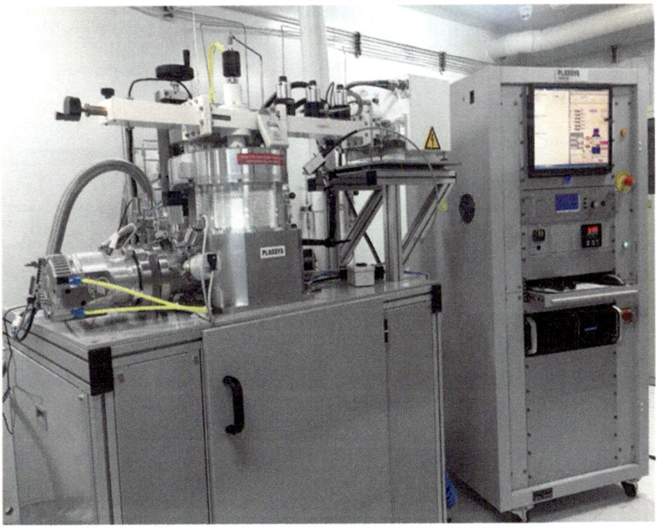

**Fig. 3.3** Diamond growth equipment MPCVD

the surface treatment employed during the homogeneous epitaxial growth process significantly impact the quality of the diamond within the CVD layer. Moreover, when the internal stress of the HPHT substrate is significant, the stress between the epitaxial layer and the substrate layer escalates as the thickness of the epitaxial layer increases. Consequently, this ultimately results in the fragmentation of the sample either during the growth process or subsequent to cutting. The utilization of the 2θ/ω mode of X-ray diffraction (XRD) has the potential to provide insights into the internal stresses present within the substrate. A high degree of symmetry in the measured 2θ peak suggests that the substrate has low internal stress and that the lattice constants in the diamond crystals are well-matched. This condition is favorable for achieving CVD homogeneous epitaxial growth. In this study, the HPHT diamond labeled as #1 was selected as the growth substrate for CVD diamond. Figure 3.4a displays the XRD 2θ outcomes of this particular sample, revealing an improved symmetry in the diffraction pattern of the substrate. Additionally, the Raman test results indicate that the calculated internal stress of the #1 sample is − 0.126 GPa, indicating a nearly stress-free state. This value is considerably smaller than the range of 0.37 to 2.21 GPa [4] reported in existing literature. In general, commercially acquired HPHT substrates often exhibit significant surface roughness and the presence of deep defect pits at the scale of hundreds of nanometers. This roughness and the formation of pits can be attributed to the polishing damage incurred during the initial rough polishing process, necessitating subsequent fine polishing to impede the propagation of dislocations induced by these surface defects. In Fig. 3.4b, the atomic force microscopy (AFM) image displays the surface topography of the sample following a meticulous polishing process. The root-mean-square roughness (Rq) of the substrate measures at 1.65 nm, indicating its suitability for achieving uniform chemical vapor deposition (CVD) epitaxial growth. However, the AFM results reveal the presence of localized dot-like bumps, which appear as white features. Consequently, additional surface pretreatment is necessary to mitigate any potential impact of lattice damage on the subsequent CVD epitaxial growth process.

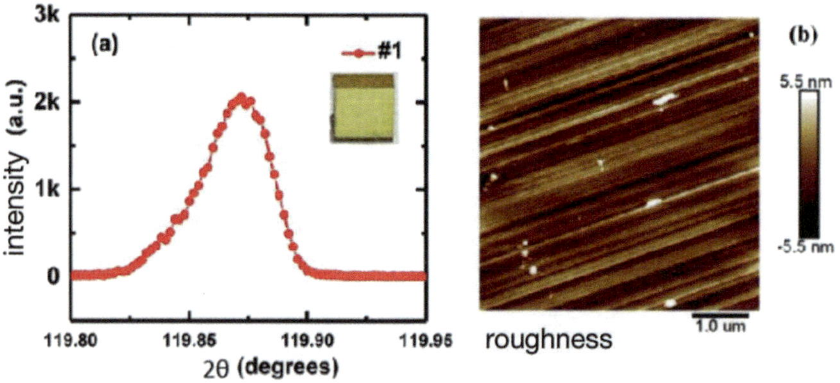


**Fig. 3.4** XRD results and surface morphology of sample #1. **a** 2θ results; **b** AFM results

(b) Substrate cleaning

Following the grinding and polishing of the HPHT substrate, the surface may retain abrasives and contamination originating from organic substances. These contaminants, if not addressed, can contribute to cavity contamination subsequent to the etching of MPCVD plasma. Consequently, the concentration of impurities in CVD single crystals may rise. Therefore, it is imperative to perform an organic cleaning of the HPHT substrate after the selection of a suitable substrate and prior to commencing growth. A 1:1 molar ratio of $HNO_3$ to $H_2SO_4$ was prepared. The resulting mixture was then subjected to a temperature of 250 °C and used to clean the substrate for a duration of 20 min. Following the initial cleaning process, the substrate was immersed in deionized water to eliminate any residual acidic solution on the surface. Subsequently, the substrate underwent a standard semiconductor organic cleaning procedure, involving sequential immersion in acetone, alcohol, and deionized water. Each immersion was accompanied by ultrasonic cleaning for a duration of 15 min.

(c) Pre-etching of substrate surface

As previously examined, it has been observed that damage to the substrate surface extends into the chemical vapor deposition (CVD) layer in the direction of growth. To mitigate the subsequent rise in dislocation density within the CVD layer resulting from surface damage, Achard et al. [5] proposed a technique involving the removal of polishing damage. This technique aims to suppress the extension of dislocations at the growth interface by employing surface etching through the utilization of $H_2/O_2$ plasma. In this study, we employ a method for the surface pretreatment of the substrate, as well as a method for the secondary selection of the substrate. The specific process is outlined as follows: During the etching process, the following parameters were set: the cavity pressure was adjusted to 295 mbar, the microwave power was set to 3.8 kW, the $H_2$ flux was maintained at 196 sccm, the $O_2$ flux was maintained at 4 sccm, and the etching time was set to 30 min. It is important to note that the etching process may induce surface polishing damage, vacancies, and dislocations on the substrate. During the etching procedure, the preferential removal of substrate surface imperfections such as polishing damage, vacancies, dislocations, and impurity inclusions, including substrate defects, occurs. Following the completion of etching, the sample surface is examined using an optical microscope, and the surface exhibiting a reduced number of pits is chosen for further growth (Fig. 3.5).

(d) Inhibition of material impurities

The aforementioned procedure was employed to enhance the surface quality of the substrate, ultimately leading to the selection of an HPHT substrate with reduced surface dislocation damage. Subsequently, the optimization of impurities within the chemical vapor deposition (CVD) epitaxial layer is pursued. Frequently, the introduction of a small amount of nitrogen as a dopant proves to be highly efficient in accelerating the growth process. Nevertheless, this approach is accompanied by the drawback of compromising the crystalline material's quality. The presence of nitrogen impurities in diamond crystals primarily arises from multiple sources,

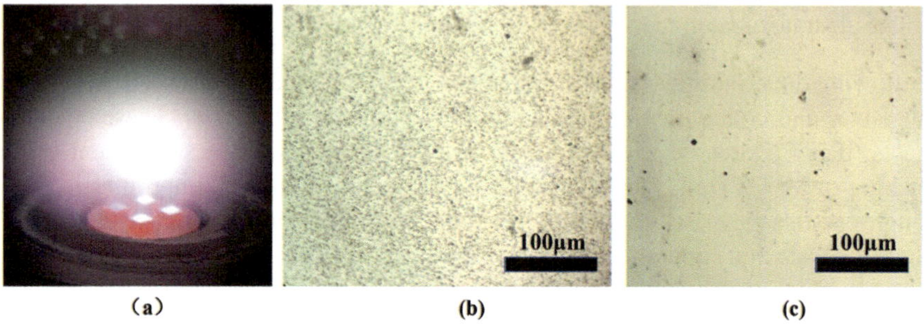


**Fig. 3.5** Etching process of substrate surface with surface optical microscopy photos of different substrates

including leakage from cavities, impurities in the reaction gas, contamination of substrates used in high-pressure, high-temperature (HPHT) processes, and coupling effects resulting from the crystal growth process. The presence of silicon impurities can be attributed to the microwave plasma etching process applied to the quartz viewing bellows. In contrast to their typically low atmospheric concentrations and the challenges associated with their detection in CVD diamonds without deliberate doping, boron and phosphorus do not serve as significant sources of impurities in Type IIa diamonds. Hence, the primary objective in the growth of high-purity diamonds is the mitigation of nitrogen and silicon impurities. The MPCVD chamber employed in this investigation achieves a vacuum level of $2 \times 10^{-7}$ mbar, which is already significantly high. Consequently, the mitigation of nitrogen impurities is primarily addressed by examining the reaction gas and the growth process. The utilization of a hydrogen purifier proves to be a highly effective method for enhancing the purity of chemical vapor deposition (CVD) diamonds, particularly in the context of the CVD growth procedure. This is due to the fact that hydrogen serves as the primary reaction gas during CVD, and impurities present in the hydrogen gas play a dominant role in affecting the purity of the resulting diamonds. The growth equipment was altered by incorporating a 9N (9 out of 9) hydrogen purifier with the purpose of preventing parasitic nitrogen impurities. However, employing a low rate of growth is also a viable approach for improving the purity of diamonds. This is attributed to the fact that a high rate of growth can result in a greater coupling rate of nitrogen impurities [6]. The occurrence of quartz bellows etching by microwave plasma is an inevitable consequence of the presence of silicon impurities. The literature suggests that the presence of silicon impurities in diamond can be mitigated by employing a reaction chamber made of stainless steel. However, it is important to note that in order to prevent the plasma from coming into contact with the metal reaction chamber, a larger reaction chamber is necessary. Additionally, it is worth mentioning that the financial implications associated with replacing such a reaction chamber are substantial. Due to the aforementioned factors, a higher reaction pressure is employed in order to mitigate the expansion of the plasma ball and minimize the etching of the quartz bellows.

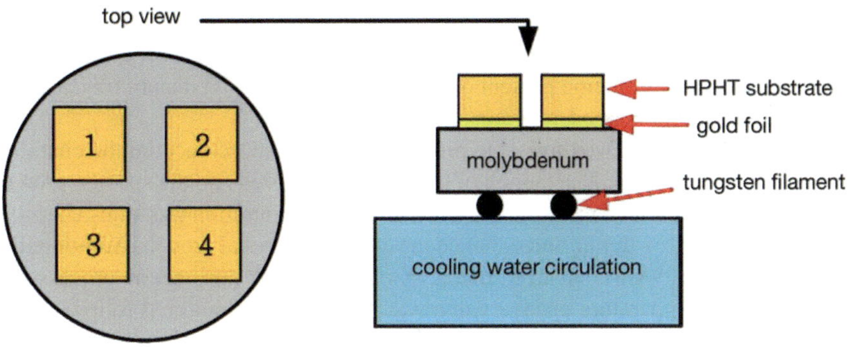


**Fig. 3.6** Schematic diagram of substrate placement before growth

3. Growth process

(a) Substrate placement

Once the substrate was chosen through the substrate optimization process described earlier, it was introduced into the MPCVD reaction chamber, as depicted in Fig. 3.6. The gold foil has a thickness of 25 μm. Subsequently, the MPCVD heating process is employed to liquefy the gold foil and join the molybdenum tray together. This is done to establish a consistent heat transfer between the HPHT substrate and the molybdenum tray, as well as to maintain temperature equilibrium within the substrate. To optimize temperature control during the growth process, a transition layer is established between the cooling water circulation table and the molybdenum tray, achieved by employing a tungsten filament with a diameter of 150 μm. The molybdenum rest has a thickness of 3 mm, while the HPHT substrate measures $3 \times 3 \times 1.5$ mm$^3$ in size. Additionally, the crystal grain orientation is aligned along the (100) direction.

(b) Growth process and parameters

Step 1: the substrate was positioned, followed by the reduction of the chamber vacuum to a level below $1 \times 10^{-6}$ mbar.

Step 2: Begin by activating the hydrogen valve and adjusting the hydrogen flow rate to 200 sccm. Concurrently, establish the chamber pressure at 20 mbar. Once the chamber air pressure has reached a state of equilibrium, initiate the microwave source and configure the power output to 1500 W. Currently, the hydrogen within the reaction chamber undergoes excitation, transitioning into a plasma state, thereby initiating the emission of luminous radiation.

Step 3: The hydrogen flow rate should be maintained at a constant level. The cavity pressure should be set to 200 mbar, while the microwave source power should be adjusted to 3500 W. Simultaneously, the oxygen should be introduced into the system at a flow rate of 4 sccm. The substrate should be subjected to an etching process for a duration of 15 min. Following this, the microwave source should be deactivated, and the valves controlling the flow of hydrogen and oxygen should be closed. Finally, the flow rate of both gases should be set to zero. Following this, the process of evacuation starts, and it is subsequently repeated three times in order to eliminate any contamination present on the substrate, the molybdenum tray, and the cavity resulting from exposure to air.

Step 4: Once the above three-step process is completed, re-vacuum the chamber to below $1 \times 10^{-6}$ mbar. The launch of hydrogen gas started and the specific growth parameters are presented in Table 3.1. The methane concentration was adjusted to 4% to minimize the growth rate and enhance the quality of crystal growth. Additionally, the hydrogen purity was enhanced to a level of 9N through purification processes.

Step 5: the temperature change value was consistently monitored following the initiation of growth. The height of the internal lifting mechanism was adjusted in response to the temperature readings, ensuring that the temperature remained within the range of 910 to 930 °C. In Fig. 3.7, the CVD diamond growth process is depicted. Initially, the individual HPHT substrates were appropriately spaced. Following 100 h of growth, the epitaxial layer experienced a substantial increase in thickness. However, simultaneous lateral expansion also took place. After 164 h of growth, the growth process concluded, leading to the termination of growth. Subsequently, the substrates were separated from the molybdenum rests, resulting in the acquisition of four pieces of CVD diamond attached to the substrates.

Step 6: Following the completion of the growth process, the measurements of the total thicknesses were conducted to ascertain the thickness of the CVD diamond that had been grown, along with the growth rate. These measurements are presented in Table 3.2. Ultimately, the thickness of the CVD layer that was cultivated ranged from 0.44 to 0.50 mm, while the growth rate fell within the range of 2.68 to 3.05 μm/h. The observed variations can be attributed to minor discrepancies in both the temperature and plasma fields.

**Table 3.1** MPCVD growth reaction parameters

| Temperature | Pressure | Microwave power | $H_2$ flow | $CH_4$ flow | Reflected power | $H_2$ purity |
|---|---|---|---|---|---|---|
| 915 °C | 295 mbar | 3800 W | 192 sccm | 8 sccm | 6 W | Greater than 9N |

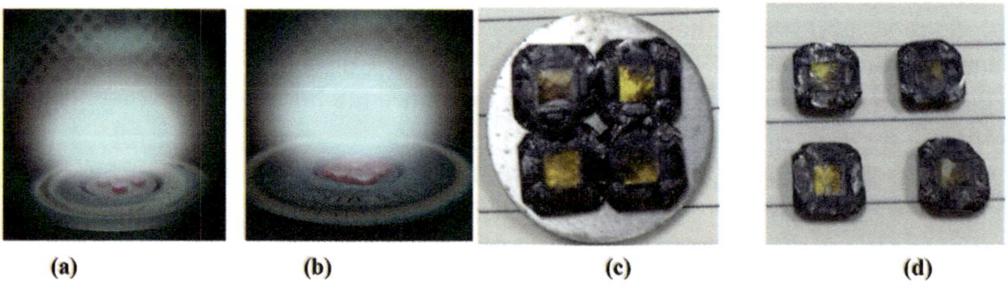


**Fig. 3.7** CVD diamond growth process. **a** start of growth; **b** after 100 h of growth; **c** end of growth for a total of 164 h; **d** after separation from the molybdenum rest

**Table 3.2** Sample thickness and growth rate after growth completion

| | Slice #1 | Slice #2 | No. 3 piece | Slice #4 |
|---|---|---|---|---|
| Total thickness before growth (mm) | 4.59 | 4.59 | 4.58 | 4.61 |
| Total thickness after growth (mm) | 5.03 | 5.08 | 5.02 | 5.11 |
| CVD layer thickness (mm) | 0.44 | 0.49 | 0.44 | 0.50 |
| Growth rate (μm/h) | 2.68 | 2.98 | 2.68 | 3.05 |

Step 7: After the growth is complete, a high-energy laser with a wavelength of 532 nm is utilized to cut the CVD layer away from the substrate and then polished, ultimately yielding a CVD single crystal diamond that can be used to fabricate devices.

4. Characterization of growth materials

To assess the quality of the CVD diamond upon completion of growth, non-destructive characterization techniques were employed to analyze its material properties.

In the context of nuclear detectors, the utilization of diamond typically necessitates the use of thick film materials, specifically epitaxial layers with thicknesses exceeding 100 μm. This requirement arises from the need to maximize the absorption of ray energy, which can only be achieved by employing detector materials of sufficient thickness. Incomplete energy deposition in α-particle detection results in heightened randomness in particle detection and diminished energy spectral properties. Conversely, in photon detection, the utilization of a suitably thick material facilitates greater energy absorption, thereby enhancing the strength of the signal.

In order to meet the demands of nuclear detectors regarding material thickness, the objective of CVD diamond growth is to achieve a substantial film of CVD diamond epitaxial layer. However, the growth of thick films of CVD diamond results in significant surface undulations, as depicted in Fig. 3.8. These undulations are readily observable to the naked eye or through the use of an optical microscope. The surface exhibits prominent step flow and undulation heights reaching the μm level. If not addressed, these surface irregularities would pose challenges to the desired functionality of the detectors. The preparation of electronic devices is contingent upon the surface being finely polished. Hence, it is imperative to elucidate the surface roughness of diamond subsequent to meticulous polishing.

Atomic force microscopy (AFM) is a valuable technique for investigating the surface characteristics of solid materials, including those that possess insulating properties. The investigation of solid matter's surface structure and properties involves the detection of the subtle interatomic forces between the sample under examination and a small force-sensitive component. This enables the acquisition of nanoscale surface topography and the measurement of surface roughness. In this study, the DIMENSION ICON AFM device from Bruker was employed as the instrument to assess the surface roughness of diamond. During the experimentation, the Atomic Force Microscope (AFM) was utilized in tap mode to examine the samples. Following the

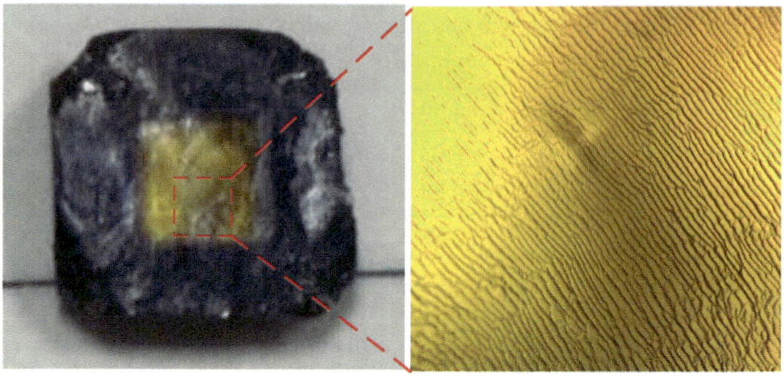

**Fig. 3.8** Pre-polishing morphology of grown CVD diamond

growth of the CVD diamond, it is customary to perform surface polishing on one or both sides of the processed material in order to satisfy the demands of device manufacturing. Typically, following the process of polishing, a degree of undulation is observed, albeit to a lesser extent than the surface of the diamond in its initial state of growth. The surface morphology of the diamond after fine polishing is depicted in Fig. 3.9. It is observed that the CVD epitaxial diamond exhibits a distinct oblique linear polishing texture, which is evenly distributed across the surface. Additionally, the Rq value of the CVD epitaxial layer measures 1.15 nm.

The FTIR results of CVD single crystal diamond, as depicted in Fig. 3.10a, indicate that the diamond grown under optimal conditions exhibits the characteristic features of a Type IIa diamond. Notably, no additional impurity peaks were observed. The results of the Raman spectrum are depicted in Fig. 3.10b, wherein the Raman peak is observed at a position of 1332.2 $cm^{-1}$ and exhibits a half-height width of 3.67 $cm^{-1}$. Upon comparing this observation with the other samples presented in Table 3.3, it becomes evident that the Raman peak exhibits a significantly

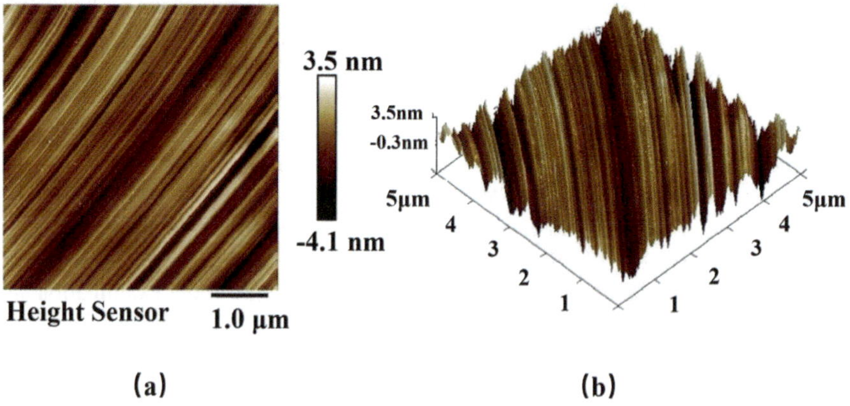


**Fig. 3.9** Diamond surface roughness after fine polishing. **a** 2D morphology; **b** 3D morphology

smaller deviation from the theoretical value (1332.3 $cm^{-1}$). The findings presented in Fig. 3.10c illustrate the XRD rocking curves of the cultivated diamond. The half-height width of the rocking curve is measured to be 46.3 arcsec, which is comparatively smaller than the half-height width of the HPHT sample (51.84 arcsec) utilized as the growth substrate. This discrepancy suggests that the dislocation density of the epitaxial CVD layer can be mitigated to some extent through the surface treatment process applied to the substrate. Consequently, this reduction in dislocation density enhances the quality of crystallization and leads to a decrease in the half-height width observed in the XRD rocking curves. When comparing the samples listed in Table 3.3, it is observed that the half-height width of the rocking curves is marginally greater in comparison to ordinary Type IIa single crystal diamonds. This is due to the fact that while the pretreatment process of the substrate surface can partially decrease the presence of extended dislocation defects, the overall quality of the CVD layer is primarily influenced by the quality of the substrate. In order to further decrease the density of dislocations in the epitaxial CVD diamond and enhance its crystallization quality, it remains imperative to undertake additional measures. To enhance the crystallization quality and minimize dislocation density in epitaxial CVD diamond, it is imperative to meticulously evaluate the HPHT substrate by considering the lower half-height width of the XRD rocking curve. Figure 3.10d–f present the outcomes of PL spectra analysis conducted on both the synthesized CVD single crystal diamond and the conventional Type IIa single crystal diamond at ambient temperature. The PL measurements were performed using three distinct laser excitations at wavelengths of 325 nm, 514 nm, and 532 nm, respectively. These different excitation wavelengths were employed to investigate the diverse luminescent intervals associated with impurity peaks. It is evident that the impurity luminescence intensity of high-quality CVD single-crystal diamonds decreases significantly following the process of normalization, in contrast to ordinary Type IIa diamonds. Moreover, the grown CVD single-crystal diamonds exhibit no discernible luminescence peaks associated with known impurities when subjected to different laser excitations. The results presented above demonstrate that the impurity content of CVD diamond can be significantly reduced through the implementation of various techniques. These techniques include the utilization of the 9N hydrogen growth process, the adoption of a slow growth method, and the application of a high-pressure and strong method to prevent plasma ball etching on the quartz window. By employing these approaches, it is possible to obtain high-purity CVD single-crystal diamond. Furthermore, the aforementioned processes effectively decrease the concentration of background impurities in the CVD diamond, with the exception of impurities introduced during the sealing of the cavities. However, it is important to acknowledge and address these impurities, which necessitates ongoing enhancements in the growth equipment.

The primary objective in the development of high-quality CVD single-crystal diamond is to prioritize the mitigation of “trace” impurities and the minimization of crystal defects within the epitaxial CVD layer. The enhancement of growth conditions, such as the utilization of high-purity growth atmosphere gases, optimization of the growth process, and selection of a superior HPHT substrate, can yield

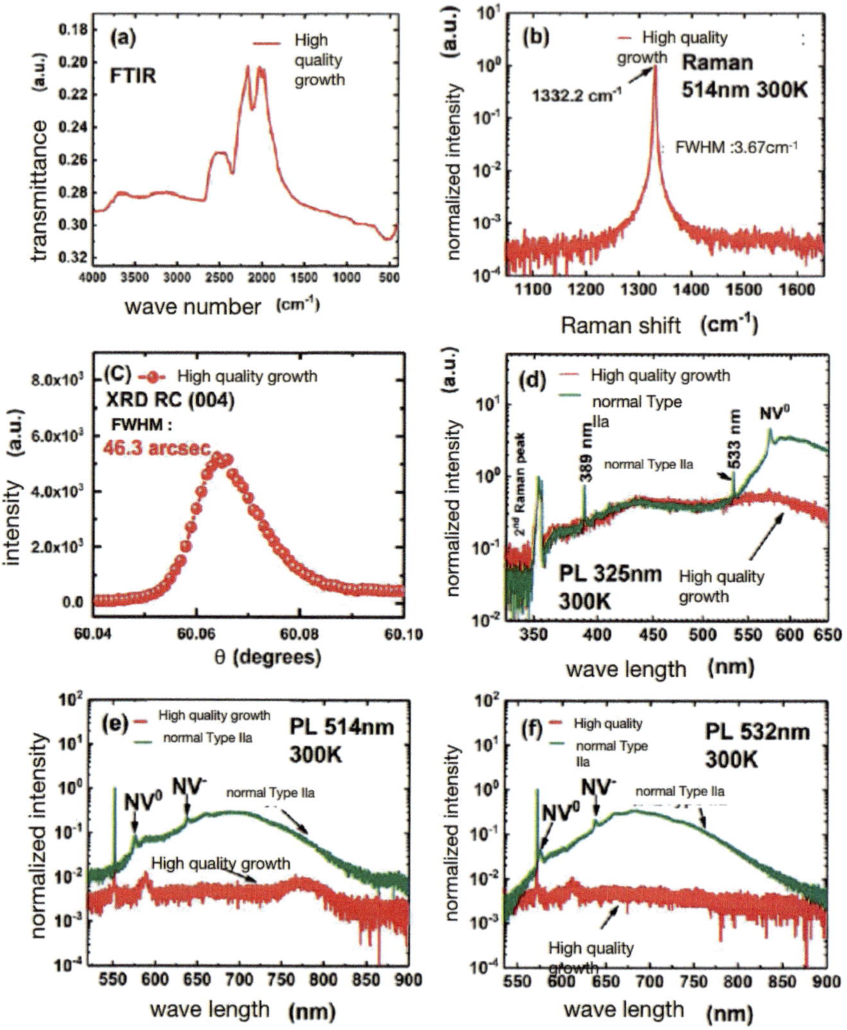


**Fig. 3.10** Non-destructive characterization of high quality CVD single crystal diamond. **a** FTIR; **b** Raman spectroscopy; **c** XRD; **d** PL spectrum—325 nm laser excitation; **e** PL spectrum—514 nm laser excitation; **f** PL spectrum—532 nm laser excitation

substantial improvements in the quality of a CVD single-crystal. These improvements can facilitate the production of diamond material suitable for nuclear detector applications.

**Table 3.3** Comparison of high quality grown CVD diamond and normal Type IIa CVD diamond samples

| | FTIR | Raman peak position (half height and width) | XRD swing curve (arcsec) |
|---|---|---|---|
| HPHT substrates | Type Ib | 1332.54 $cm^{-1}$ $(3.96\ cm)^{-1}$ | 51.84 |
| Growing CVD diamond | Type IIa | 1332.2 $cm^{-1}$ $(3.67\ cm)^{-1}$ | 46.3 |
| General Type IIa | Type IIa | 1332.61 $cm^{-1}$ $(2.97\ cm)^{-1}$ | 39.02 |

## 3.2 Characterization Analysis of Nuclear Radiation Detector Grade Diamond Materials

### *3.2.1 Diamond Sample Selection Strategy*

Based on the analysis presented in the preceding section, it becomes evident that the carrier mobility and lifetime of the detector material constitute crucial factors that impose limitations on the charge collection performance of nuclear detectors. The carrier lifetime and mobility of diamond material can be influenced by various factors related to its material properties, such as impurities, crystallization quality (including dislocations), and the presence of non-diamond phases (such as graphite phases). Hence, employing diamond material characterized by a reduced impurity concentration and dislocation density holds promise as a viable approach to augment the performance of nuclear detectors. When approaching the fabrication of diamond nuclear detectors, the initial step involves the careful selection of appropriate diamond materials. This selection process entails conducting a thorough characterization analysis of the materials. The purpose of this analysis is to provide guidance for the growth strategy of CVD diamond materials and the subsequent development of high-performance diamond nuclear detectors.

Within the framework of the diamond material classification system, the utilization of HPHT diamond as a substrate for homogeneous epitaxial growth is a common practice. It is worth noting that this particular type of diamond, typically categorized as Type Ib, exhibits a distinct yellow color. The most superior quality of diamond is classified as Type IIa, exhibiting a significant nitrogen impurity concentration of less than 5 ppm. The material properties of CVD diamond exhibit significant variations depending on the growth process employed. These variations manifest as a range of colors, from slightly brownish tinted to colorless and transparent. Furthermore, CVD diamond can be classified into two categories based on crystal class, namely monocrystalline and polycrystalline CVD diamond. Consequently, we opted to examine one specimen of HPHT Type Ib substrate and five specimens of the most superior grade Type IIa CVD diamond for the purpose of investigating their material properties. These specimens are detailed in Table 3.4. Samples #5 and #6 consist of "electronic grade" diamonds sourced from Element Six [7]. The product information provided by Element Six indicates that these diamonds have nitrogen impurities of less than 5 ppb and boron impurities of less than 1 ppb.

Despite the significant reduction in diamond material costs through the utilization of CVD for synthesis, diamonds remain relatively expensive compared to other semiconductor materials. In order to achieve Type IIa CVD diamond of superior quality, it is typically necessary for the epitaxial material size to closely match that of the substrate. Furthermore, the non-splicing method poses challenges in obtaining CVD diamond single crystals larger than 12 mm in size. Furthermore, the presence of inconsistencies within the slices impedes the ability to accurately characterize the entire diamond wafer material solely through slicing and sampling methods. In the pursuit of manufacturing nuclear detectors, there exists a consistent preference for achieving a substantial sensitive area while preserving the integrity of the material structure. Due to the aforementioned rationales, employing non-destructive analysis techniques holds significant significance in characterizing the quality of CVD diamond-based materials and evaluating diamond materials that meet the prerequisites for nuclear detector fabrication.

**Table 3.4** Diamond samples for material characterization

| Sample number | Category | Color tones | Category |
|---|---|---|---|
| #1 | HPHT |  | Type Ib |
| #2 | CVD single crystal |  | General Type IIa |
| #3 | CVD single crystal |  | General Type IIa |
| #4 | CVD single crystal |  | General Type IIa |
| #5 | CVD single crystal |  | Type IIa electronic grade |
| #6 | CVD polycrystalline |  | Type IIa electronic grade |

### 3.2.2 Analysis of Non-destructive Characterization of Diamond

1. FTIR-diamond type screening

The primary classification criterion utilized for diamonds is infrared spectroscopy, specifically Fourier Transform Infrared Spectroscopy (FTIR), which is employed to ascertain the "Type" of various diamonds. The diamond type classification system is extensively employed in diamond research due to its efficacy in categorizing diamonds based on their chemical and physical attributes. The identification of diamond types and the characterization of trace element impurities can be efficiently and expeditiously achieved through the utilization of FTIR. FTIR analysis is a non-destructive and cost-effective technique that offers a substantial amount of valuable information pertaining to impurities present within the diamond lattice.

The process of FTIR analysis involves the emission of an infrared radiation beam through a diamond and subsequent measurement of the absorbed amount at specific wavelengths. The infrared region of the electromagnetic spectrum exhibits distinct characteristics due to the influence of nitrogen and boron impurity configurations on the surrounding carbon atoms. Each nitrogen and boron impurity of a specific type is associated with the formation of one or more distinct absorption bands. The diamond lattice exhibits distinctive absorption characteristics, enabling FTIR spectroscopy to not only ascertain the presence of diamond in a sample but also disclose the nature and quantity of impurities. The Bruker Vertex70 micro-infrared spectrometer was conducted using the transmission method. The spectrometer had a scanning wave number range of 400 to 4000 $cm^{-1}$ and a resolution of 4 $cm^{-1}$. The calibration process involved averaging 16 scans over 16 scans, and it was performed at a room temperature of 24 °C and a humidity level of 38%. The calibration was carried out using air as the medium.

Diamond exhibits pronounced absorption properties in the mid-infrared spectral region, specifically within the wave number range of 400 to 4000 $cm^{-1}$. The three regions are contained within the structure, as depicted in Fig. 3.11 [8]. The initial phonon region spans from approximately 1332 to 400 $cm^{-1}$ wavenumber and is associated with the distinctive absorption peaks attributed to nitrogen impurities. In the case of Type IIa diamonds, the first phonon region typically lacks any discernible characteristic peaks. The spectral region spanning from 2665 to 1332 $cm^{-1}$ wave number, commonly referred to as the two-phonon region, represents an intrinsic characteristic of diamond. The aforementioned characteristic region is observable across all types of diamond. The aforementioned characteristics arise as a result of the oscillations of the carbon–carbon bonds within the diamond lattice when subjected to infrared radiation. The Three-phonon region is situated within the wave number range of approximately 4000 to 2665 $cm^{-1}$. The vibrational peaks associated with boron impurities are observed in both the second and third phonon regions, with notable absorption peaks occurring at approximately 2458 $cm^{-1}$, 2803 $cm^{-1}$, and 2930 $cm^{-1}$. These peaks exhibit strong and distinct absorption characteristics, indicative of the presence of boron impurities. Additional impurities, such as hydrogen

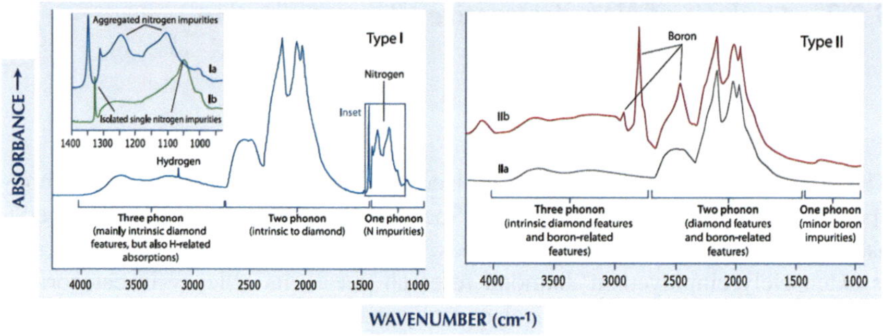


**Fig. 3.11** Diamond classification using FTIR

(3107 $cm^{-1}$) [3], are also present within the three-phonon region. The phonon region exhibits characteristic spectra that allow for the identification of various nitrogen state types. Specifically, substitutional soliton nitrogen is typically observed at frequencies around 1344 and ~ 1130 $cm^{-1}$, type A aggregated nitrogen at ~ 1282 $cm^{-1}$, and type B aggregated nitrogen at ~ 1175 $cm^{-1}$.

The FTIR characteristic spectrum of the HPHT #1 sample, representing a yellow-toned HPHT diamond, is depicted in Fig. 3.12. This spectrum clearly exhibits the presence of substitutional lone nitrogen at 1344 and ~ 1130 $cm^{-1}$, which is identified as the primary type of nitrogen impurity. Additionally, a minor presence of Type A polyazide can be observed at approximately 1282 $cm^{-1}$. However, it does not constitute the primary impurity type. Therefore, it can be concluded that the yellow-tinted HPHT substrate identified as #1 corresponds to Type Ib diamond.

Figure 3.13 displays the FTIR characteristic spectra of CVD single crystal diamonds. While there may be a subtle variation in the color of CVD diamonds, ranging from a slightly brownish hue to complete colorlessness and transparency,

**Fig. 3.12** FTIR characteristic spectrum of the HPHT #1 sample

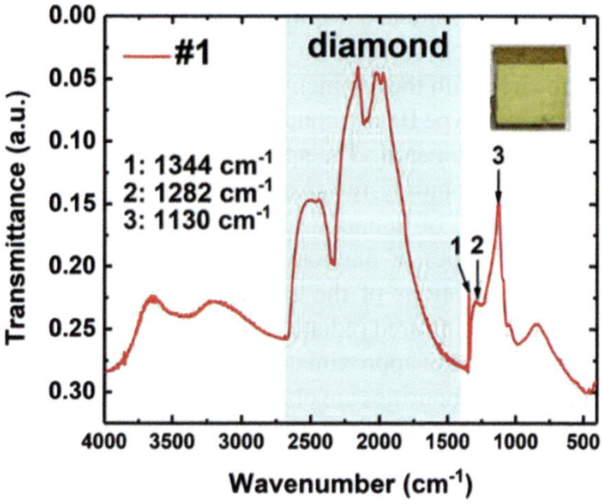

the fundamental characteristics observed in the FTIR spectra remain largely consistent. Notably, samples #2, #3, #4, and #5 exhibit no discernible characteristic peaks associated with nitrogen impurities in the first phonon region, nor any characteristic peaks related to boron in the second phonon region. Furthermore, no additional impurity peaks are detected in the third phonon region. No more peaks indicating impurities were seen inside the third sonority zone. Hence, in accordance with the conventional gemmological categorization of the aforementioned specimens, it may be inferred that these CVD diamonds can be classified as Type IIa diamonds.

The FTIR characteristic spectrum of Sample #6, a CVD polycrystalline diamond, is displayed in Fig. 3.14. Just like single crystal diamond used in CVD, polycrystalline diamond also possesses grain boundaries. However, analysis of its FTIR characteristic spectrum reveals that it belongs to the Type IIa diamond category. This also shows that the FTIR method does not have the ability to distinguish between polycrystalline and monocrystalline diamonds.

The aforementioned findings indicate that the chosen CVD diamonds can be classified as Type IIa diamonds of the highest quality based on their gemological classification using FTIR characteristic spectrum analysis. Additionally, the impurity level of these CVD diamonds is found to be below 5 ppm. The diamonds exhibiting a

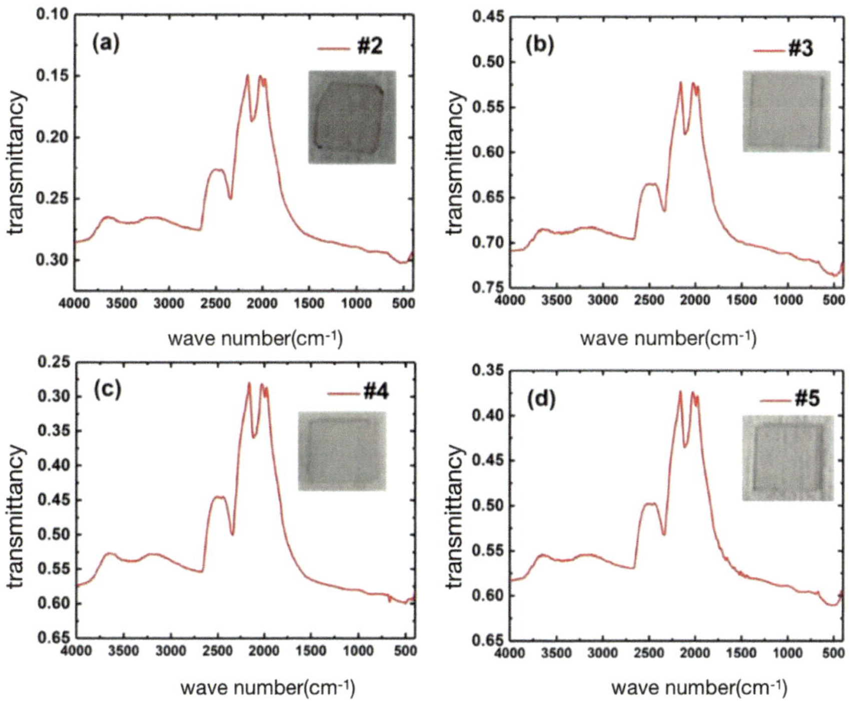


**Fig. 3.13** FTIR characteristic spectrum of CVD single crystal diamond. **a** sample #2; **b** sample #3; **c** sample #4; **d** sample #5

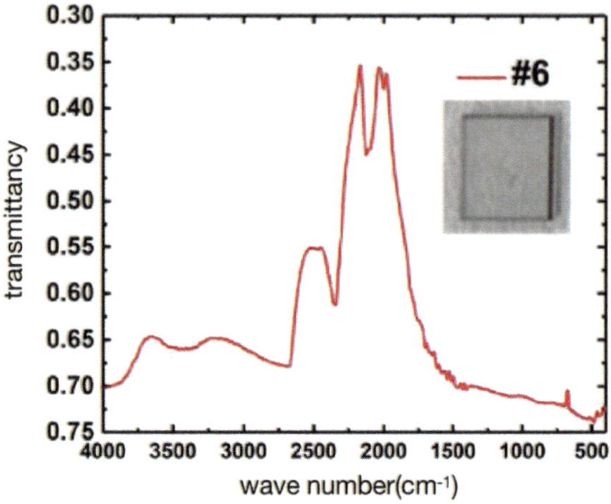


**Fig. 3.14** FTIR spectra of CVD polycrystalline diamond

yellow hue due to HPHT treatment can be classified as Type Ib diamonds, characterized by a notable presence of substitutional solitary nitrogen inside their crystal lattice.

2. Raman Spectroscopy

Raman scattering is classified as a form of inelastic scattering, when photons undergo scattering due to the influence of polarization fluctuations associated with different excitations within a material. The Raman frequency shift in Raman scattering is governed by the energy of the meta-excitation involved in the scattering process and is indicative of the material properties. This shift refers to the difference in frequency between the scattered light and the incident light. The utilization of Raman scattering of phonons in semiconductors allows for the observation of the lattice vibrational spectra of the material, as well as the localized phonon spectra associated with impurity defects, in order to determine the properties of the material. The utilization of Raman scattering in the characterization of diamond materials facilitates the assessment of their growth quality and stress profile. Additionally, Raman spectroscopy allows for the identification of diamond and non-diamond phases, as carbon isotopes originating from non-diamond phases are frequently incorporated into diamond crystals during their growth [9]. The wave number ($cm^{-1}$) is commonly employed as a unit for characterizing spectral measurements in Raman spectroscopy. The reciprocal of the wavelength, known as the wave number, has a proportional relationship with the frequency. If the wavelength of the incident light is designated as $\lambda_0$ and the wavelength of the scattered light is designated as $\lambda_1$, the Raman shift (expressed in units of $cm^{-1}$) may be mathematically represented as:

$$\frac{1}{\lambda_0} - \frac{1}{\lambda_1} = Raman\ shift \tag{3.3}$$

According to the statistics presented in reference [4], it is acknowledged that the theoretical Raman frequency shift of diamond at ambient temperature is 1332.3 ± 0.2 $cm^{-1}$, whereas the Raman peak half-height breadth measures 1.8 $cm^{-1}$. In this study, the theoretical value of diamond Raman frequency shift is taken as 1332.3 $cm^{-1}$. The internal stress $\sigma$ in diamond can be mathematically represented as the disparity $\Delta\omega$ between the theoretical Raman frequency shift $\omega_0$ and the observed Raman frequency shift $\omega$, ie:

$$\Delta\omega = \omega - \omega_0 = -\sigma \tag{3.4}$$

Here κ is the diamond Raman stress factor, which has been determined to be 1.9 $cm^{-1}$/GPa [4]. In this context, a negative value of the diamond stress $\sigma$ signifies the presence of compressive stress within the diamond material. In the Raman spectrum of diamond, in addition to the characteristic peaks of diamond, there will be a D peak at 1350 $cm^{-1}$ if there is a lattice defect of carbon atoms [10], and a G peak at 1580 $cm^{-1}$ if there is a graphite phase with $sp^2$ hybridization [11]. The Raman spectrometer used in the experiments was a Jobin Yvon LavRam HR800 spectrometer, which can test Raman spectra, and for the Raman spectra, the 1800 g/mm grating was selected and the laser wavelength was 514.5 nm (514 nm). The Raman spectrum of HPHT diamond (sample #1) is shown in Fig. 3.15, with the Raman frequency shift peak at 1332.54 $cm^{-1}$ and a half-height width of 3.96 $cm^{-1}$, and no non-diamond phases were found in this sample. The crystalline quality of Type Ib diamond closely approximates the Raman shift peak of an ideal diamond, hence exhibiting a high degree of crystallinity. The selection of Type Ib diamond as the CVD epitaxial substrate is primarily motivated by its exceptional crystalline quality, which facilitates the creation of more desirable CVD epitaxial diamond through the CVD process.

The Raman spectra of the four CVD single crystal diamond samples are shown in Fig. 3.16, the Raman peaks are located at 1332.54 $cm^{-1}$, 1332.73 $cm^{-1}$,

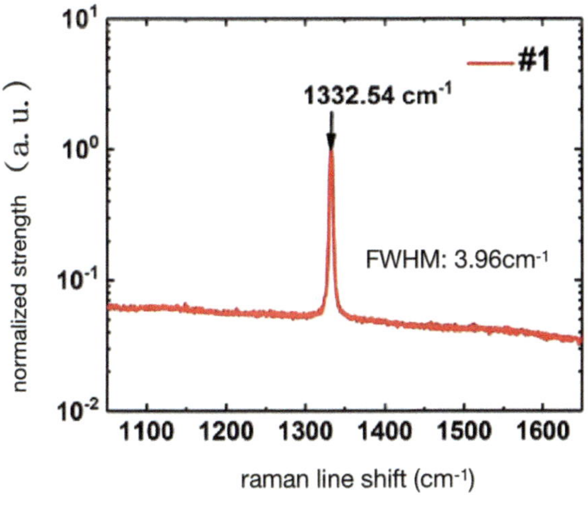


**Fig. 3.15** Raman spectrum of HPHT diamond (sample #1)

1332.61 cm$^{-1}$, 1332.40 cm$^{-1}$, and the half-height and widths are 3.59 cm$^{-1}$, 3.56 cm$^{-1}$, 2.97 cm$^{-1}$, 3.70 cm$^{-1}$, respectively, and no D peaks were found and G peaks, similar to the results of Raman frequency shift peaks of HPHT diamond. The CVD single crystal diamond samples in this study are classified as Type IIa diamonds, which have a significantly lower nitrogen impurity level compared to Type Ib HPHT diamonds. However, the analysis of Raman spectroscopy data indicates that the impurity content has minimal impact on the Raman main peaks and the half-height width.

Figure 3.17 displays the Raman spectrum of CVD polycrystalline diamond, specifically referring to sample #6. Based on the findings from the Raman spectral analysis, it is observed that polycrystalline CVD diamond exhibits a significantly high level of crystalline quality, as evidenced by a Raman frequency shift of 1332.88 cm$^{-1}$ and a half-height width of 3.70 cm$^{-1}$. This is noteworthy considering the presence of numerous grain boundaries and variations in crystallographic orientations among the diamond grains in polycrystalline diamond.

The aforementioned findings indicate that both CVD Type IIa diamonds and HPHT Type Ib diamonds exhibit a notable degree of crystalline quality, as seen by

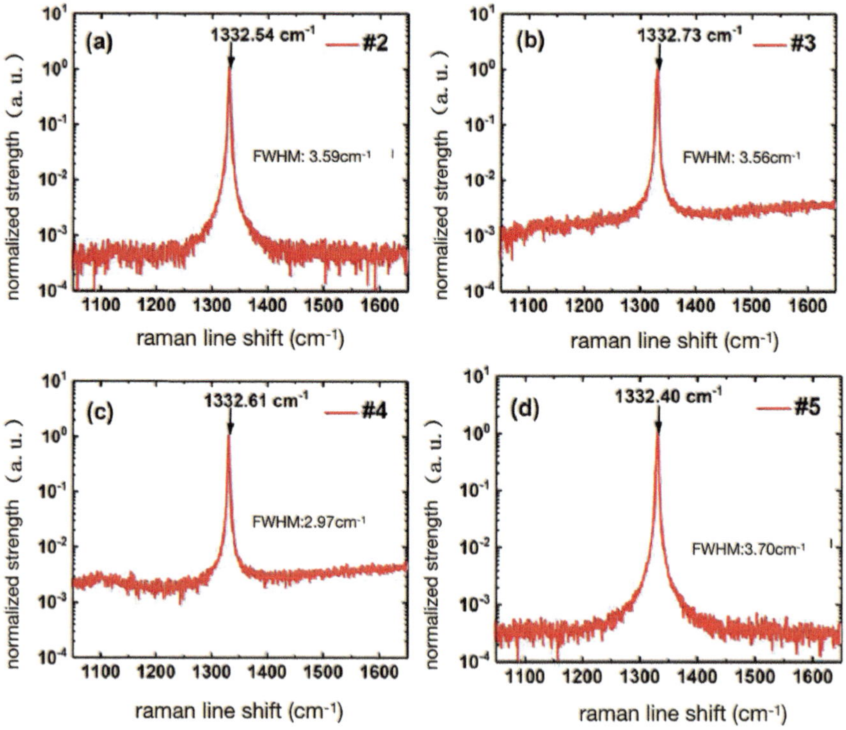


**Fig. 3.16** Raman spectra of CVD single crystal diamond. **a** sample #2; **b** sample #3; **c** sample #4; **d** sample #5

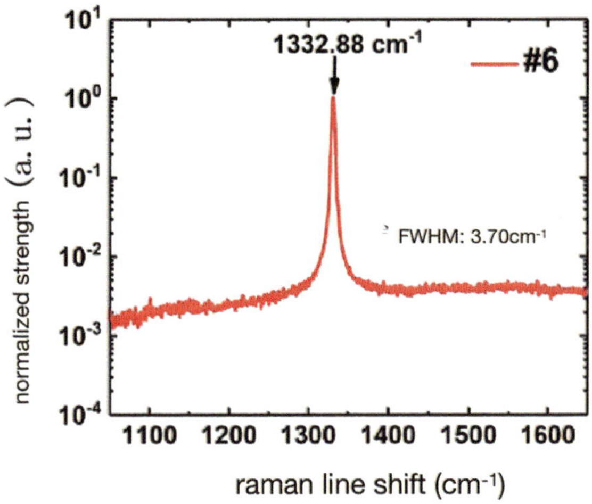


**Fig. 3.17** Raman spectrum of CVD polycrystalline diamond

the absence of D and G peaks. Raman spectroscopy can serve as a valuable tool in the process of material selection for diamond core detectors, as it enables the confirmation of diamond phases and assessment of crystal quality. Nevertheless, the utilization of Raman spectroscopy in isolation is inadequate in efficiently discerning diamonds of excellent quality.

3. PL spectrum

Photoluminescence (PL) spectroscopy refers to the analysis of the phenomenon in which a semiconductor material emits photons as a result of electrons transitioning from a high-energy state to a low-energy one upon laser irradiation. The examination of photoluminescence spectra offers insights on the material's structural and physical characteristics, as well as the interplay between radiative and non-radiative components. In order to quantify PL spectra, a preset excitation light wavelength is employed, and a detector is utilized to systematically examine the material for the wavelength and intensity of the produced photons that accompany the process. PL spectroscopy is a valuable technique employed for investigating the radiative interaction between energy bands within diamond material and the energy levels associated with impurities. It serves to validate the luminescence energy levels of various impurities present in the diamond material, as well as identify the specific types of impurities. Additionally, PL spectroscopy offers a semi-quantitative analysis capable of detecting exceedingly faint impurity luminescence.

The experimental setup employed a Jobin Yvon LavRam HR800 spectrometer, capable of measuring both Raman and PL spectra. Specifically, a 600 g/mm grating was selected for the acquisition of PL spectra. In order to mitigate interference between Raman peaks and neighboring impurity luminescence peaks, it is necessary to employ lasers of varying wavelengths for PL spectroscopy. In this study, lasers with wavelengths of 325 nm, 514 nm, 532 nm, and 633 nm were utilized.

(a) Common impurity luminescence defects in diamond

Photoluminescence can be induced by certain lattice defects present in diamond. While these defects are not inherently linked to the categorization of diamond, investigating them is valuable for material characterization and the advancement of diamond core detectors with enhanced performance capabilities. The luminescence resulting from impurities and imperfections is commonly referred to as diamond color centers. The primary forms of color centers found in diamonds include [3, 8]:

N3 (415 nm): This defect consists of three nitrogen atoms surrounding a vacancy and contributes to the yellowing of the diamond.
N2 (478 nm): This broad spectral band is associated with N3 and is part of many yellow diamond spectra. It is also associated with nitrogen impurities.
480 nm: The current origin of this color center is unknown, but it often arises from Type Ia diamonds and is associated with the yellow spectrum in diamonds.
H4 (496 nm): This defect consists of four nitrogen atoms with two vacancies, and is produced by the combination of the vacancies in the diamond and the B-type poly-nitrogen impurity, which also produces a yellow luminescence in the diamond.
H3 (503.2 nm) This defect is uncharged and consists of two nitrogen atoms bonded to a vacancy, with the two nitrogen atoms separated by the vacancy, i.e., the $[N–V–N]^0$ structure, which is also associated with yellow luminescence.
3H (503.5 nm) is associated with interstitial carbon atoms, originating from radiation damage, or GR1 color centers.
550 nm: from plastic deformation of diamond and other causes.
$NV^0$ (575 nm): consists of a nitrogen atom with a vacancy next to it, usually in the neutral state.
595 nm: This luminescence is associated with a nitrogen defect, but the exact structure is unknown.
$NV^-$(637 nm): this defect also consists of a nitrogen atom with a vacancy next to it, but is negatively charged.
$SiV^-$(737 nm): consists of a silicon atom with a vacancy next to it, usually negatively charged overall.
GR1 (742 nm): This luminescent defect consists of an uncharged vacancy in the diamond, usually due to radiation exposure of the diamond.
$SiV^0$ (946 nm) consists of a silicon atom with a vacancy next to it, usually in a neutral state overall.
H2 (986 nm): this luminescent defect consists of a vacancy inserted in the middle of two nitrogen atoms and is negatively charged overall, i.e., the $[N–V–N]^-$ structure.

This list does not include several other forms of defects. Additionally, due to the diverse methods of diamond formation, the spectra of impurities in diamond exhibit significant variability. The examination of color centering in diamond is currently a specialized topic of study, which will not be extensively addressed in this discussion.

(b) PL spectrum of HPHT diamond

The classification of Sample #1 has been conducted using FTIR spectroscopy, leading to its categorization as a Type Ib diamond. The primary impurity present in this sample is substitutional lone nitrogen. However, additional investigation into the luminescence centers of the impurity is necessary. Based on the photoluminescence (PL) spectrum observed at room temperature (300 K) as depicted in Fig. 3.18, it can be inferred that when subjected to 325 nm laser excitation, two distinct peaks at 473 nm and 478 nm are present. This phenomenon can be attributed to the presence of Fe, Co, and Ni catalytic solvents typically found in diamond grown through the HPHT technique. Notably, Ni exhibits inherent optical luminescence properties, thus suggesting that the aforementioned peaks correspond to the S2 and S3 centers associated with Ni and nitrogen. The broadband emission spectra, exhibiting peak values within the wavelength range of 500–550 nm, have also been proposed to have a correlation with the S2 and S3 centers [12]. The PL spectrum obtained from the excitation of a 14 nm laser reveals the presence of a single Raman peak at a wavelength of 552.5 nm, with no other color centers observed. In a similar vein, the PL spectra obtained from laser excitations at wavelengths of 532 nm and 633 nm exhibit solely the first-order Raman peak, with the absence of any additional color centers.

(c) PL spectrum of CVD single crystal diamond (#2, #3, #4, #5)

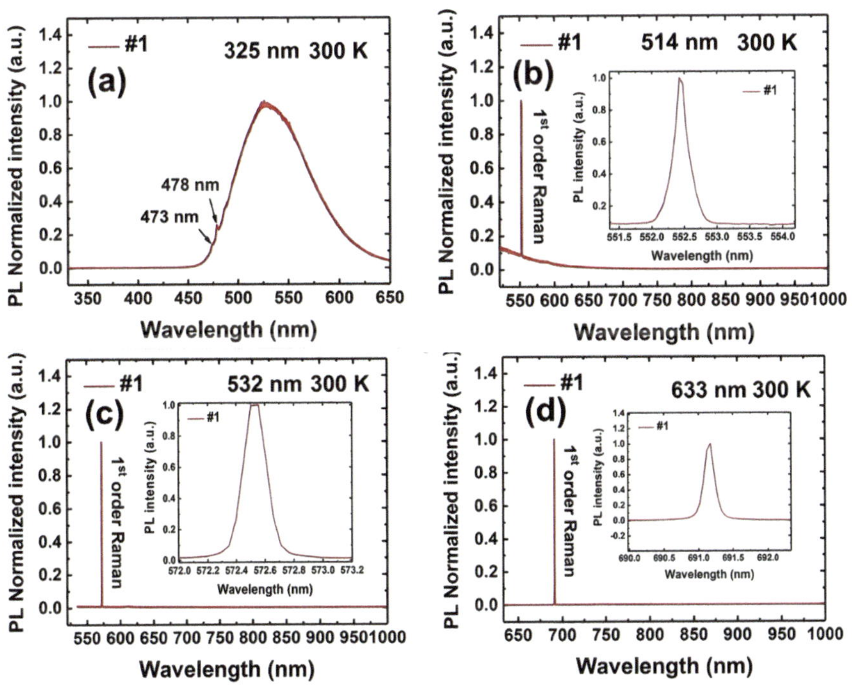


**Fig. 3.18** PL spectra of sample #1 at 300 K at room temperature at different excitation wavelengths. **a** 325 nm laser; **b** 514 nm laser; **c** 532 nm laser; **d** 633 nm laser

Figure 3.19 displays the PL spectra of single crystal diamond samples grown using CVD at ambient temperature (300 K). Figure 3.19a illustrates the PL spectra obtained through laser excitation at a wavelength of 325 nm. It is observed that the selected samples do not exhibit any impurity peaks, except for the presence of a first-order Raman peak at 340 nm. This Raman peak arises due to the presence of nitrogen impurities in the growth environment of CVD diamond. Typically, these nitrogen impurities are bonded to vacancies with substitutional nitrogen atoms and do not tend to form nitrogen pairs or multi-nitrogen centers [3]. In Fig. 3.19b, the PL spectrum is presented, obtained through laser excitation at a wavelength of 514 nm. The peaks labeled as "1" and "2" correspond to the first-order Raman peaks and the second-order Raman peaks, respectively. It is worth noting that samples #2 and #5 exhibit no discernible impurity peaks. Conversely, sample #3 displays peaks at 575 nm, 637 nm, and 737 nm, which can be attributed to the presence of $NV^0$, $NV^-$, and $SiV^-$ defect centers. Similarly, sample #4 contains $NV^0$ and $NV^-$ defect centers. The PL spectrum under 532 nm laser stimulation is depicted in Fig. 3.19c. It is observed that sample #5 does not exhibit any impurity peaks, however sample #2 has higher intensity $NV^0$ and $NV^-$ defect peaks. Additionally, samples #3 and #4 exhibit a PL spectrum consistent with the 514 nm laser. Figure 3.19d displays the PL spectrum obtained using a 633 nm laser excitation. Analysis of the spectrum reveals that the $NV^0$ and $NV^-$ defect centers are absent in all samples, while only the $SiV^-$ defect center is visible in sample #3.

The variation observed in PL spectra when subjected to varied laser excitations can be attributed to the optical interference resulting from impurity peaks induced by intense laser peaks or first-order Raman peaks. These interference effects are not adequately accounted for in the resulting PL spectra. Under 514 nm laser illumination, the first-order Raman peaks are observed at a wavelength of 552 nm. This particular wavelength exhibits a significant disparity in the energy centers of the $NV^0$, $NV^-$, and $SiV^-$ defects. Consequently, it is less susceptible to interference and masking effects caused by the first-order Raman peaks. Under laser irradiation at a wavelength of 532 nm, the first-order Raman peak is observed at 572 nm. This peak significantly impacts the center of $NV^0$ defects, which are positioned at 575 nm. Consequently, the first-order Raman peak effectively masks the presence of the $NV^0$ defect center. At an excitation wavelength of 633 nm, the excitation energy of the NV0 defect center surpasses the energy of the laser, rendering it undetectable. Conversely, the 637 nm luminescence center, where the NV-defect center is situated, is already in close proximity to the excitation light. Consequently, the intensity of luminescence is significantly diminished by the laser, thereby enhancing the efficacy of detecting SiV-defect centers at this particular excitation wavelength.

Based on the aforementioned results of PL analysis, it is observed that Type IIa diamonds, which represent the highest grade in the gemmological classification system, exhibit notable variations in impurity luminescence characteristics within their PL spectra. Notably, the PL spectra of "electronic grade" single-crystal diamonds do not exhibit detectable impurity luminescence at room temperature.

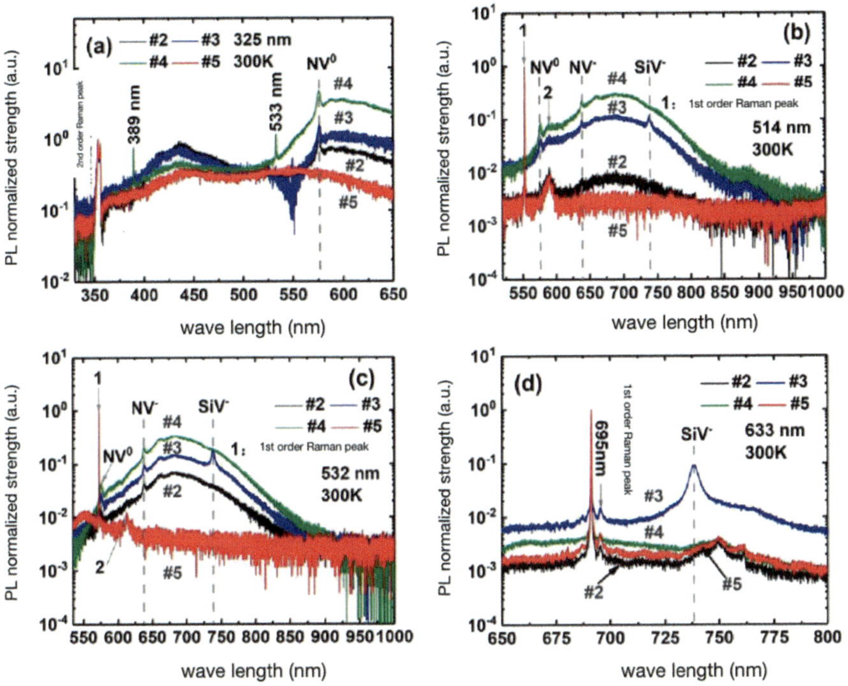

**Fig. 3.19** PL spectra of CVD single crystal diamond samples (#2, #3, #4, #5) at different laser wavelengths. **a** 325 nm laser PL spectrum; **b** 514 nm laser PL spectrum; **c** 532 nm laser PL spectrum; **d** 633 nm laser PL spectrum

This absence of impurity luminescence can be attributed to the extremely low impurity content in these diamonds, which distinguishes them from ordinary Type IIa diamonds. The utilization of PL spectroscopy as a means to assess diamond crystals for their suitability in the production of nuclear detectors with exceptional performance characteristics has shown to be a very efficient approach.

(d) PL spectrum of CVD polycrystalline diamond sample (#6)

The polycrystalline diamond sample (#6) was analyzed for its room temperature PL spectrum, as depicted in Fig. 3.20. Notably, no luminescence peaks associated with impurities were observed across the range of laser excitation wavelengths. This finding suggests that the material exhibits promising characteristics for serving as a diamond nucleation detector with high performance capabilities. However, it is important to conduct additional experiments to validate and ascertain the actual performance of the material.

4. XRD-crystalline quality analysis

X-ray diffraction (XRD) is a widely used technique for the non-destructive analysis of crystalline materials, enabling the determination of several characteristics

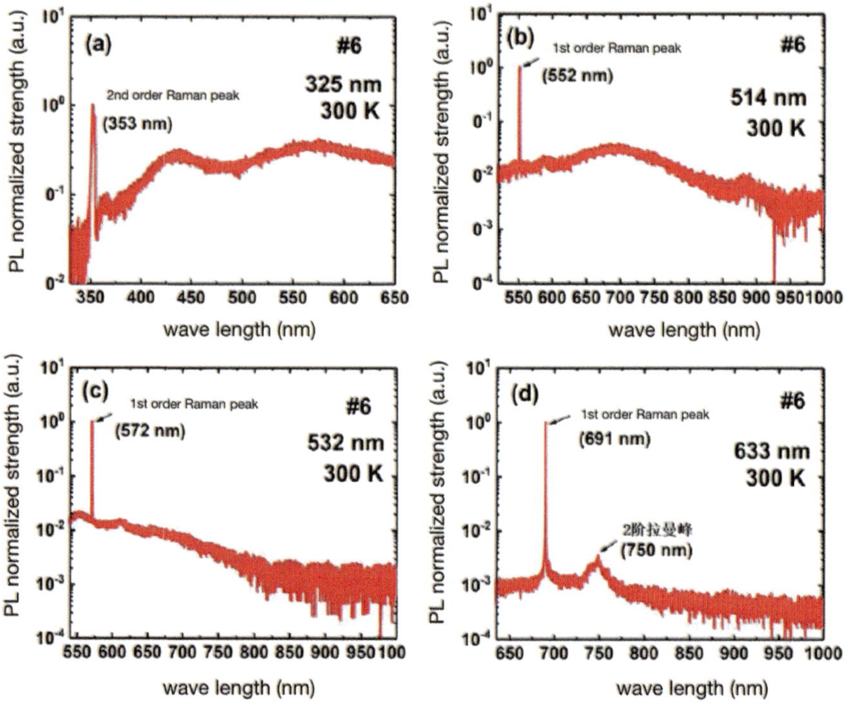


**Fig. 3.20** PL spectra of CVD polycrystalline diamond samples at different laser wavelengths. **a** 325 nm laser PL spectrum; **b** 514 nm laser PL spectrum; **c** 532 nm laser PL spectrum; **d** 633 nm laser PL spectrum

including crystalline quality, stress, and lattice constants. When X-rays impinge upon the surface of a specimen, they undergo diffraction in relation to a particular set of crystalline planes. In the context of crystalline materials, diffraction phenomena arise when there is a correspondence between the spacing of crystal planes (denoted as d) and the wavelength of received X-rays. This correlation is described by Bragg's equation, a widely recognized formula in the field. The equation can be stated as follows:

$$2d \sin \theta = n\lambda \tag{3.5}$$

In this context, 'n' represents the quantity of diffraction levels, whereas 'θ' denotes the diffraction angle. In the case of diamond structures, it is observed that facets do not exhibit diffraction when the facet family (hkl) consists of a combination of odd and even values, or when all the values of (hkl) are even but the sum of h, k, and l does not equal a multiple of 4. In this context, "n" represents the quantity of diffraction levels, whereas "θ" denotes the diffraction angle. In diamond structures, the crystallographic families (hkl) exhibit a combination of odd and even values, or alternatively, exclusively even values. However, no diffraction occurs from this crystal plane when $h + k + l \neq 4n$. The $2\theta$ angle is commonly employed for the purpose of ascertaining the crystal orientation and lattice constant of diamond. This technique is applicable to both polycrystalline and powder samples. In the assessment of crystallization quality for high-quality single crystal diamonds, the half-height width of the X-ray diffraction (XRD) rocking curve is commonly employed. A smaller half-height width indicates a higher crystallization quality and a lower dislocation density in the diamond. The present work used a Bruker D8-Discover high-resolution X-ray single crystal diffractometer, together with a D8-Advance type polycrystalline diffractometer, to assess the crystalline quality of the samples. The X-rays utilized in the experiment were Cu $K\alpha_1$ rays with a wavelength of 0.154 nm.

The XRD results of the selected samples are presented in Fig. 3.21. In order to assess the quality of the crystal structure, XRD rocking curves were used for the characterization of the single crystal samples. Specifically, the half-height width of the XRD rocking curve for the (004) face of the #1 HPHT sample is measured to be 51.84 arcsec, as depicted in Fig. 3.21a. It is worth noting that this broadening is not attributed to a high concentration of nitrogen impurities, which typically lead to an increased half-height width in comparison to other Type IIa diamonds. This discrepancy can be attributed to the fact that the HPHT process emulates the natural formation of diamonds, thereby facilitating the release of internal stresses and minimizing interlattice deformation. Conversely, nitrogen impurities typically exist as point defects, which do not significantly contribute to the increase in dislocation density.

Among the Type IIa single-crystal samples, sample #4 exhibits the most prominent luminescence resulting from its PL impurities. However, it is noteworthy that this sample also demonstrates the narrowest half-height breadth of the XRD rocking

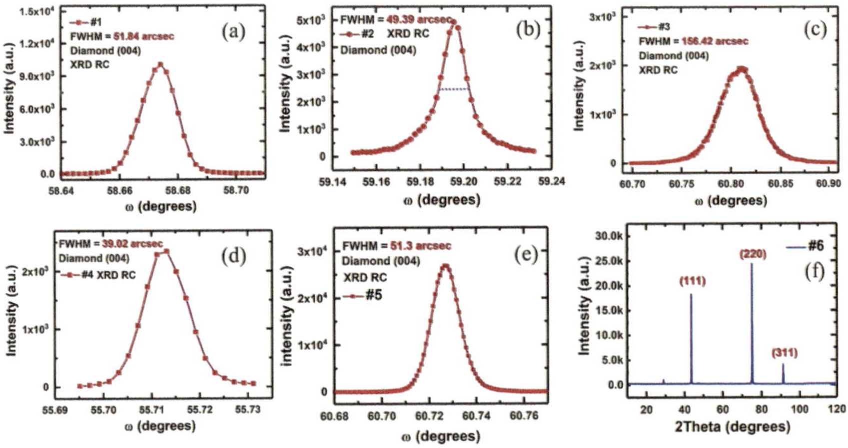


**Fig. 3.21** XRD results of diamond samples. **a** sample #1; **b** sample #2; **c** sample #3; **d** sample #4; **e** sample #5; **f** sample #6

curve, measuring merely 39.02 arcsec (as depicted in Fig. 3.21d). The XRD result for sample #5, which has the lowest impurity content, is not optimal at 51.3 arcsec (Fig. 3.21e); There was a notable disparity between the results of sample #2 and sample #3, both of which exhibited PL luminescence intensity within the intermediate range. Specifically, sample #2 displayed a measurement of 49.39 arcsec (Fig. 3.21b), whereas sample #3 exhibited a measurement of 156.42 arcsec (Fig. 3.21c). The aforementioned findings indicate that the quality of Type IIa single-crystal diamond is not directly correlated with the concentration of impurities. Additionally, the XRD wobble curves of this diamond exhibit significant variations in terms of half-height and width. This discrepancy can be attributed to the presence of dislocations in the CVD epitaxial layer, which typically originate from the interface between the substrate and the growth. These dislocations tend to propagate in the direction of growth. Furthermore, trace impurities primarily exist as point defects within the interior of the diamond. Consequently, in order to achieve high-quality single-crystal diamond growth, it is necessary to optimize the selection of the HPHT substrate and implement appropriate treatments at the growth interface.

The polycrystalline diamond sample labeled as #6 in Fig. 3.21f exhibits structural properties that prevent the utilization of XRD wobble curve analysis for assessing its crystalline quality. Instead, XRD analysis is employed to investigate its crystallographic orientation. The results reveal that the dominant crystallographic orientations of the sample are the (111), (220), and (311) faces.

5. Orthogonal polarizing microscopy-diamond anomalous birefringence

Birefringence arises from the transmission of light through anisotropic crystals, which exhibit anisotropic body. In these crystals, the velocity and refractive index of light vary depending on the direction of its vibration. Consequently, the light can be separated into two polarized components, each with distinct propagation speeds and refractive indices, oriented perpendicular to one another. Diamond is classified within the cubic crystal system and typically exhibits isotropy, resulting in a lack of birefringence. Consequently, its refractive index remains consistent regardless of the direction in which it is measured. When observing a diamond under an orthogonal polarized microscope, the polarized light beam passing through the diamond will cause the detector positioned at a 90-degree angle to the polarized light to be unable to transmit the light. As a result, the eyepiece will seem completely dark. Nevertheless, the presence of impurities, dislocations, and stresses within the diamond can result in deformation of its lattice structure. Consequently, anisotropy emerges in specific regions, causing alterations in the rate of light propagation along certain lattice directions. This phenomenon manifests as anomalous birefringence when light traverses through the diamond. There exists a positive correlation between the degree of birefringence and the degree of lattice strain, refractive index, and thickness. Orthogonal polarized light microscopy is commonly employed for the purpose of observing the anomalous birefringence phenomenon in diamond. This principle is visually depicted in Fig. 3.22. Birefringence offers a visual method for observing the stress distribution in diamond, allowing for a clearer understanding of the spatial distribution of stress. This approach is capable of detecting faults at the micron scale

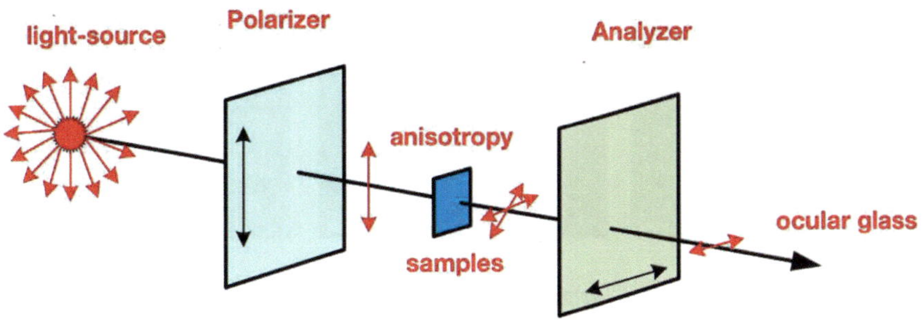


**Fig. 3.22** Principle of anomalous birefringence of diamond observed by orthogonal polarizing microscopy

due to the ability of the strain field associated with these defects to stretch over tens of micrometers. The technique's non-destructive characteristics and quickness make it a valuable adjunct to both Raman spectroscopy and XRD. The microscope employed in this investigation was a ZEISS Scope. The A1 microscope was utilized to capture images, which were subsequently captured using a CCD camera.

The birefringent morphology of various diamond samples is depicted in Fig. 3.23. It is evident from the figure that diamonds produced through distinct growth techniques and varying procedures within the same growth method exhibit discernible disparities in their birefringent morphology. In order to enhance the mobility lifetime product of high-performance diamond core detectors, it is imperative to acquire material samples that possess minimal endogenous defects and do not exhibit notable anomalous birefringence. Specifically, samples falling under type (d) or (e) in Fig. 3.23 are preferred for this purpose. Conversely, other samples exhibit substantial birefringence, which is commonly attributed to crystallographic defects like dislocations and stress. Such defects are highly detrimental to the improvement of detector performance.

## 3.3 Preparation Process Method and Electrical Properties of Diamond Nuclear Radiation Detector

### *3.3.1 Device Structure Design and Preparation*

The primary structural parameters of nuclear detectors encompass the sensitive area, non-sensitive area thickness, and sensitive area thickness.

The measurement of sensitive area is a crucial physical parameter in the context of nuclear detectors. An increased surface area enhances the count rate of the detector, particularly in the identification of sources of low intensity, where precise counting measurements are crucial. Nevertheless, it is important to note that a substantial sensitive area gives rise to a significant parasitic capacitance, so elongating the impulse

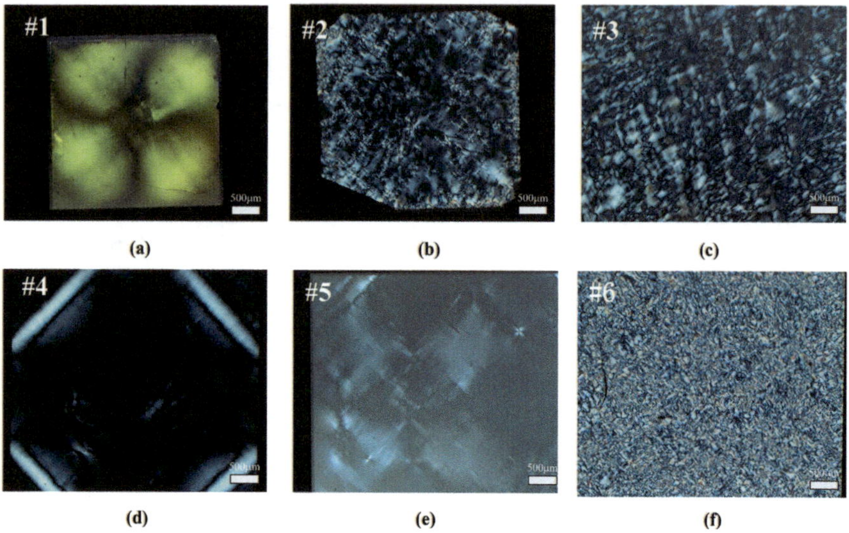


**Fig. 3.23** Diamond anomalous two-ray morphology. **a** HPHT sample; **b** sample #2; **c** sample #3; **d** sample #4; **e** sample #5; **f** sample #6—CVD polycrystal

response. Conversely, a larger sensitive area also leads to a greater leakage current, resulting in a diminished energy resolution of the detector. Consequently, it becomes imperative to strike a balance while designing the construction of the detector. The non-sensitive region, known as the incident window or dead layer, is characterized by its thickness. This region is traversed by the ray prior to entering the sensitive region, but it does not contribute to the output signal. However, the incident particles lose their energy within this region, leading to a decrease in the amplitude of the output pulse. Consequently, this reduction in amplitude negatively impacts the CCE and energy resolution. In the context of detectors, it is common for the metal electrode to serve as a non-sensitive region. Consequently, it is desirable to maintain a minimal thickness for the electrode layer, while yet ensuring effective electrical connectivity with the detector. Another commonly observed non-sensitive region pertains to the non-exhaustion region of junction detectors. In this context, when the width of the depletion region of the p–n junction is smaller than the geometrical thickness of the device, particularly in the area of the incident plane, energy loss occurs, resulting in a decrease in the amplitude of the output pulse.

The sensitive zone thickness refers to the extent of the effective detection region, which directly influences the amplitude of the output pulse generated by the detector. In the event of a ray's incidence, if the particle's range is smaller than the sensitive zone's thickness, the energy is fully absorbed, resulting in the maximization of the output pulse amplitude. Consequently, the detector's output pulse amplitude exhibits a linear relationship with the energy of the incident particle. Hence, in order to optimize the detection capability of the detector, it is imperative to maximize the sensitivity thickness for the detection of energetic particles.

The structure of a diamond nucleus detector can usually be designed in one of the following four ways, as shown in Fig. 3.24. As previously discussed, the preparation of diamond nucleus detectors of p–n junction, Schottky junction, and p–i–n junction types is challenging due to the complexity of diamond doping. Achieving excellent performance in these detectors is also difficult. Additionally, the sensitive region of the detector, which is formed by the barrier region, typically has a thickness of less than 10 μm. On the other hand, the non-depletion region and the dead zone formed by the metal electrodes have a relatively large thickness. This results in a significant depletion of the incident particles' energy and leads to increased fluctuations in the electron–hole pairs, ultimately degrading the energy resolution. In a study conducted by Shimaoka et al. in [13], a p–n junction type, p–i–n junction type diamond nucleus detector was developed. The detector exhibited an energy resolution of 30% and an energy deposition of less than 10% for an α-particle with an energy of 5.486 MeV. These results indicate that the performance of CVD diamond nucleus detectors is insufficient for practical energy spectroscopy measurements. Hence, the prevailing diamond detector structures that are feasibly applicable continue to be predominantly characterized by MiM structures (as depicted in Fig. 3.24d).

The SRIM code can be utilized to calculate the range of non-relativistic α particles within a diamond material. The SRIM programme use Monte Carlo methods to compute the energy loss and distribution of ions upon interaction with a target surface. It is extensively utilised in several domains such as particle physics, semiconductor technology, and other related fields. SRIM refers to a collection of software programmes designed to compute the energy dissipation and spatial dispersion of ions subsequent to their interaction with a target surface. The utilisation of Monte

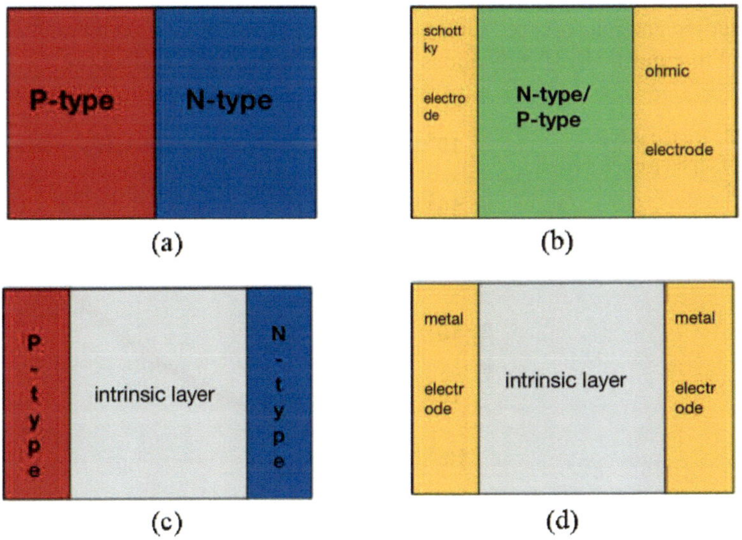


**Fig. 3.24** Structure of CVD diamond core detector

Carlo Fig. 3.25 is used to illustrate the relationship between the range and energy of α-particles in diamond, as determined using SRIM simulations. CVD diamond layer must be at least 1.73 μm in order to achieve complete energy deposition for α-particles with an energy of 1 MeV. Similarly, for α-particles with an energy of 5 MeV, the thickness of the CVD diamond layer must meet certain requirements. When α-particles possess an energy of 5 MeV, it is necessary for the thickness of CVD diamond to be equal to or more than 11.79 μm. However, as the energy of α-particles increases to 100 MeV, the thickness of CVD diamond must exceed 2 mm. The method, which is currently extensively employed in the fields of particle physics, semiconductor research, and other related domains. Hence, it is imperative to appropriately optimise the thickness of the intrinsic layer in the MiM structure based on the specific energy of the observed particles.

Diamond core detectors are commonly manufactured using conventional semiconductor fabrication techniques. The initial step involves the sequential execution of cutting, grinding, and polishing procedures for the CVD epitaxial layer. The HPHT diamond single crystal substrate exhibits a notable presence of impurities, rendering it unsuitable for the purpose of detector development. Consequently, it becomes imperative to separate the CVD diamond epitaxial layer from the aforementioned substrate. Diamond, being a highly rigid material, is commonly subjected to laser cutting, resulting in an uneven cut surface that is often marred by the presence of amorphous carbon and imperfections. The growth process of a single crystal at the periphery of the growth surface is susceptible to the formation of a ring of polycrystalline inclusions. Consequently, it becomes imperative to engage in grinding and polishing procedures to refine the diamond single crystal material as it undergoes growth. The primary objective of the grinding and polishing process is to minimise the presence of non-monocrystalline components, minimise surface damage, enhance both macroscopic and microscopic surface flatness, and obtain diamond monocrystalline material that meets the specifications necessary for detector production.

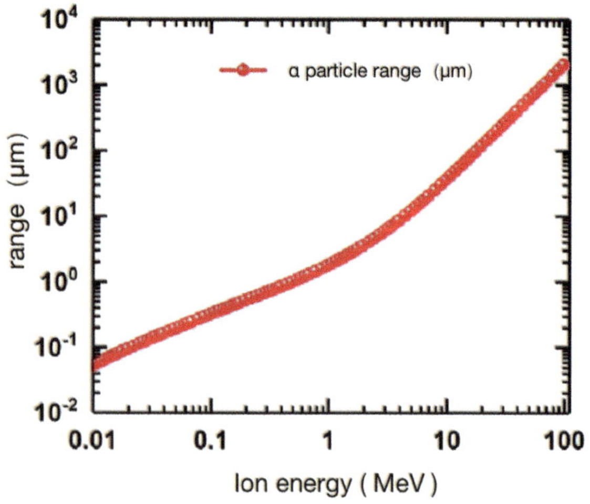


**Fig. 3.25** Range versus energy of alpha particles in diamond

The subsequent step involves the fabrication process of the detector core. Prior to production, the surface of the diamond single crystal undergoes a cleaning process. This typically involves degreasing using substances such as acetone, chromic acid, and sulfuric acid. Subsequently, the crystal is rinsed with a mixture of alcohol, $NH_4OH$, and $H_2O_2$, followed by a final rinse with deionized water. Subsequently, the process of preparing the ohmic contact electrode is undertaken. Diamond has a substantial electron affinity energy, rendering the establishment of an ohmic contact with diamond challenging for conventional metals. According to research reports, the application of hydrogen terminal treatment on the surface of diamonds can induce a negative electron affinity energy, facilitating the formation of ohmic contact when in contact with a high work function metal. Additionally, the utilisation of an all-carbon electrode structure and the annealing of Ti gold to form a TiC alloy structure are other methods that can be employed to achieve ohmic contact electrodes. Currently, the investigation of ohmic contact in diamond devices has emerged as a prominent area of research. In the case of the hydrogen terminal structure, this process involves conducting hydrogen treatment on the diamond surface using MPCVD, followed by photolithography patterning, and ultimately depositing metal electrodes on the surface through thermal evaporation or electron-beam evaporation. Ultimately, in order to mitigate the surface leakage of the detector, it is important to perform oxygen termination on the surface of the detector core, excluding the electrodes. This process serves to eliminate the presence of surface conductive ions and effectively reduce the occurrence of dark current. The aforementioned process methodology is depicted in Fig. 3.26.

To facilitate the observation of electrical signals generated by radiation-induced electron–hole pairs in the diamond detector core, it is imperative to encapsulate the core in accordance with the structural design requirements of the detector. Additionally, establishing a reliable connection between the electrically-contacted metal layer of the core and the external circuits is necessary. This ensures the protection and support of the diamond detector core, while also facilitating its practical utilisation. The successful encapsulation of the detector necessitates the absence of any material damage during the encapsulation procedure, the maintenance of cleanliness and effective insulation of the material, and the provision of convenient accessibility to the detector casing and leads for its use and installation. At present, the generally

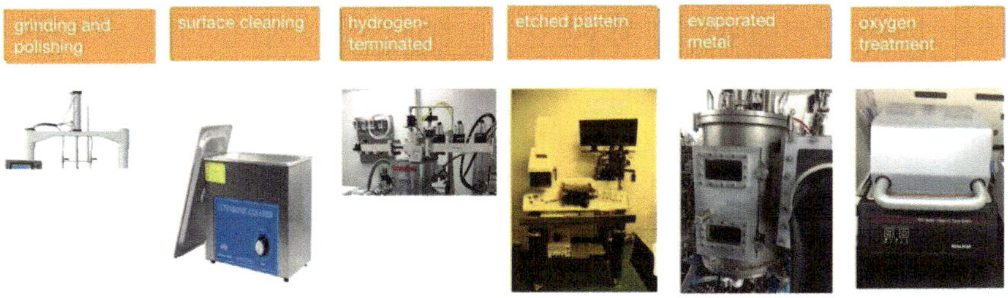


**Fig. 3.26** Diamond core detector fabrication process (hydrogen terminal ohmic contact method)

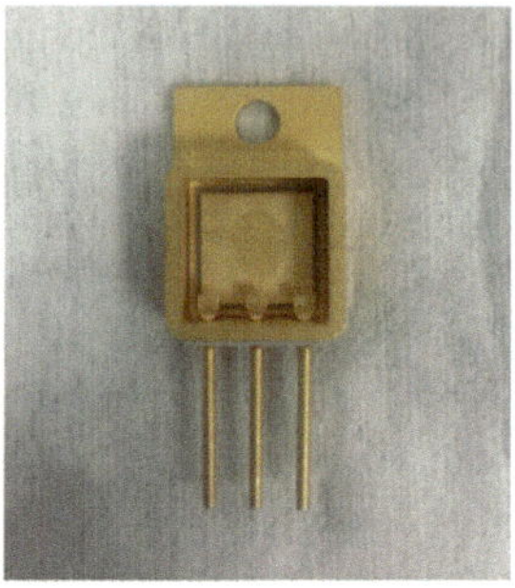

**Fig. 3.27** Common TO ceramic package shells

used package shell for the plastic shell, PCB package, ceramic package. Figure 3.27 illustrates the ceramic shell commonly used in TO packages, which exhibits superior insulation and lower leakage compared to the other two packaging materials. The ceramic shell is capable of withstanding voltages above 1000 V, making it an ideal choice for robust and lasting detector packaging.

### *3.3.2 Preparation Process and Electrical Properties of Hydrogen–Oxygen Terminated CVD Diamond Core Detectors*

This section describes a process for the preparation of a high-performance hydrogen–oxygen terminated CVD diamond core detector with ohmic contact characteristics and dark current.

1. Introduction

High-performance CVD diamond core detectors often necessitate exceptional CCE, energy resolution, and I–V properties. Traditionally, the material qualities of diamonds, the quality of electrode contact, and the condition of the surface have been identified as the primary determinants affecting these properties. The parameters of CCE, as determined by Hecht's equation, are influenced solely by three factors: the carrier mobility-lifetime product, the electric field strength, and the material thickness. The impact of electrode contact and the influence of surface condition on CCE have not been taken into account. The metal-diamond interface often exerts a significant influence on the operational efficacy of nuclear detectors. Various electrode manufacturing procedures in MiM structures lead to the formation of diverse forms, including ohmic connections, Schottky contacts, and other variations. The presence of an exceptional ohmic contact implies the absence of substantial extra impedance. This characteristic not only enhances the output current of the diamond core detector while operating in current mode, but also leads to a rapid time response. When incident rays interact with a diamond core detector, the ensuing nonequilibrium carriers are caught in traps that are formed by defects present at the interface between the electrode and the diamond. The trap-capture phenomenon has a negative impact on the collection of carriers and can result in substantial discrepancies between the output signals determined by electrons and holes in the energy spectrometer. An additional concern arises when polarisation arises due to inadequate electron extraction or injection by the electrode contact during the passage of current through the diamond core detector. In this scenario, following the cessation of ionising radiation, the nuclear detector fails to promptly return to its neutral state within two consecutive time intervals. Consequently, charges accumulate within the diamond material, resulting in the creation of an inherent electric field that opposes the applied electric field. This phenomenon subsequently diminishes the amplitude of the output signal and hampers the collection of charges by the nuclear detector [14]. In the past, the phenomenon of polarisation was primarily associated with the presence of significant flaws in natural and early CVD diamonds. However, in contemporary high-quality "electron-grade" CVD single-crystal diamonds, polarisation is now understood to be influenced by the characteristics of the metal-diamond interface [14]. In order to solve these problems, multilayer metallic Ti-C phase/Pt/Au electrode structures [15], Diamond-like-carbon (DLC) layer structures [14], pulsed laser deposition (PLD) contacts [16], and graphene oxide/Au electrode structures [17] for CVD diamond nucleation detectors have been shown to overcome the above drawbacks to a certain extent.

The utilisation of surface modulation characteristics in diamond has been employed to enhance the efficiency of diamond-based electrical devices. The hydrogen and oxygen terminals, which are the primary surface attributes of diamond surfaces, exhibit distinct physical characteristics, including electron affinity energy, figure of merit, and surface conductance [18]. The surfaces of diamond that are terminated with hydrogen exhibit a lack of dangling bonds and possess a structurally stable surface. The electron affinity energies of hydrogen-terminated diamonds with (001) facets have been reported to be in the range of $-$ 1.3 to $-$ 0.4 eV [19, 20]. Exposure of the hydrogen terminals to air results in the occurrence of surface conductance, while the transfer doping-induced creation of interfacial dipoles causes a marginal enhancement in its electron affinity energy relative to the figure of merit [21]. The energy required to remove one electron from a hydrogen-terminated diamond surface exposed to air is approximately 5.3 eV [22]. One notable advantage of gold (Au) as a metal electrode material is its resistance to oxidation, coupled with its work function of 5.1 eV. The Au potential barrier height while in contact with the hydrogen-terminated diamond surface has been measured to be $0.29 \pm 0.13$ eV [23]. Additionally, the contact resistivity of this interface ranges from $1 \times 10^{-3}$ to $1 \times 10^{-4}$ $\Omega \cdot cm^2$ [24]. Therefore, the interaction between Au and hydrogen-terminated diamond can be considered a viable ohmic electrode contact. The utilization of an oxygen termination as a surface isolation and passivation layer is facilitated by its notable characteristics, including a remarkably high surface resistance and a positronic affinity energy of 1.7 eV [18]. Consequently, the reduction of the dark current in a diamond nuclear detector is a crucial milestone in the effective production of this detector.

2. Device process preparation

Figure 3.28a illustrates the sequential process involved in device manufacture. The initial stage involved selecting a commercial single-crystal diamond with the least luminescence properties of material impurities, as determined by prior research on the material properties of CVD diamond. This choice was made to ensure that the fabrication of the diamond core detector would not be affected by any defects present in the material itself, thereby preserving the device's performance. The dimensions of the sample are $4.6 \times 4.6 \times 0.3$ mm$^3$, and it has a crystallographic orientation of (100) in order to mitigate the impact of material flaws. During the second step of the experiment, the diamond samples experienced a series of cleaning procedures. Initially, the samples were immersed in $H_2SO_4$ for a duration of 30 min to eliminate any amorphous carbon and organic pollutants. Subsequently, the samples were subjected to an ultrasonic cleaning process at 60 W power for 15 min each, using a sequence of acetone, alcohol, and deionized water. During the third step, the samples' upper and lower surfaces underwent hydrogen plasma treatment to generate a hydrogen-terminated surface using MPCVD. The MPCVD treatment process involved the following parameters: a hydrogen flow rate of 700 sccm, a cavity pressure of 100 mbar, a temperature set at 850 °C, and a microwave power of 2 kW. During the fourth phase, a layer of Au with a thickness of 150 nm was applied as electrodes onto both the upper and bottom surfaces of the sample. This was achieved by the process of electron beam evaporation, utilizing a metal mask configuration with electrodes measuring $4 \times 4$ mm$^2$. During the fifth phase of the experiment, following the establishment of electrode contacts, the specimen was introduced into a reactive ion etching (RIE) chamber. Within this chamber, the exposed hydrogen terminals, not shielded by the Au electrodes, underwent a conversion process to become oxygen terminals. This transformation occurred under an oxygen atmosphere, with a flow rate of 80 sccm within the RIE chamber. The chamber pressure was maintained at 8 Pa, while a reaction power of 150 W compared was applied. The processing time for this step was set at 10 min. Subsequently, the contrived nuclear detection apparatus was enclosed within the TO tube casing, and the process of lead bonding was carried out utilizing silicon and aluminum wires. The completed constructed device is depicted in Fig. 3.28b, while the device structure is illustrated in Fig. 3.28c.

3. Ohmic contact characteristics

The objective of this study was to determine the contact resistivity of the hydrogen-terminated diamond surface with the Au electrode. To do this, the circular transmission line model (C-TLM) was employed for the calculation of this parameter. The schematic representation of the C-TLM concept is shown in Fig. 3.29. This structure comprises a conducting circular inner zone of radius r, and a conducting outer zone, with the radius from the inner zone to the outer zone being R, and the spacing between them being R–r. Here, the radius of r is fixed at 30 μm, and the radius of R is increased from 35 μm, at intervals of 5 μm, all the way up to 60 μm. In contrast to the linear transmission line model (TLM), the C-TLM model effectively mitigates

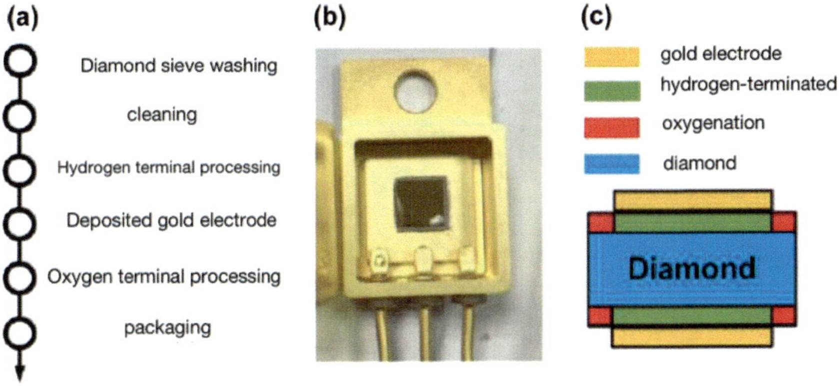


**Fig. 3.28** **a** Device fabrication flow; **b** photo of the finished device; **c** device structure

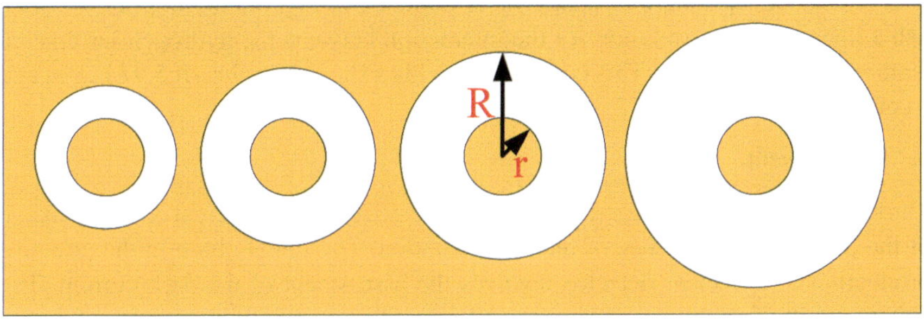


**Fig. 3.29** C-TLM transmission line model

the issue of significant computational inaccuracies arising from current transmission dispersion resulting from inadequate isolation of the external table.

According to the C-TLM model, the total resistance $R_T$ can be expressed by the following equation [25], i.e.:

$$R_{\mathrm{T}} = (R_{\mathrm{S}}/2\pi)[\ln(R/r) + L_{\mathrm{T}}(1/R + 1/r)] \tag{3.6}$$

Here $R_{\mathrm{S}}$ is the resistance of the square on the surface of the hydrogen-terminated diamond, $R$ is the outer diameter length, and $r$ is the inner diameter length. $L_{\mathrm{T}}$ is the transfer length, which indicates the distance over which the most current is transferred from the semiconductor to the metal or from the metal to the semiconductor. Equations (3.6) can be derived as a linear fit of $R_{\mathrm{S}}$ to $L_{\mathrm{T}}$ with a slope of $R_{\mathrm{S}}/2\pi$ and an intercept of $R_{\mathrm{S}}L_{\mathrm{T}}/\pi r$, so that the characteristic contact resistivity is $L_{\mathrm{T}}{}^2R_{\mathrm{S}}$. The I–V relationship for the transition of $R$-$r$ from 5 to 30 μm is depicted in Fig. 3.30a. The observed I–V relationship exhibits characteristics consistent with an ohmic contact at various distances, wherein the current progressively rises with decreasing $R$–$r$. Determine the gradient of the linear function at various intervals, representing the

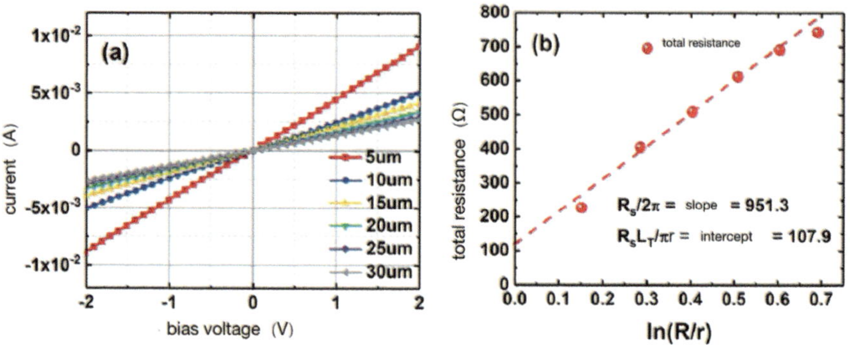


**Fig. 3.30** Characteristic contact resistivity of hydrogen-terminated diamond and Au electrodes

cumulative resistance $R_T$ at each interval. The Eq. (3.6) is utilized to incorporate the total resistance $R_T$ at various spacings, as depicted in Fig. 3.30b, in order to establish a linear fitting correlation for the interaction between the hydrogen-terminated diamond surface and Au. This correlation yields a sheet RS value of 5.97 kΩ/sq and a contact resistivity of $11.73 \times 10^{-4}\ \Omega\cdot\text{cm}^2$.

4. Dark current

The black current of a nuclear detector is characterized as the inherent current leakage of the device in the absence of incident radiation. The initial phase in the effective production of a nuclear detector involves the assessment of the dark current. The enhancement of signal-to-noise ratio and detection sensitivity in a nuclear detector can be achieved by efficiently reducing the dark current, given that the current signal caused by the rays in the external circuit is now of a diminutive magnitude. Dark currents in detectors typically comprise body leakage, surface leakage, and leakage caused by series resistance or inadequate contact. These leakage currents can emerge as scattering noise, undulation noise, and other similar phenomena, hence impacting the energy resolution of nuclear detectors.

The dark current parameter was assessed using a KEITHLEY 6517B Electrostatic Meter in the conducted investigation. The reason for this is that nuclear detectors often function at bias voltages in the range of several hundred volts, while the dark current typically falls within the nA or even pA range. This relationship is characterized by high voltages and low currents. The 6517B Electrostatic Meter has the capacity to attain bias voltages of up to 1000 V, enabling the testing of leakage currents at levels on the order of fA. Typically, dark current measurements are conducted under controlled ambient conditions at standard room temperature and in a light-protected environment.

Figure 3.31 illustrates the dark current characteristics of the hydroxide-terminated diamond core detector at ambient temperature. The graph depicts the standard ohmic contact characteristics commonly observed in the hydroxide-terminated diamond core detector. The formation of a favorable ohmic contact between hydrogen-terminated diamond and Au is facilitated by the matching figure of merit between

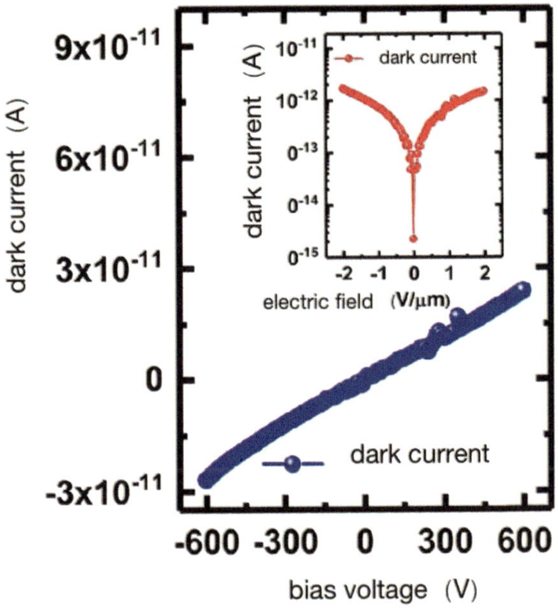


**Fig. 3.31** Dark current characteristics of the hydrogen–oxygen terminal diamond core detector

the two materials, which is attributed to the negative electron affinity energy on the surface of hydrogen-terminated diamond. At a bias voltage of 0 V, the dark current attains a minimum value of $2.22 \times 10^{-15}$ A/mm$^2$. However, when the electric field strength is raised to 1 V/μm, the dark current exhibits a linear increase, reaching $7.46 \times 10^{-13}$ A/mm$^2$. This phenomenon contributes to the achievement of superior surface insulation properties for the oxygen-terminated surface in the non-electrode region, which is formed through the RIE process. Kumar [26] observed a dark current exceeding $10^{-9}$ A at an electric field strength of 1 V/μm for the device following the implementation of chemical oxidation treatment. The dark current properties of the hydroxide-terminated diamond core detector exhibit notable improvements as compared to the chemical oxidation treatment approach utilized in the existing literature for comparison.

## 3.4 Response of Diamond Nuclear Radiation Detectors to α Particles

Obtaining alpha particles from isotopic radioactive sources is a reasonably straightforward process. Alpha particles exhibit a limited range within nuclear detectors, thereby enabling the evaluation of transport properties pertaining to non-equilibrium carriers within these detectors. The resulting alpha particle spectra can be utilized to assess the charge collection efficiency and energy resolution of the detection process, which represents a crucial performance parameter for nuclear detectors. The conversion layer in thermal neutron detectors serves the purpose of converting thermal neutrons into alpha particles for indirect detection. Similarly, other charged particles such as fission fragments and proton beams follow a detection principle akin to that of alpha particles. Therefore, it is of utmost importance to investigate the response of diamond nuclear detectors to alpha particles.

### *3.4.1 Energy Spectrum Characteristics Parameters*

1. Charge collection efficiency

The charge collection efficiency (CCE) of a detector is the ratio of the charge collected at the electrode tip, $Q_c$, to the charge produced by the rays in the body of the detector, $Q_{in}$, viz:

$$\mathrm{CCE} = \frac{Q_c}{Q_{in}} \tag{3.7}$$

The output pulse amplitude of the detector is directly influenced by the quantity of collected charge, assuming the external circuit parameters remain constant. As a result, enhancing the CCE characteristics of the detector has consistently been a significant area of investigation in the field of diamond nucleus detection.

The charge collection efficiency of a particular semiconductor nuclear detector can be characterized by the Hecht equation [1]. This equation assumes a homogeneous distribution of the electric field within the semiconductor and establishes the correlation between the charge collection efficiency and various parameters, including the thickness ($d$) of the sensitive region of the nuclear detector, the nonequilibrium carrier mobility ($\mu$), the carrier lifetime ($\tau$), and the electric field strength ($E$). This equation theoretically establishes a connection between the charge collecting efficiency and the transport properties of the semiconductor material.

Figure 3.32 presents a schematic representation of the Hecht equation model, which describes the charge collection efficiency in a nuclear detector. In this model, an alpha particle with energy $E_\alpha$ is directed towards the cathode plane. Upon interaction, an electron with charge $Q_{e0}$ and a hole with charge $Q_{h0}$ are generated at position $z$, representing a nonequilibrium electron–hole pair.

$$Q_{e0} = Q_{h0} = e\frac{E_\alpha}{w} \tag{3.8}$$

The variable "$e$" represents the elementary charge, while "$w$" is the average ionization energy of the nuclear detector material. When subjected to an applied electric field denoted as $E$, the movement of electrons is directed towards the anode, whereas the movement of holes is directed towards the cathode. Consequently, the temporal and spatial decay of these non-equilibrium electron–hole pairs can be mathematically

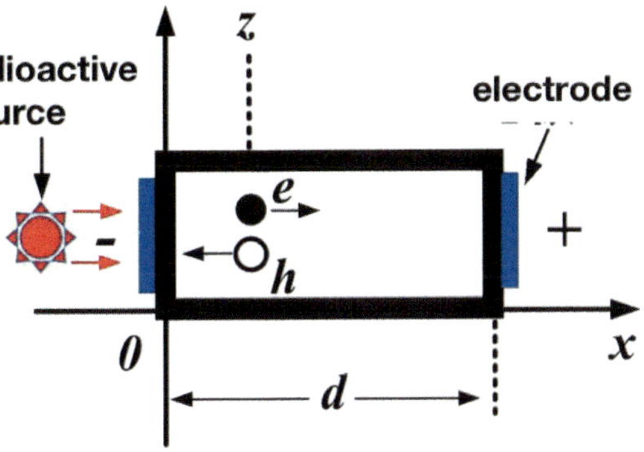


**Fig. 3.32** Hecht equation model of charge collection efficiency schematic diagram

represented as:

$$Q_e = Q_{e0} \cdot e^{-\frac{t}{\tau_e}} = Q_{e0} \cdot e^{-\frac{\mu_e E t}{\mu_e E \tau_e}} = Q_{e0} \cdot e^{-\frac{(x-z)}{\mu_e E \tau_e}} \tag{3.9}$$

$$Q_h = Q_{h0} \cdot e^{-\frac{t}{\tau_h}} = Q_{h0} \cdot e^{-\frac{\mu_h E t}{\mu_h E \tau_h}} = Q_{h0} \cdot e^{-\frac{(x-z)}{\mu_h E \tau_h}} \tag{3.10}$$

When an electron–hole pair is transported inside the nuclear detector, combined with Eq. (3.1), the charge collected on the electrode is:

$$Q_{e\_c} = \int_z^d Q_{e0} \cdot e^{-\frac{(x-z)}{\mu_e E \tau_e}} \frac{dx}{d} \tag{3.11}$$

$$Q_{h\_c} = \int_z^d Q_{h0} \cdot e^{-\frac{(x-z)}{\mu_h E \tau_h}} \frac{dx}{d} \tag{3.12}$$

Then the charge collection efficiency is:

$$CCE = \frac{\mu_e E \tau_e}{d}\left(1 - e^{-\frac{(d-z)}{\mu_e E \tau_e}}\right) + \frac{\mu_h E \tau_h}{d}\left(1 - e^{-\frac{(d-z)}{\mu_h E \tau_h}}\right) \tag{3.13}$$

Equation (3.13) is the full equation for the charge collection efficiency, which can be simplified to a carrier effect based on the direction of the electric field when the ray is incident on the detector and generates electron–hole pairs at positions $z << d$, i.e.:

$$CCE = \frac{\mu_{e/h} E \tau_{e/h}}{d}\left(1 - e^{-\frac{d}{\mu_{e/h} E \tau_{e/h}}}\right) \tag{3.14}$$

2. Energy resolution

The energy resolution of nuclear detectors is a crucial quantity in the context of energy spectrum applications, as it signifies the capacity to distinguish and separate comparable energy spectral lines. The energy resolution, denoted as $\Delta E_0$, is determined experimentally at the maximum peak of a spectral line. It is quantified as the half-height width, which represents the energy value. Additionally, the energy resolution can be stated as the ratio of the half-height width to the energy value in the center of the peak of the spectral line, denoted as $\Delta E_0/E_0$, and is typically presented as a percentage. The energy resolution of a nuclear detector is influenced by several factors. These factors include the statistical fluctuation of electron–hole pairs generated by incident radiation within the detector, the decrease in the CCE caused by the trapping of nonequilibrium carriers in composite materials and traps, the electronic noise inherent in the testing system, the energy loss in the non-sensitive region of the detector structure, and the energy deficit resulting from collisions between incident particles and the atomic nucleus. The energy resolution limit achievable by a nuclear detector is controlled by the statistical fluctuations in the generation and recombination of nonequilibrium carriers resulting from the energy dissipation of the radiation within the semiconductor's sensitive zone. If the mean value of the nonequilibrium carriers generated by each ionizing radiation particle surpasses several hundred orders of magnitude, the statistical distribution can be characterized by a Gaussian distribution. The intrinsic energy resolution of the detector is denoted by the spectral line spread $\Delta E_0/E_0$, which arises from stochastic fluctuations in the rise and fall of the signal.

$$\frac{\Delta E_0}{E_0} = 2.355\sqrt{\frac{w}{E_0}} \tag{3.15}$$

In practical applications, it has been observed that the measured energy resolution of a detector can be lower than its intrinsic energy resolution. This discrepancy arises from the assumption that the independent events of the statistical distribution are not entirely independent when they occur within the detector [27]. Consequently, for nuclear detectors, the actual energy resolution is determined as follows:

$$\frac{\Delta E_0}{E_0} = 2.355\sqrt{\frac{wF}{E_0}} \tag{3.16}$$

The Fano factor, denoted as $F$, is a fundamental parameter of intrinsic semiconductor materials. It characterizes the statistical fluctuations in the generation and recombination of electron–hole pairs within a detector, resulting in nonequilibrium conditions. In the context of a specific perfect semiconductor material, the parameter $F$ remains constant. However, in theoretical analyses, variations in the ionization model lead to differences in the value of $F$. Typically, when a ray interacts with the semiconductor material, a portion of the ray's energy is scattered by lattice excitation of phonons, without resulting in ionization or the creation of electron–hole pairs. This energy loss is directly proportional to the magnitude of $F$, with greater energy losses corresponding to higher values of $F$. In the context of nuclear detectors, the energy resolution is influenced by various factors. While the statistical rise and fall effects have been discussed earlier, the primary factor that significantly impacts the energy resolution is the reduction of CCE. This reduction is attributed to both the compound effect and the trapping effect of the trap. Consequently, incomplete charge collection occurs, resulting in a decrease in the output pulse amplitude and a consequent degradation of the energy resolution. Hence, enhancing the performance of energy core detectors in terms of CCE can significantly enhance the energy resolution parameter. The F-factor of Shimaoka et al. has been experimentally determined to be $0.44 \pm 0.04$ [28] for a high-performance diamond detector.

The energy resolution of a nuclear detector measured in an experiment is usually the result of the superposition of a number of factors, which can be described using the root-mean-square expression, viz:

$$\left(\frac{\Delta E_0}{E_0}\right)_{\text{all}} = \sqrt{\left(\frac{\Delta E_0}{E_0}\right)^2_{\text{detector}} + \left(\frac{\Delta E_0}{E_0}\right)^2_{\text{noise}} + \left(\frac{\Delta E_0}{E_0}\right)^2_{\text{electronics}} + \ldots} \quad (3.17)$$

To achieve enhanced energy resolution, it is important to undertake several endeavors such as optimizing device performance and employing advanced electronics measurement equipment. These efforts are essential for attaining improved performance outcomes.

3. Time response characteristics

The primary factors influencing the time response characteristics include the rise time of the impulse response and the duration required for charge collection.

The rise time ($T_r$) of the impulse response is typically influenced by the time constants of the detector and measurement system. In comparison to other semiconductor materials like silicon, diamond exhibits a significantly lower dielectric constant. Consequently, diamond generally enables faster impulse response times than silicon. In practical applications, the presence of the non-sensitive zone and the series resistance caused by contact might contribute to an elevation in the rise time during the development of detectors. The pulse rise time, as determined by experimentation, refers to the duration necessary for the amplitude of a pulse to increase from 10 to 90%.

For the charge collection time $T$, which is determined by the thickness of the detector's sensitive region versus the average velocity of the nonequilibrium carriers, it can be expressed as:

$$T = \frac{d}{v} = \frac{d}{\mu E} \quad (3.18)$$

### 3.4.2 Measurement Methods

1. Energy spectrum characteristics

The investigations involved the utilization of an energy spectroscopy electronic system to examine the energy spectral response characteristics of diamond nucleus detectors towards alpha particles. The Pulse Height Spectrum (PHS), as depicted in Fig. 3.33, is a widely used electronic system for energy spectroscopy. It comprises several components, including a bias power supply, a charge sensitive preamplifier (Ortec 142), a main amplifier (Ortec 672), a multichannel analyzer (Ortec 927), and a vacuum cavity with a pressure of $10^{-3}$ mbar. The test was conducted at room temperature.

The pulse height spectrometer is a system used for measuring linear energy spectra. The spectrogram's horizontal axis represents the channel address, which corresponds to the charge collected by the detector. This charge is amplified by the charge sensitive amplifier and the main amplifier, resulting in a voltage value. The voltage value is then quantized into a digital representation using an ADC in the multi-channel analyzer. The number of channels in the system's performance determines the number of addresses. In the case of the same gain, a higher amount of collected charge leads to a greater voltage value. Additionally, an increase in the number of addresses is directly proportional to the amount of collected charge. The energy resolution can be derived by calculating the ratio between the half-height breadth of the spectrum and the channel site. The y-axis of the graph depicts the frequency of events taking place at a specific channel location, demonstrating the statistical characteristics of the incident particles as they produce an electric charge within the detector and undergo movement and accumulation. The peak value on the graph corresponds to the greatest likelihood of an event occurring and the largest number of events at that particular channel site, as depicted in Fig. 3.34.

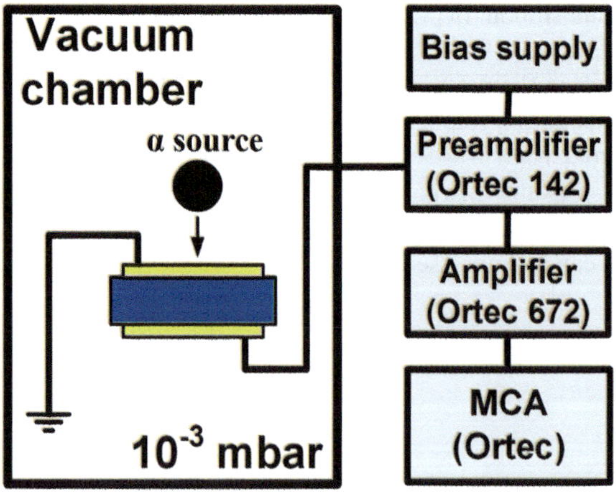


**Fig. 3.33** Pulse height spectrometer

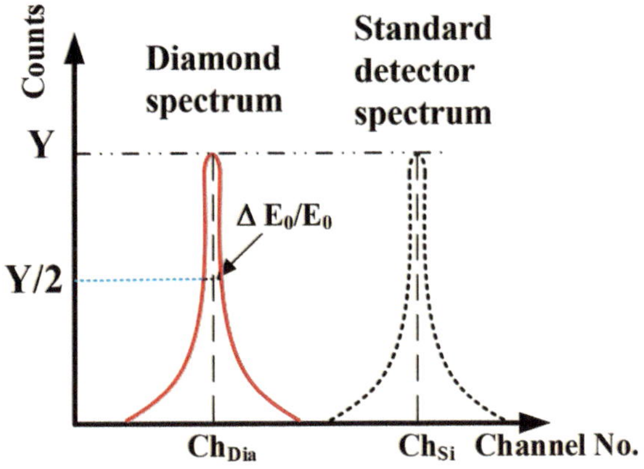


**Fig. 3.34** Diamond core detection CCE and energy resolution calibration method

The calibration process for determining the CCE of diamond core detectors often involves the utilization of a standard silicon detector, as depicted in Fig. 3.34. In this particular case, Ortec's CU-014-050-100 gold–silicon surface barrier detector is employed for calibration purposes. The $CCE_{\text{diamond}}$ under investigation can be determined using the following procedure: Firstly, the diamond nucleus detector under investigation will undergo irradiation by alpha particles with an energy of $E_\alpha$ for the purpose of measuring the energy spectrum. The amplifier's gain during this process, denoted as $G_{\text{diamond}}$, will determine the channel site, $Ch_{\text{dia}}$. Subsequently, the standard silicon detector will be measured at a gain of $G_{\text{Si}}$ to determine the channel site, $Ch_{\text{Si}}$. This relationship can be expressed as:

$$Ch_{\text{Si}} = \frac{G_{\text{Si}} \cdot E_\alpha \cdot CCE_{\text{Si}}}{w_{\text{Si}}}, \quad Ch_{\text{dia}} = \frac{G_{\text{diamond}} \cdot E_\alpha \cdot CCE_{\text{diamond}}}{w_{\text{diamond}}} \tag{3.19}$$

The average ionization energy of silicon, denoted as $w_{\text{Si}}$, and the average ionization energy of diamond, denoted as $w_{\text{diamond}}$, are being referred to in this context. The CCE of a typical silicon detector is commonly observed to be 100%. Additionally, the energy spectrometer typically maintains a constant magnification. Consequently, the $CCE_{\text{diamond}}$ can be mathematically represented as:

$$CCE_{\text{diamond}} = \frac{Ch_{\text{dia}} \cdot w_{\text{diamond}}}{Ch_{\text{Si}} \cdot w_{\text{Si}}} \tag{3.20}$$

2. Time response

The evaluation of a detector's temporal performance involves analyzing its response time, which refers to the breadth of the temporal response waveform at half of its maximum height, as well as the rise time, among other factors. The characterization of the detector's temporal response speed is accomplished by the measurement of its response time. In the present era, the nuclear detection domain necessitates the temporal responsiveness of detectors to operate at the nanosecond scale

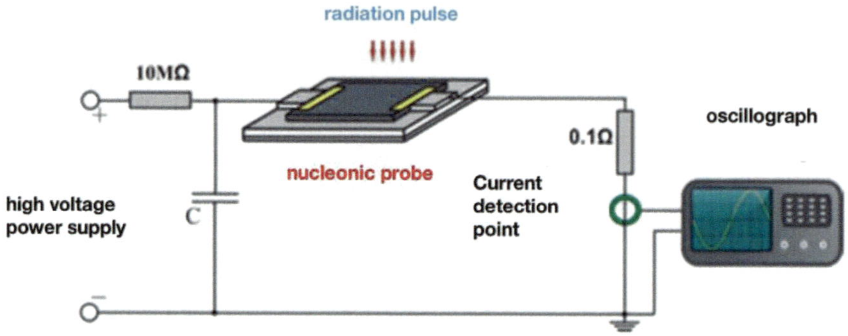


**Fig. 3.35** Diamond core detector impulse response test schematic

or even below. However, it is worth noting that the relative dielectric constant of diamond single-crystal material stands at a mere 5.7, significantly lower than that of conventional silicon–germanium material. Consequently, detectors constructed from diamond exhibit reduced capacitance and exhibit enhanced speed. The temporal behavior of the detector is also influenced by factors such as the distance between electrodes, the intensity of the electric field, the velocity of carrier drift, the resistance of the load, the nature of the radiation, and the depth of penetration. These aspects necessitate further investigation. The investigation of the temporal performance of the detector in an experimental setting typically involves the utilization of α-particles as a source of radiation pulses. However, due to the low intensity of the signal, it is often necessary to employ a high-speed amplifier in conjunction with the detector. Alternatively, other sources of irradiation such as pulsed electron beams or pulsed X-rays can also be used. The resulting output pulse waveforms can be observed using an oscilloscope, as depicted in the test principle illustrated in Fig. 3.35.

### *3.4.3 Energy Spectral Response of Single Crystal Diamond Core Detectors to α Particles*

The assessment of the diamond nuclear detector's charge collection efficiency (CCE) was conducted by utilizing a $^{241}$Am source possessing a radioactivity level of 1.1 $\times$ $10^4$ Bq. This source releases alpha particles with an energy of 5.486 MeV. The nuclear detector will be integrated into the pulse height spectrometer following the test method outlined in Fig. 3.33. As the voltage is gradually increased, the peak position of the detector's response will eventually reach a plateau, indicating the saturation of the charge collection efficiency. To ensure the stable operation of the detector, it is necessary to maintain a certain voltage margin. In this particular case, the operating voltage is set at 200 V, applied to the bottom electrode of the detector. Consequently, electron–hole pairs are generated near the top electrode, with the electrons drifting towards the bottom electrode and the holes drifting towards the top electrode.

Based on the carrier drift theory, it can be observed from Fig. 3.36a that the depth of incidence of the α-particle with an energy of 5.486 MeV inside the diamond, as calculated using SRIM, is approximately 14 μm. This value is derived using the sensitive zone thickness of the detector, which is determined to be 300 μm. Hence, it is possible to employ a simplified Hecht equation to elucidate the underlying physical mechanism, wherein the impact of holes on the CCE can be disregarded in the electron collection mode. The non-sensitive region of the electrode causes a loss of energy when the incident alpha particle passes through it, resulting in an absence of a useful output signal. According to SRIM calculations, the collisional ionization energy loss rate for 5.486 MeV alpha particles within the Au atomic layer is approximately 434 keV/μm, as depicted in Fig. 3.36b. Despite the Au layer being only 150 nm thick, the energy loss incurred while passing through it amounts to approximately 65 keV, which accounts for merely 1.18% of the total energy of the incident particles. Consequently, the impact of the gold layer is disregarded in the calculation of CCE in this context.

The pulse height spectrum of the hydroxide-terminated diamond core detector in electron collection mode is depicted in Fig. 3.37. The spectrum was obtained with a measured channel address number of No. 332. Additionally, the standard silicon detector was subjected to the same test conditions and placed in the pulse height spectrometer, resulting in a channel address number of No. 1219. The average ionization energy of silicon is determined to be 3.62 eV, while that of diamond is found to be 13.1 eV. By utilizing Eq. (3.20), it is determined that the CCE for electrons is 98.6%. At this point, the half-height width of the energy spectrum peak of the diamond nucleus detector, denoted as $\Delta E_0$, measures 5.72 lanes. Consequently, the energy resolution in the electron collection mode is calculated to be 1.7% of $\Delta E_0/E_0$. In a similar vein, when a bias voltage of − 200 V is applied to the lower electrode, the device operates in hole collecting mode. In this mode, the observed charge collection efficiency is 99.01%, accompanied with an energy resolution of 1.5%.

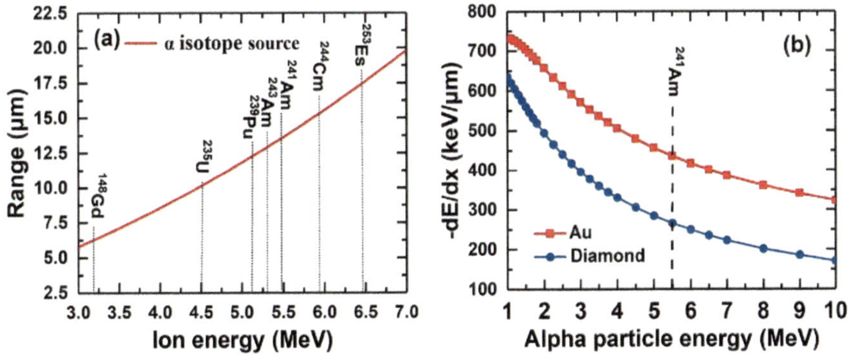


**Fig. 3.36** SRIM calculated values. **a** Common alpha source range within diamond; **b** collisional ionization energy loss rate of alpha particles of different energies within Au and diamond

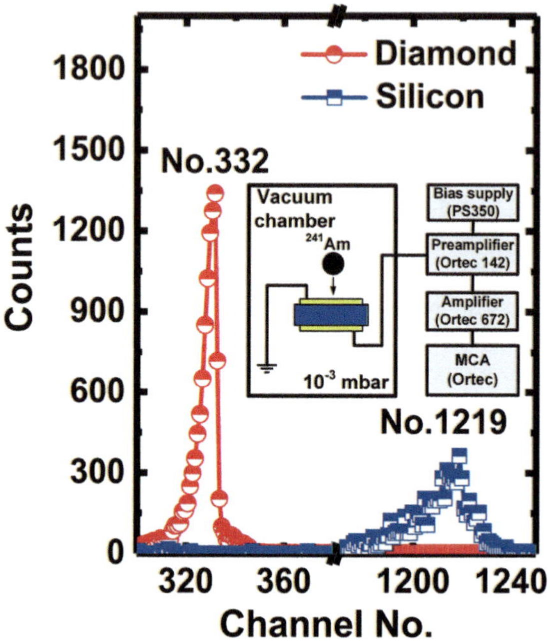


**Fig. 3.37** Calibration of diamond core detector CCE

The calculation of the CCE for the diamond nucleus detector may be performed using Eq. (3.13) when the distances of electrons from the anode and holes from the cathode are of similar magnitudes. Equation (3.13) represents the complete expression for the CCE. In the current configuration, where the range of α-particles (z = 14 μm) is significantly lower than the thickness of the detector (d = 300 μm), Eq. (3.13) can be simplified to Eq. (3.14). Consequently, the CCE can be seen as being influenced by a single nonequilibrium carrier. The measurement of the CCE at various bias voltages can be conducted in accordance with Eq. (3.14). By employing a fitting procedure, the electron/hole mobility-lifetime product ($\mu\tau$) can be determined independently. This process is illustrated in Fig. 3.38a. When the lower electrode of the nuclear detector is subjected to negative pressure, the CCE is mostly influenced by the presence of holes. The value of the $\mu_h\tau_h$ product, which represents the product of the mobility and lifetime of holes, is determined to be $1.2 \times 10^{-4}$ $cm^2 \cdot V^{-1}$ using fitting analysis. When a positive pressure is applied to the bottom of the nuclear detector, the CCE is mainly determined by the electrons, whose $\mu_e\tau_e$ product is close to $6.0 \times 10^{-5}$ $cm^2 \cdot V^{-1}$ for $V_{bias} \leq 40$ V, and $1.2 \times 10^{-4}$ $cm^2 \cdot V^{-1}$ for $V_{bias} > 40$ V. The CCE is also determined by the electrons. The higher voltage bias ($V_{bias} > 40$ V) reveals a greater alignment between the CCE characteristics of electrons and holes, as depicted in Fig. 3.38a. This finding suggests that the interface between Au and hydrogen-terminated diamond exhibits a reduced density of traps. Additionally, there is no notable disparity in the energy spectrum qualities of electrons and holes, indicating that both species can achieve more favorable energy spectral characteristics.

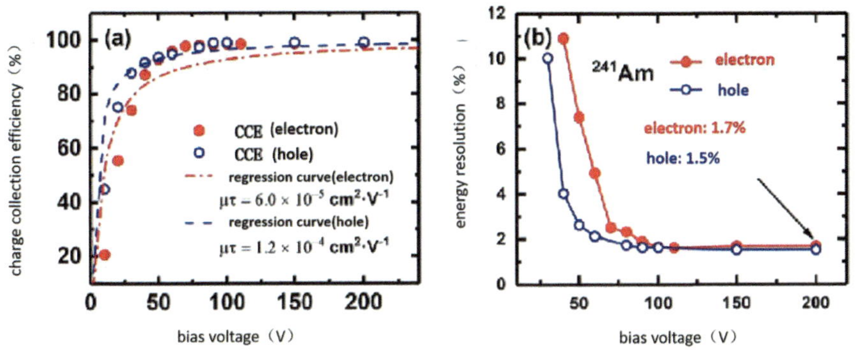


**Fig. 3.38** Energy spectrum characteristics under irradiation from the $^{241}$Am source. **a** CCE versus bias voltage; **b** energy resolution versus bias voltage

Figure 3.38b depicts the energy resolution of the nuclear detector when subjected to irradiation from a $^{241}$Am radioactive source at various bias voltages. It is evident that the energy resolution of the electron/hole decreases as the bias voltage increases. This phenomenon can be attributed to the notable improvement in the CCE of the electron/hole as the bias voltage rises. Consequently, the ratio $\Delta E_0/E_0$ begins to decline, indicating a decline in the energy resolution. This implies that enhancing the CCE of nucleus detectors might be a viable approach to enhance the effectiveness of diamond nucleus detectors. When $V_{\text{bias}} > 100$ V, the CCE of the nuclear detector hits saturation. At this point, the energy resolution also starts to saturate. Eventually, the energy resolution in the collection mode of electrons and holes, when exposed to radiation from the $^{241}$Am radioactive source, reaches its minimal value.

To enhance the validation of the energy spectral resolution of the hydroxide-terminated CVD single-crystal diamond nucleus detector, we conducted energy spectral measurements utilizing a composite $^{243}$Am-$^{244}$Cm source exhibiting radioactivity levels spanning from 1000 to 1600 Bq. The primary energy peak of the α-particles discharged from the $^{243}$Am source was measured to be 5.275 MeV, whereas the corresponding peak energy for the α-particles ejected from the $^{244}$Cm source was determined to be 5.805 MeV. The bias voltage applied to the detector was ± 200 V. The energy spectrum characteristics, as depicted in Fig. 3.39, demonstrate the detector's ability to effectively differentiate between the several energy levels of α-particle energy spectra. These spectra exhibit prominent spectral features. Conversely, in the energy range outside of α-particle energies, no discernible energy peaks are observed. The energy resolution of the detector in the electron collection mode is 1.04% for the $^{243}$Am source and 1.05% for the $^{244}$Cm source. In the hole collection mode, the energy resolution of the detector is 0.76% for the $^{243}$Am source and 0.87% for the $^{244}$Cm source. The energy resolution of the spectral characteristics exhibited by the hybrid source consisting of $^{243}$Am-$^{244}$Cm is found to be superior in comparison to the energy spectrum of the $^{241}$Am source. This observation leads us to propose that the discrepancy may be attributed to minor variations in contamination resulting from the

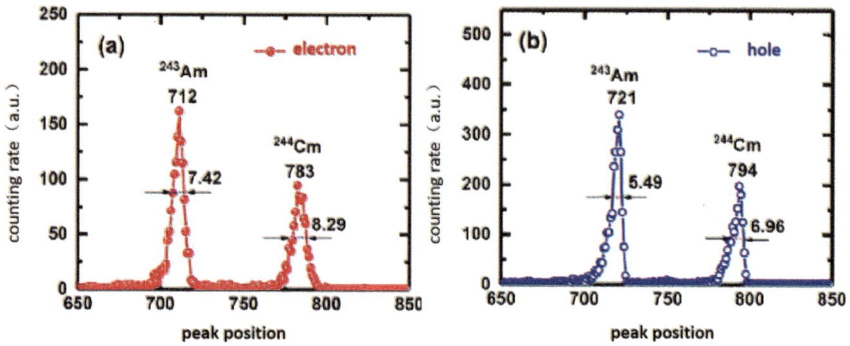


**Fig. 3.39** Energy resolution characteristics of the detector irradiated by a mixed source at $^{243}$Am-$^{244}$Cm. **a** electrons; **b** holes

hygroscopic nature of the radioactive source following extended periods of storage. Furthermore, it should be noted that the energy ratio of the hybrid source, specifically 5.805/5.275, exhibits equivalence to the peak ratios of 783/712 for electron collection and 794/721 for hole collection. This correspondence can be attributed to the exceptional CCE properties and the symmetrical collection of electrons and holes in the hydroxide-terminated CVD single-crystal diamond core detector.

### *3.4.4 Time Response of Single Crystal Diamond Core Detectors*

The time response properties of the detector were evaluated by subjecting it to a pulsed electron source with an energy of 40 MeV and a pulse width of 1 ps, while maintaining a bias voltage of 200 V. Figure 3.40 displays the pulse response, revealing that the rise time of the pulse response measures 347.4 ps, the pulse width spans 2.9 ns, and the pulse fall time amounts to 980.5 ps. Due to the favorable ohmic contact qualities exhibited by the Au and hydroxide terminals, with a contact resistivity of $1.73 \times 10^{-4}\ \Omega\cdot\text{cm}^2$, the time response does not exhibit a gradual decay component. However, it is worth noting that a ringing signal, attributed to the inductive connection of the cable is observed.

### *3.4.5 Comparison Results*

The present study involves the fabrication of hydroxide-terminated CVD single-crystal diamond nucleation detectors with enhanced performance. This was achieved through the implementation of a diamond surface manipulation technique. The experimental findings indicate that employing oxygen plasma etching on the non-electrode

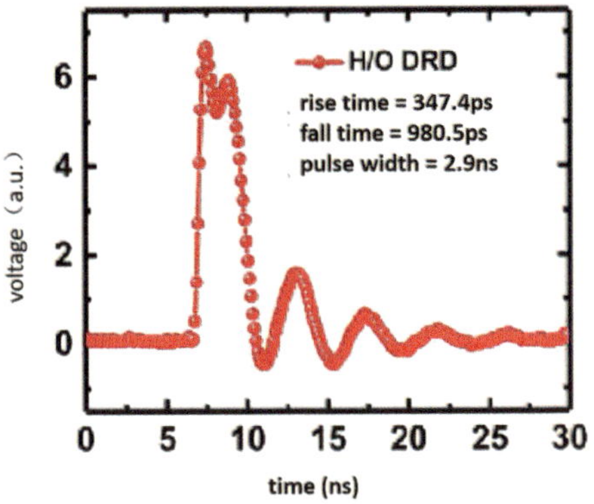


**Fig. 3.40** Time response of the hydrogen–oxygen terminated CVD single crystal diamond core detector

area of diamond to create an oxygen-terminated surface leads to a decrease in surface leakage and the creation of a well-functioning surface passivation region. Under an electric field of 1 V/μm, the dark current measures only $7.46 \times 10^{-13}$ A/mm$^2$. The utilization of hydrogen plasma is employed to create hydrogen terminals on the surface of diamond. This process results in the establishment of favorable ohmic contact qualities with the gold electrode, a reduction in surface trap density, and the attainment of efficient charge collection. The energy spectrum measurements obtained from the alpha source indicate that the nuclear detector exhibits a charge collecting efficiency of 99.01% for holes and 98.6% for electrons. Additionally, the energy resolution for holes and electrons is determined to be 0.76% and 1.04% respectively. The impulse response rise time of the detector was measured to be 347.4 ps when subjected to a time response test utilizing a picosecond pulsed electron source. The hydroxide-terminated CVD single-crystal diamond nucleus detector exhibits superior performance when compared to the findings shown in Table 3.5 of various published literature sources, both domestically and globally. The gadget has the potential to be utilized in charged particle spectroscopic measurements, as well as in the screening of high-speed pulses.

### *3.4.6 Energy Spectral Response of Polycrystalline Diamond Core Detectors to Alpha Particles*

Based on the preceding study, it is evident that the presence of impurities and dislocations in single-crystal materials exerts a notable influence on the CCE features of diamond nuclear detectors. Nevertheless, the fabrication of large-area nuclear detectors is challenging as a result of the size constraints imposed by CVD single-crystal materials. CVD polycrystalline diamond is a frequently encountered variant of diamond material. Polycrystalline diamond has been effectively employed in the fabrication of diamond pixel array nuclear detectors [36] as well as diverse field effect transistor devices [37, 38]. Polycrystalline materials offer distinct advantages

**Table 3.5** Comparison of hydrogen–oxygen terminated CVD single crystal diamond core detector with some literature results

| | CCE (%) | $\Delta E_0/E_0$ (%) | Structure |
|---|---|---|---|
| Domestic | – | 3.3 | DLC/Pt/Au [29] |
| | – | 3.5 | Graphitic [30] |
| | 97 | 3.7 | Al [31] |
| | – | 2.25 | Cr/Au [32] |
| Foreign | ~ 100 (e)/89 (h) | 3(e)/8.5 (h) | PLD-C/Ni [16] |
| | 97.7 (e)/99.4 (h) | 2.3(e)/1.5 (h) | TiC/Pt-Ru [33] |
| | 94 (e)/94 (h) | 4.5 | Printed silver [34] |
| | 92 (e)/98 (h) | 3.7 (e)/3.8 (h) | Ti/Au-Al [35] |
| This work | 98.6 (e)/99.01 (h) | 1.04 (e)/0.76 (h) | Hydroxide terminal/Au |

due to their substantial dimensions and comparatively reduced expenses. To investigate the spectrum characteristics of polycrystalline diamond nucleus detectors, we employ high purity "electron-grade" polycrystalline diamond with a designation of #6 to mitigate the potential impact of extraneous factors, such as impurities.

Based on the findings of the preceding study on material characterization, it is evident that there is no discernible distinction between high-purity "electron-grade" polycrystalline diamond and "electron-grade" single-crystalline diamond based on the non-destructive characterization analysis. Furthermore, the analysis of PL spectra did not reveal any impurity peaks, and the outcomes obtained from Raman spectroscopy exhibited a high degree of similarity. The primary distinction is observed in the grain boundary of polycrystalline diamond, as depicted in Fig. 3.21f. It is evident that #6 polycrystalline diamond comprises numerous crystal orientations, namely (111), (220), and (311). To ascertain the grain size of the sample, a scanning electron microscope (SEM) was employed for a subsequent examination, as illustrated in Fig. 3.41. The SEM analysis revealed distinct grain boundaries on the surface, exhibiting varying grain sizes characterized by a significant degree of randomness. The determined grain size range for this particular polycrystalline diamond specimen falls within the approximate range of 80 to 150 μm.

The CVD polycrystalline diamond core detectors were prepared by the same process as described above, and the sample information is shown in Table 3.6.

The energy spectral properties of the CVD polycrystalline diamond nucleus detector were examined by experimentation involving a $^{241}$Am source that generated α-particles with an energy of 5.486 MeV. The resulting energy spectral properties are depicted in Fig. 3.42. The figure illustrates that there is negligible alteration in the energy spectrum properties of the detector when subjected to bias voltages of 100 V, 200 V, and 300 V. The highest point on the energy spectrum is associated with a CCE value of 4%. The energy spectrum observed by the detector exhibits an exponential decay curve rather than a Gaussian distribution. Consequently, the polycrystalline diamond core detector lacks the capability to effectively discriminate energy for the detection of charged particles. The maximum value for the

**Fig. 3.41** SEM test image of polycrystalline diamond

**Table 3.6** CVD polycrystalline diamond core detector sample information

| Category | Sample number | Thickness | Electrode size | Remarks |
|---|---|---|---|---|
| Polycrystalline | #6 | 0.50 mm | 4 × 4 $mm^2$ | Electronic grade polycrystalline with low impurities |

CCE is approximately 35%. However, it is worth noting that the particle counts have already dropped to levels below 20. The energy spectrum characteristic, also known as the pulse height-amplitude spectrum, is a statistical distribution that represents the number of pulse signals detected at a specific channel site, which has been converted to CCE in this case. The figure illustrates that the CCE exhibits the highest number of pulse signals at 4%. This indicates that the incident alpha particles produce a charge within the detector, and the majority of the pulse signals are concentrated at the 4% level after being collected through the transport process. Due to the inherent variability in size and orientation of polycrystalline diamond grains, the distribution of incident α particles exhibits a discontinuous nature, wherein only a fraction of these particles yield a signal of CCE that can attain a maximum value of 35%. Based on the SEM findings, it is evident that the shooting region exhibits a maximum grain size of 150 μm. Considering the sample's thickness of 500 μm, the resulting quotient from dividing these values is 0.3. This aligns closely with the maximum CCE (35%) derived from the spectroscopic characterization. Hence, in the context of high-purity "electron-grade" polycrystalline diamond core probing, the maximum CCE is constrained by the upper limit of grain size. This limitation arises from the presence of grain boundaries, which act as traps and composite centers. When carriers are transported towards these grain boundaries, the strong composite effect hinders the carrier transport process, leading to a decrease in the CCE value. Furthermore, if we account for the impact of impurities and dislocations in polycrystalline diamond,

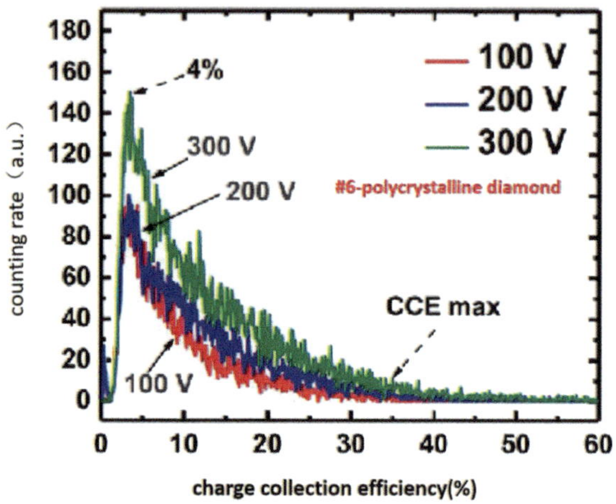


**Fig. 3.42** Energy spectral characteristics of CVD polycrystalline diamond core detectors

the CCE value is further diminished. Incorporating the influence of impurities and dislocations in polycrystalline diamond will further diminish the CCE value.

## 3.5 Response of Diamond Nuclear Radiation Detectors to X-Rays

In the context of high-throughput radiometric conditions, the application of pulse-height spectroscopy for measuring energy-spectrum properties becomes impractical. This is primarily due to the significantly brief time intervals between incident rays, resulting in overlapping pulse spectra. Consequently, the prevalent approach for detection is through current mode. One advantage of diamond nuclear detectors in X-ray detection is their similarity in atomic number to the human body model, enabling precise evaluation of absorbed dose within the human body. Additionally, their compact size enhances spatial resolution. In the context of X-ray detection applications, it is common for diamond nucleus detectors to function in current mode. This operational mode necessitates the use of nucleus detectors that possess several key characteristics, including high output current and sensitivity, excellent linearity and repeatability, as well as rapid response time. This section discusses the reaction of diamond nucleus detectors to X-rays, based on the aforementioned reasons.

### 3.5.1 Performance Parameters and Test Methods

1. Performance parameters

(a) Dark current and signal-to-noise ratio

Dark current, as previously defined, refers to the inherent leakage current of a device in the absence of external radiation. Typically, the measurement of dark current involves placing the device in a dark room. However, due to diamond's lack of response to visible light, it is also feasible to conduct the test in a natural light environment.

The signal-to-noise ratio (SNR) is defined as the ratio of the light current value to the dark current value.

(b) Photoconductive gain and responsiveness

The gain factor is defined as the ratio of the photogenerated test current $I_{\mathrm{m}}$ to then photogenerated theoretical current $I_{\mathrm{g}}$:

$$G_{\mathrm{ain}} = \frac{I_{\mathrm{m}}}{I_{\mathrm{g}}} \tag{3.21}$$

The incident light irradiating the detector material produces a theoretical current value of:

$$I_{\mathrm{g}} = \frac{D\rho ev}{w} \tag{3.22}$$

Here, $D$ is the absorbed dose rate, $\rho$ is the material density, $e$ is the unit charge, $v$ is the sensitive volume of the material, and $w$ is the average ionization energy of the material to produce a pair of electron–hole pairs, which is 13.1 eV for diamond.

(c) Linearity

The incident light irradiating the detector material produces a theoretical current value of:

$$I = I_{\mathrm{dark}} + \mathrm{A}D^{\Delta} \tag{3.23}$$

$\Delta$ is the linearity index, which responds to the linearity of the device, and A is a constant.

(d) Sensitivity and characteristic sensitivity

The sensitivity of the detector is defined as the linear scaling factor of the integrated charge generated by the detector to the total dose of irradiation, as shown in the following equation:

$$Q = \mathrm{a} + bDt \tag{3.24}$$

Here, $b$ is the linear coefficient, i.e., sensitivity, $t$ is the irradiation time, and a is a constant.

The characteristic sensitivity is defined as the sensitivity divided by the detector sensitivity volume.

(e) Repeatability

Repeatability is defined as the standard deviation *SD* of the response current divided by the average current value to determine:

$$\text{Reproducibility} = \left[\left(\frac{SD}{\text{Average current}}\right)\%\right] \tag{3.25}$$

The standard deviation of the current usually consists of a combination of inconsistencies in carrier transport of the detector device, errors in the measurement equipment, and random fluctuations in the radiation source, and the International Atomic Energy Agency (IAEA) recommends that the repeatability should be less than 0.5%.

(f) Time response

The time response mainly includes the rise time and fall time, which refers to the time for the current signal to rise from 10 to 90% of the maximum value and to fall from 90 to 10%, respectively.

2. Test method

The diamond's response to X-rays was evaluated using a MultiRad 160 X-ray irradiator. This irradiator is equipped with a maximum tube voltage of 160 kV and a maximum tube current of 30 mA. Additionally, it features an integrated dosimeter that allows for real-time monitoring of the absorbed dose of irradiation. The test's principle is illustrated in Fig. 3.43. During the experiment, a 6517B electrostatic meter was utilized to establish a voltage bias and quantify the response current. Simultaneously, the dosimeter positioned in the center was employed to measure the absorbed dosage rate in real-time. The irradiator possessed an X-ray exit angle of 37.5°, while the detector was positioned at a distance of 44.5 cm from the X-ray source. Within a diameter of 31 cm at this height, the distribution of X-ray absorbed dose rate remains stable.

The experimental setup involved setting the tube voltage to 120 kV while varying the tube current to manipulate the absorbed dose rate. The exact tube currents corresponding to the absorbed dose rates are presented in Table 3.7. It is important to note that all tests were conducted under room temperature conditions.

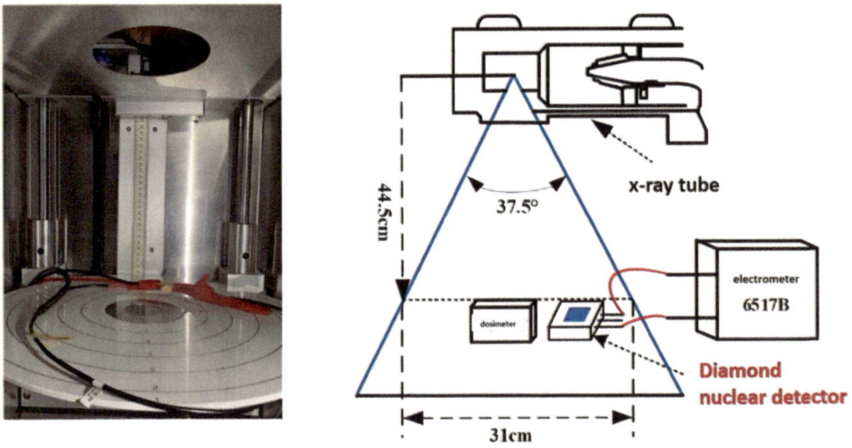


(a) (b)

**Fig. 3.43** X-ray response test method. **a** Real products; **b** test principle

**Table 3.7** Absorbed dose rates corresponding to different tube currents at 120 kV tube voltage

| Tube current (mA) | 30 | 20 | 15 | 10 | 5 | 1 | 0.5 |
|---|---|---|---|---|---|---|---|
| D (Gy/min) | 6.157 | 4.159 | 3.138 | 2.110 | 1.063 | 0.214 | 0.108 |

### 3.5.2 *X-Ray Response of Single Crystal Diamond Nuclear Radiation Detectors*

The previous study examined the reaction of hydroxide-terminated CVD single-crystal diamond nuclear detectors to α-particles. The findings demonstrated that establishing effective ohmic contacts on nuclear detectors composed of high-purity "electron-grade" diamonds greatly improves the charge collection capabilities of these detectors. When the detector is utilized in current mode, the presence of an excellent ohmic contact will not introduce a substantial additional impedance, hence facilitating an increase in the output current of the detector. Hence, this part proceeds to examine the response characteristics of the nuclear detector in question with respect to X-rays.

1. Device fabrication and current–voltage characteristics

The device fabrication process is the same as in Chap. 4, and the single crystal number chosen is as well as the specific device dimensions as shown in Table 3.8, with the sensitive size being the product of the electrode coverage size and the sample thickness.

The CCE of the device was determined to be 99% using the methodology outlined in the preceding chapter. Figure 3.44 illustrates the relationship between the dark

**Table 3.8** Device feature dimensions

| Number | Sensitive size | Remarks |
|---|---|---|
| #5–1 | $3 \times 3 \times 0.3$ mm$^3$ | X-ray response |

current of the device and the current–voltage characteristic curve when subjected to X-irradiation. The dark current of the device has a value of $2.14 \times 10^{-11}$ A under a bias voltage of 300 V, corresponding to an electric field strength of 1 V/μm. Conversely, when the bias voltage is reduced to 0 V, the dark current decreases to $2.35 \times 10^{-14}$ A. This observation suggests that the diamond nucleus detector exhibits a remarkably minimal leakage current, hence enabling the attainment of a much reduced measurement threshold. When the X-ray tube voltage is set to 120 kV and the tube current is set to 1 mA, it is observed that the SNR of the device experiences a minor rise as the bias voltage is adjusted. Specifically, the SNR of the device is measured to be $3.2 \times 10^4$ when the bias voltage is set to 200 V, and $1.8 \times 10^4$ when the bias voltage is set to $1.8 \times 10^4$. In the same way, when the X-ray tube voltage is set at 120 kV and the tube current is 30 mA, the SNR of the device is found to be $1 \times 10^6$ at a bias of 200 V, and $5 \times 10^5$ at a bias of 50 V. When the dosage rate of the input X-rays is augmented by a factor of ten, the resulting output current experiences a rise, thus leading to an amplification of the SNR. The International Atomic Energy Agency (IAEA) recommends SNR > 1000 [16] for X-ray measurements. This level of SNR is deemed enough for achieving accurate detection outcomes. Insufficient switching ratio can lead to inaccuracies in dosimetry and have an impact on further parameters, including the linearity and sensitivity of the instrument.

2. Current time response characteristics of the device

The evaluation of a device's photoconductivity gain and responsivity, linearity, sensitivity, and characteristic sensitivity at a certain voltage can be achieved by measuring its current time response characteristics to X-rays. Figure 3.45 shows the current

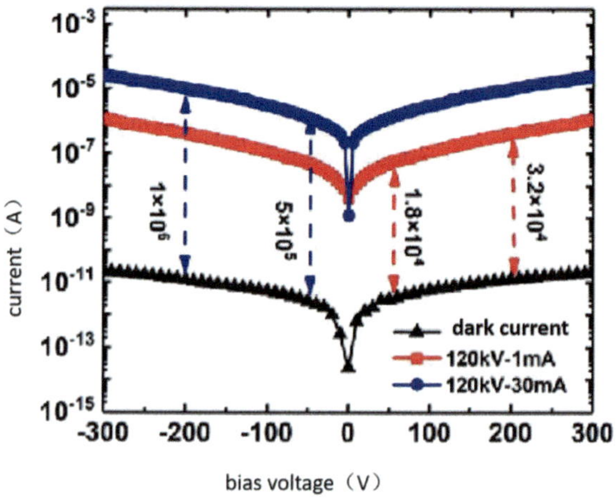


**Fig. 3.44** Current–voltage characteristics of the device response to X-rays

response of the device to different dose rates of X-rays at different bias voltages. Prior to activating the X-ray source, the gadget exhibits a dark current on the order of $10^{-11}$ A. Upon activation of the X-ray source, the device exhibits a prompt escalation in response current, with rise times below 0.6 s observed at various dosage rates. This behavior is observed when the bias voltage is set at 50 V as opposed to 200 V. Figure 3.45a, b depict the rising time of X-rays when subjected to a dosage rate of 1.063 Gy/min, which measures around 0.56 to 0.58 s. When comparing the bias voltages of 50 V and 200 V, it is observed that the response current at various dose rates exhibits an increase proportional to the dose rate. This observation is made under the condition of a constant irradiation time of 3 min for each dose rate. Additionally, upon turning off the X-rays, the device current promptly reverts back to the dark current state. The fall-off time for this transition is measured to be 0.17 s and 0.36 s for the bias voltages of 50 V and 200 V, respectively. At varying levels of dosage, the persistent photocurrent phenomena [39] does not exhibit any discernible priming effect [39].

Based on the device current time response, the gain of the device can be calculated from Eq. (3.21), as shown in Table 3.9.

The gains at different dose rates were averaged to obtain an average gain of 19 times for the device at 50 V bias voltage and 151.83 times for the device at 200 V bias voltage. The analysis reveals that the device's gain exhibits a standard deviation of 0.111 when subjected to a bias voltage of 50 V. However, this standard deviation rises to 3.84 when the bias voltage is increased to 200 V. Our hypothesis posits that the observed rise in the fluctuating float of the gain of the device at high bias voltages can be attributed to the amplified impact of collisional ionization on the carriers.

The linearity of the device was determined by employing the Fowler model and fitting it to the data, as illustrated in Fig. 3.46. The current values were acquired by calculating the average current over a period of 3 min during testing at various dose rates, as presented in Table 3.9. The figure illustrates that the device exhibits an $\Delta$ value of $1.014 \pm 0.001$ at a bias voltage of 50 V, and an $\Delta$ value of $1.033 \pm 0.014$ at a bias voltage of 200 V. This observation suggests a nearly linear correlation between the device and the measured range of dosage rates. The observed trend indicates that the linear exponent $\Delta$ value exhibits a positive correlation with the bias voltage, which has also been documented in previous studies [40, 41]. However, it is worth noting that the literature reports low CCE values for diamond nucleus detectors. The increase in the $\Delta$ value resulting from the enhancement voltage can be attributed to the reduction in carrier crossing time, leading to a decrease in the number of carriers captured by the trap and consequently an increase in the linear exponent. Nevertheless, the CCE of our detector exhibits a remarkably high value, nearing 100%. Additionally, the $\mu\tau$ derived from the computation of Hecht's equation yields an approximate value of $10^4$ cm$^2$/V. Based on the carrier mobility values documented in the literature, ranging from 1000 to 4500 cm$^2$/V·s, it can be inferred that the carrier lifetime is approximately on the order of magnitude of $10^2$ ns. Consequently, for a detector with a thickness of 300 μm, the carrier transport time is estimated to be a few nanoseconds. It is evident that the carrier lifetime significantly exceeds the transit time. Under these circumstances, at higher voltages, the collisional ionization of

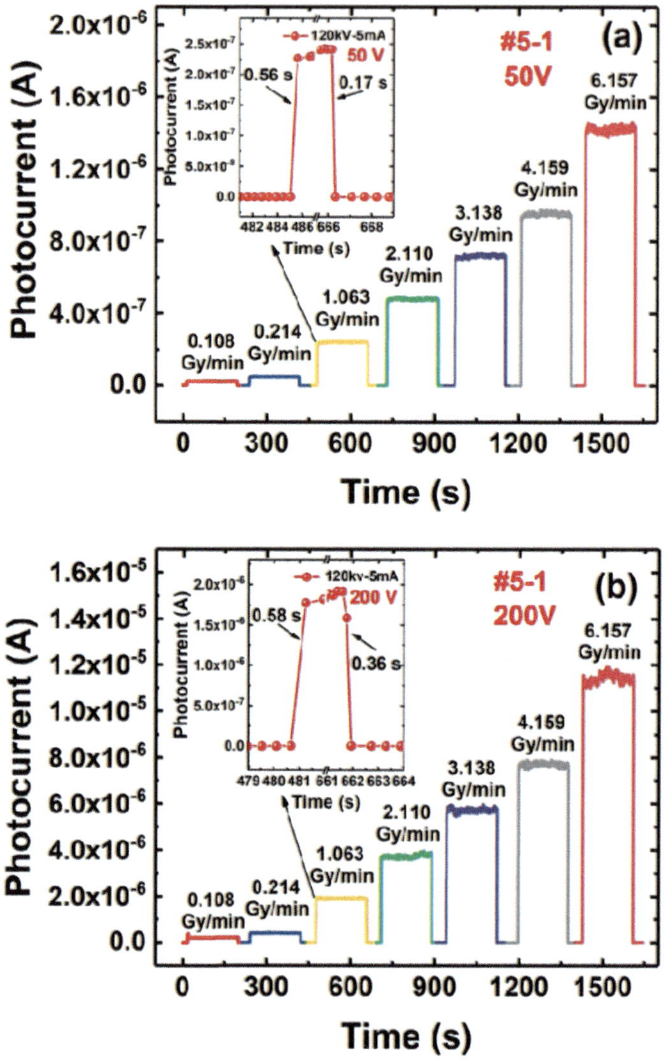


Fig. 3.45 Current response of the device to different dose rates of X-rays at different bias voltages. **a** 50 V; **b** 200 V

carriers within the lattice becomes more prominent, leading to an enhanced level of randomness and subsequently increasing the value of Δ. Figure 3.45b demonstrates that the stochastic enhancement at a bias voltage of 200 V induces a more prominent fluctuation in the current time profile, as opposed to the fluctuation observed in Fig. 3.45a. Despite a marginal rise in the value of Δ at 200 V, the linearity index of the device remains in close proximity to 1. According to the guidelines set forth by the International Commission on Radiological Units (ICRU), it is recommended that

**Table 3.9** Average response current of the device at 50 V versus 200 V

| Gy/min | 0.108 | 0.214 | 1.063 | 2.110 | 3.138 | 4.159 | 6.157 |
|---|---|---|---|---|---|---|---|
| Average current (50 V) | $2.461 \times 10\ A^{-8}$ | $4.914 \times 10\ A^{-8}$ | $2.409 \times 10\ A^{-7}$ | $4.793 \times 10\ A^{-7}$ | $7.168 \times 10\ A^{-7}$ | $9.529 \times 10\ A^{-7}$ | $1.418 \times 10\ A^{-6}$ |
| Gain (50 V) | 18.95 | 19.10 | 18.85 | 18.89 | 18.99 | 19.05 | 19.16 |
| Average current (200 V) | $2.051 \times 10\ A^{-7}$ | $3.881 \times 10\ A^{-7}$ | $1.893 \times 10\ A^{-6}$ | $3.722 \times 10\ A^{-6}$ | $5.702 \times 10\ A^{-6}$ | $7.707 \times 10\ A^{-6}$ | $1.139 \times 10\ A^{-5}$ |
| Gain (200 V) | 158.02 | 150.86 | 148.15 | 146.73 | 151.13 | 154.13 | 153.85 |

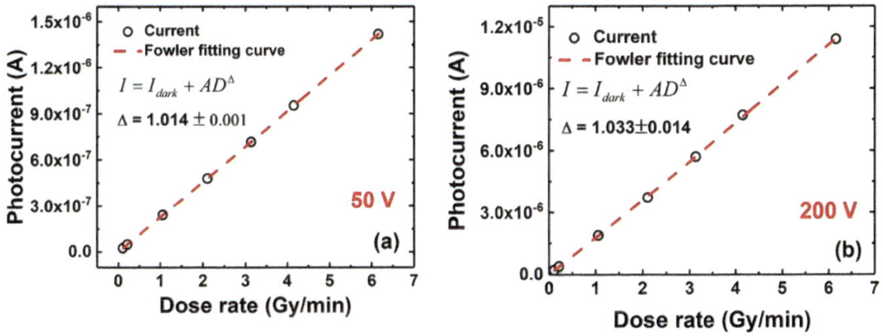


**Fig. 3.46** Linearity of the device Fowler model fit. **a** 50 V; **b** 200 V

fluctuations in the value should be maintained at a level not exceeding 5% [42]. In the present context, the maximum fluctuation in the Δ value of the device is found to be less than 2%.

In this study, the sensitivity of the device is determined by subjecting it to X-ray irradiation for a duration of 3 min at various dose rates. The output photocurrent is then integrated over the same 3-min period to obtain the charge. Subsequently, a linear regression analysis is performed to establish the relationship between the accumulated dose and the obtained charge. This approach is employed due to the high linearity exhibited by the device. The sensitivity of the device at bias voltages of 50 V and 200 V is depicted in Fig. 3.47. The sensitivity of the device is determined by the slope of the linear fit observed in the figure. The sensitivity of the device is observed to be 13.858 μC/Gy at a bias of 50 V and 111.890 μC/Gy at a bias of 200 V. The dimensions of the sensitive volume of the device are $3 \times 3 \times 0.3\ mm^3$. To determine the characteristic sensitivity of the device, the sensitivity is divided by the sensitive volume. Specifically, at a bias of 50 V, the characteristic sensitivity is calculated to be 5.133 μC/Gy $mm^3$, while at a bias of 200 V, the characteristic sensitivity is calculated to be 41.441 μC/Gy $mm^3$.

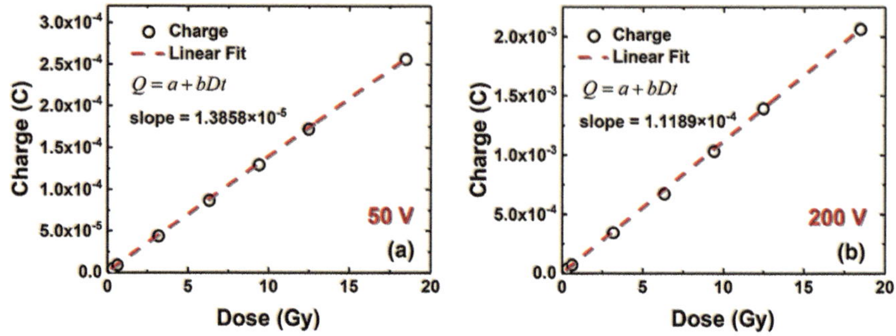


**Fig. 3.47** Sensitivity of the device at different voltages. **a** 50 V; **b** 200 V

The repeatability test of the device is conducted utilizing the square wave scanning mode of the 6517B electrostatic meter, as depicted in Fig. 3.48. Initially, the X-ray source is typically in an active state, while the electrostatic meter administers varying levels of high and low voltage to the devices. Subsequently, the meter records the resulting current response of the device at the high voltage level. Statistical computations are then conducted to derive the mean and standard deviation of the device's response current, thereby facilitating the assessment of the device's repeatability. The duration of measurement in this context spans 10 s.

The current values of the device under X-ray irradiation at a tube voltage of 120 kV and tube current of 30 mA are depicted in Fig. 3.49. The measurements were conducted across numerous cycles, specifically cycles exceeding 35. It is observed that the current response of the device fluctuates by 0.759% when the bias voltage of the device is set to 50 V. When the bias voltage is 200 V, the current response fluctuates 0.929%. This yields a device repeatability of 0.759% at 50 V bias and a device repeatability of 0.929% at 200 V bias. Based on the aforementioned findings, it is evident that the device's repeatability diminishes with an increase in bias voltage. This trend is analogous to the amplification of current gain fluctuations observed at

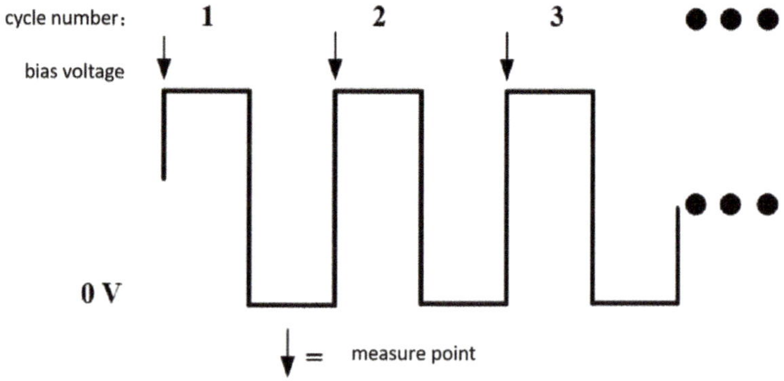


**Fig. 3.48** Device repeatability test method

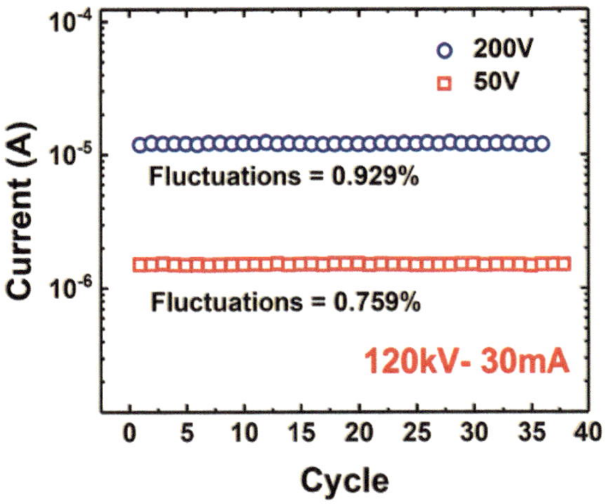


**Fig. 3.49** Repeatability of the device at 50 V bias versus 200 V bias

higher bias voltages. The underlying cause for this phenomenon is likely associated with the intensified collisional ionization effect induced by the elevated voltage. To get a consistent output current, it is necessary to correctly decrease the bias of the device. However, this adjustment will lead to a decline in gain. Therefore, a balance must be struck between the desire for repeatability and the need for gain.

3. Analysis and Discussion

The aforementioned test findings demonstrate that the CVD single crystal diamond nucleation detector with a hydroxide terminal/gold electrode (H/O/Au) structure has several notable advantages, including a significant increase in current gain, heightened sensitivity to features, and a favorable Δ-index. The significant increase in gain is primarily attributed to the combined effects of two factors. Firstly, the low defect density of the bulk material properties leads to a reduction in traps and composite centers within the diamond body. This reduction subsequently enhances the output current. Secondly, for X-ray sources with tube voltages exceeding 50 kV, the energy deposition across the diamond layer, which is less than 1 mm thick, approaches unity. This behavior differs significantly from the α-particle detection model. Consequently, the model of the detector can be regarded as a photoconducting model. The current gain of this model can be mathematically expressed as:

$$G_{\mathrm{ain}} = \frac{I_{\mathrm{m}}}{I_{\mathrm{g}}} = \tau\left(\frac{1}{t_{\mathrm{rn}}} + \frac{1}{t_{\mathrm{rp}}}\right) \tag{3.26}$$

In this context, $\tau$ represents the average lifetime of the carriers, whereas $t_{\mathrm{rn}}$ and $t_{\mathrm{rp}}$ denote the transit durations of the electrons and holes as they traverse the two electrodes, respectively. To enhance the gain of the detector, it is necessary to extend the lives of carriers and reduce their transit durations. This assertion is supported by the observed rise in gain resulting from the augmentation of the bias voltage. Alternatively, when the electrodes of the detector exhibit favorable ohmic contact properties, the carrier possessing greater mobility will be expelled from the detector initially, while the carrier with lower mobility will remain within the body. To preserve the electroneutrality of the detector, the carrier with higher mobility will subsequently be reintroduced into the inner detector body by the electrodes, thereby enhancing the gain. There exists a notion positing that the "sensitizing effect" may also contribute to the amplification of photoconductive detectors. In the scenario where the electrode contact exhibits ideal ohmic behavior and a significant number of one type of carriers are confined within the detector's body while the other type of carriers remain unbound, it becomes necessary to preserve the electrical neutrality of the body. This is achieved by reintroducing the unbound carriers from the electrodes of the ohmic contact into the detector's body, until the trapped carriers are de-trapped [43].

In the structure of our detector, the CCE for both electrons and holes approaches 100%. As a result, the concentration of electron/hole traps within the detector body is significantly reduced, leading to a low center density. Consequently, the material body effect plays a partial role in enhancing gain and feature sensitivity. In contrast, the hydrogen terminals exhibit favorable ohmic contact properties with the Au electrodes, and the influence of carrier traps on the "sensitizing effect" is negligible. The absence of a substantial disparity in CCE between electrons and holes, as observed in the detector's performance, indicates the gain enhancement may be attributed to the re-injection of carriers resulting from the swift removal of carriers with high mobility.

In the study conducted by Galbiati et al. [14], CVD single-crystal diamond with nitrogen and boron impurity concentrations below 5 ppb was utilized. This resulted in a significant reduction of the material's bulk effect. Furthermore, the three electrode structures employed in the study, namely TiW, Cr/Au, and DLC/Pt/Au (as indicated in Table 3.10), exhibited considerably lower characteristic sensitivities compared to the current research due to their inferior ohmic contact properties. In a study conducted by Schirru et al. [44], high-purity "electron-grade" single crystals were employed to create a detector utilizing DLC/Pt/Au. The resulting detector exhibited a current gain of merely 0.8 for X-rays. Furthermore, the characteristic sensitivity of the current structure surpassed that of a commercial PTW diamond nucleus detector. In a study conducted by Abdel-Rahman et al. [45], it was shown that X-ray detectors built with Pt and Al/Pt electrodes on "electron-grade" single crystals achieved gains of just 0.722 and 2.31, respectively. In the aforementioned investigation, despite efforts to eliminate material restrictions, the electrodes utilized could not establish a satisfactory ohmic contact with the diamond material. Consequently, the electrostatic equilibrium hindered charge re-injection, posing challenges in achieving high gain. The application of pulsed laser deposition (PLD) for the formation of C electrodes on the diamond surface and subsequent redeposition of Ni metal has demonstrated notable sensitivity to features and significant amplification [16]. However, the signal-to-noise ratio and exponent of this approach are not as favorable as those observed in the current structure. The above mentioned gain can be analyzed and concluded as follows: in order to achieve a high current gain, two conditions must be satisfied. Firstly, the carriers within the detector body should possess a long carrier lifetime and a short transition time. Secondly, the detector electrode contact should exhibit favorable ohmic contact characteristics. This can be achieved by employing a hydroxide termination structure on the diamond and establishing an excellent ohmic contact with an Au electrode. Additionally, the utilization of high-purity "electronic grade" materials serves to reduce the carrier lifetime limitation. Simultaneously, the ohmic contact facilitates an increase in the detector's response to X-rays, resulting in enhanced current gain and sensitivity characteristics.

The Δ-index serves as a significant parameter in the context of X-ray dose rate assessments. When the value of Δ is equal to 1, it signifies that the response current of the detector remains constant regardless of the dose rate. This characteristic is advantageous as it enables precise evaluation of the absorbed dose. Based on the Fowler model, the value of Δ is determined to be 0.5 when the semiconductor lattice structure is in an ideal state without any traps and is dominated by composites. If the traps within the material body have the same capture cross section, $0.5 < \Delta < 1$. If the traps within the material have different capture cross sections or if the distribution of traps within the material is not uniform, $\Delta > 1$. However, interpreting the Δ-index only in terms of bulk material properties is not always valid; this is because, the Fowler model is based on the ohmic contact. The findings reported by Mohamed et al. [45] indicate that the Δ-index is influenced by both the electrode contact and the annealing process. Specifically, the lower Δ-value observed can be attributed to the presence of a non-ideal ohmic contact, which results in the accumulation of negative charge near the anode and positive charge near the cathode of the detector. This charge accumulation subsequently diminishes the electric field, ultimately leading to a Δ-value below unity. In Galbiati's study [14], it was observed that the Δ-value of the TiW electrode deviated significantly from unity, reaching a value of 1.58. This

**Table 3.10** Comparison of the performance of this work with other published single crystal diamond X-ray detectors

| Electrode type | Gain | SNR | Feature sensitivity (μC/Gy mm$^3$) | Δ |
|---|---|---|---|---|
| H/O/Au (50 V) | 19 | $10^4$ ~ $10^5$ | 5.133 | 1.014 ± 0.001 |
| H/O/Au (200 V) | 151.83 | $10^4$ ~ $10^6$ | 41.441 | 1.033 ± 0.014 |
| TiW [14] | – | ~ 6.1 × $10^4$ | 0.246 | 1.58 ± 0.04 |
| Cr/Au [14] | – | ~ 2.8 × $10^4$ | 0.153 | 1.05 ± 0.01 |
| DLC/Pt/Au [14] | – | ~ 1.1 × $10^5$ | 0.144 | 1.011 ± 0.008 |
| DLC/Pt/Au [44] | 0.8 | ~ 5.1 × $10^4$ | 0.165 | 1.014 ± 0.008 |
| PTW (Commercial) [44] | – | – | 0.058 | 0.999 ± 0.008 |
| Pt [45] | 0.722 | $10^3$ ~ $10^4$ | 0.185 μC/Gy mm$^3$ | 0.975 |
| Al/Pt [45] | 2.31 | $10^4$ ~ $10^5$ | 0.600 μC/Gy mm$^3$ | 0.968 (annealed) |
| PLD-C/Ni [16] | ~ 20 | 7.2 × $10^3$ | 4.87 μC/Gy mm$^3$ | 0.97 ± 0.02 |

deviation can be attributed to the existence of traps located at the interface between the TiW and diamond materials. Hence, the diamond core detector with a H/O/Au structure exhibits a Δ-index of $1.014 \pm 0.001$ at a bias voltage of 50 V, owing to its exceptional interfacial characteristics.

In comparison to the performance of previous diamond detectors that have been published and are commercially available, the diamond X-ray detector in the current structure exhibits a notable performance advantage. This advantage can be attributed to the hydrogen-terminated diamond, which forms a favorable ohmic contact with Au. This finding is supported by the data presented in Table 3.10. The gain, SNR, characteristic sensitivity, and Δ-index of the device at 50 V bias reached 19, $10^4$–$10^5$, 5.133 μC/Gy $mm^3$, $1.014 \pm 0.001$, respectively. The gain, SNR, characteristic sensitivity, and exponent of the device at 200 V bias, on the other hand, were 151.83, $10^4 \sim 10^6$, 41.441 μC/Gy $mm^3$, and $1.033 \pm 0.014$, respectively.

### 3.5.3 X-Ray Response of Polycrystalline Diamond Nuclear Radiation Detector

One notable benefit of utilizing polycrystalline diamond as a material for nuclear detectors lies in its comparatively greater dimensions and cost-effectiveness. Consequently, the utilization of polycrystalline diamond enables the acquisition of a more expansive sensitive region for detecting purposes. While the dark current of the detector exhibits an upward trend with an increase in the detecting area, the implementation of surface etching by the utilization of oxygen plasma can significantly mitigate the surface leakage. Prior research has indicated that the interface between the hydrogen terminals and the Au electrodes can establish effective ohmic contacts, thereby enhancing the device's response current. Consequently, this section proceeds to examine the response of the hydroxide-terminated CVD polycrystalline diamond core detector to X-rays.

1. Device fabrication and current–voltage characteristics

The fabrication process of the device followed the methodology described in Chap. 4. The polycrystalline diamond used in this study was a commercially available "electronic grade" polycrystalline diamond obtained from sample #6. The dimensions of the diamond were $5 \times 5 \times 0.5$ $mm^3$, while the electrode dimensions were $4 \times 4$ $mm^2$. An Au layer with a thickness of 150 nm was applied to the diamond. Finally, the device was packaged in a TO shell and tube package. The ultimate, contrived apparatus is depicted in Fig. 3.50.

The current–voltage characteristics of the hydroxide-terminated CVD polycrystalline diamond core detector are depicted in Fig. 3.51. The measurements were conducted within a voltage range of − 600 V to 600 V. The highest dark current value is observed at a bias of 600 V, measuring $3.08 \times 10^{-11}$ A, while the lowest dark current value is recorded at − 160 V, measuring $3.55 \times 10^{-14}$ A. The image illustrates that the minimal value of dark current does not coincide with 0 V, a phenomenon

**Fig. 3.50** Physical view of polycrystalline diamond core detector

attributed to the polarization effect [46]. In polycrystalline diamond, grain boundaries and defects typically manifest as traps and composite centers. When an external bias is applied, the limited number of carriers within the diamond will undergo drift in response to the external bias. Consequently, electron traps will accumulate near the anode of the detector, while hole traps will accumulate near the cathode. This accumulation leads to the establishment of built-in electric fields that oppose the external bias. As a result, the overall electric field within the detector is attenuated, causing a modification in the distribution of electric field throughout the detector's body, rendering it non-uniform. Thus causing an offset of the dark current minimum with respect to the zero point. Upon conducting tests on the device's current–voltage characteristics using an applied X-ray source, it is observed in the accompanying figure that the minimum current value reverts back to a bias close to 0 V. The observed alteration in the device's dark current characteristics represents a substantial departure, which can be elucidated through the lens of the priming effect. Upon putting the device to an X-ray source, the examination of its current–voltage characteristics reveals that the minimum current value converges towards a bias voltage in close proximity to 0 V, as depicted in the accompanying figure. The observed alteration in the device's dark current characteristics is a substantial departure, which can be attributed to the phenomenon known as the priming effect. The irradiation of X-rays induces the generation of a significant quantity of electron–hole pairs within the diamond material. These pairs subsequently occupy the trap state present in the material, resulting in a uniform distribution of the electric field within the in vivo detector. Consequently, this process suppresses the polarization effect and restores the symmetrical distribution of the current–voltage characteristic curve [47]. When the device was performing ray tracing at an X-ray source with a tube voltage of 120 kV and a tube current of 1 mA, the highest output current was around $10^{-7}$ A, and the SNR at a bias of 300 V compared to 80 V was approximately $2 \times 10^3$. When the tube current is augmented to 30 mA, the resultant output current of the device is around $10^{-6}$ A. The SNR is observed to be $10^5$ when a bias of 300 V is applied, whereas a bias of 80 V yields an SNR of $5 \times 10^4$.

2. Current time response characteristics of the device

Fig. 3.51 Current–voltage characteristics of polycrystalline diamond core detectors

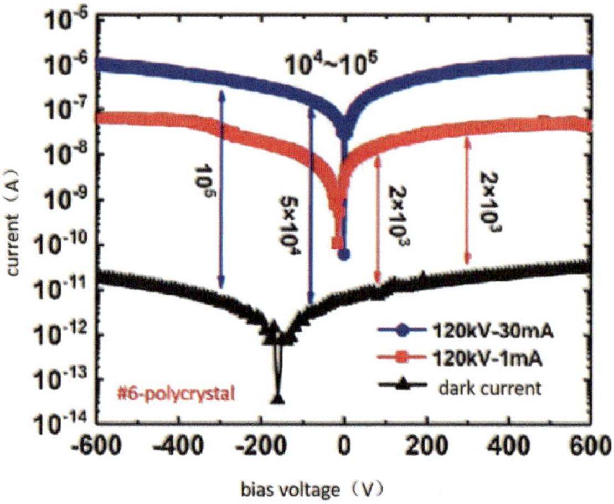


The current time response characteristics of the device for various dosage rates of X-rays at varied bias voltages are depicted in Fig. 3.52. The figure illustrates that upon activation of the X-ray source, there is a gradual increase in the current response over time. Specifically, at voltages of 80 V and 300 V, the final current values exhibit a respective increase of 1.6% and 4.8% compared to the initial value. This observed phenomenon can be attributed to the priming effect, as previously discussed [39]. The devices underwent irradiation with varying doses of X-ray radiation for a duration of 3 min each. Additionally, the devices exhibited a drop time of 1 s all around when the X-ray source was activated and deactivated. The phenomenon of persistent photocurrent was not seen in the study [39]. Nevertheless, it should be noted that the devices exhibit modest variations in their rise times, wherein the rise times exceed 300 V at 80 V for the identical dosage rate. This phenomenon can be attributed to the accelerated trapping of electron–hole pairs at elevated bias voltages, resulting in a faster filling of the trap. When exposed to a dose rate of 1.063 Gy/min, the rise time was shown to be roughly 3.14 s at 80 V and 2.35 s at 300 V.

Table 3.11 displays the mean current of the device at various bias voltages and dosages. The device's gain can be determined by utilizing Eq. (3.21) and considering the device's current time response. From the table, it is evident that the device's gain gradually diminishes as the X-ray irradiation dose increases, ultimately reaching a state of stability. The reason for the greater increase in gain at lower doses of irradiation compared to higher doses may be attributed to the elevated rate of carrier complexation at higher doses. This leads to a situation where a portion of the electron–hole pairs generated are already being complexed before the transport process is completed, resulting in a decrease in gain. The device's average gain at 80 V is 1.31, while at 300 V it is 3.92, as determined by the acquired gain values.

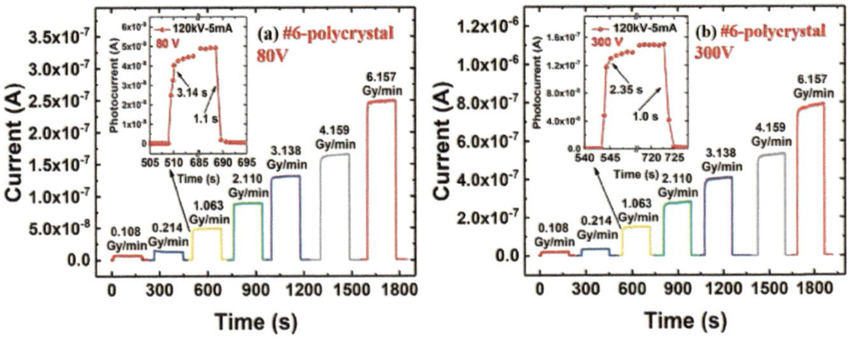


**Fig. 3.52** Current time response characteristics of polycrystalline diamond core detectors. **a** 80 V; **b** 300 V

**Table 3.11** Average response current of polycrystalline diamond core detector at 80 V versus 300 V

| Gy/min | 0.108 | 0.214 | 1.063 | 2.110 | 3.138 | 4.159 | 6.157 |
|---|---|---|---|---|---|---|---|
| Average current (80 V) | $6.466 \times 10$ $A^{-9}$ | $1.284 \times 10$ $A^{-8}$ | $4.801 \times 10$ $A^{-8}$ | $8.844 \times 10$ $A^{-8}$ | $1.299 \times 10$ $A^{-7}$ | $1.634 \times 10$ $A^{-7}$ | $2.471 \times 10$ $A^{-7}$ |
| Gain (80 V) | 1.68 | 1.67 | 1.26 | 1.17 | 1.16 | 1.10 | 1.12 |
| Average current (300 V) | $1.865 \times 10$ $A^{-8}$ | $3.404 \times 10$ $A^{-8}$ | $1.466 \times 10$ $A^{-7}$ | $2.750 \times 10$ $A^{-7}$ | $3.999 \times 10$ $A^{-7}$ | $5.239 \times 10$ $A^{-7}$ | $7.730 \times 10$ $A^{-7}$ |
| Gain (300 V) | 4.86 | 4.44 | 3.87 | 3.66 | 3.57 | 3.53 | 3.52 |

The graph in Fig. 3.53 demonstrates the linearity of the polycrystalline CVD diamond core detector, which was analyzed using Fowler's model. The obtained values of $0.981 \pm 0.032$ and $0.978 \pm 0.010$ correspond to different absorbed dose rates ranging from 0.108 to 6.157 Gy/min. These measurements were taken at bias voltages of 80 V and 300 V, respectively. Figure 3.53 demonstrates the linearity of the polycrystalline CVD diamond core detector, which was fitted using Fowler's model. The values of $\Delta$ are $0.981 \pm 0.032$ and $0.978 \pm 0.010$ when the absorbed dose rate is changed from 0.108 Gy/min to 6.157 Gy/min at a bias voltage of 80 V versus 300 V, respectively. The standard deviation of the $\Delta$ value is smaller at 300 V bias compared to 80 V bias. This is because at 300 V bias, the electric field strength inside the detector is larger, resulting in a shorter drift time for the carriers. This reduces the uncertainty caused by the random complexation of carriers. The linear exponents of the devices are discovered to be highly proximate to 1. When $\Delta$ is set to 1, the photocurrent of the devices remains unaffected by the dose rate. This characteristic is advantageous for accurately determining the dose rate.

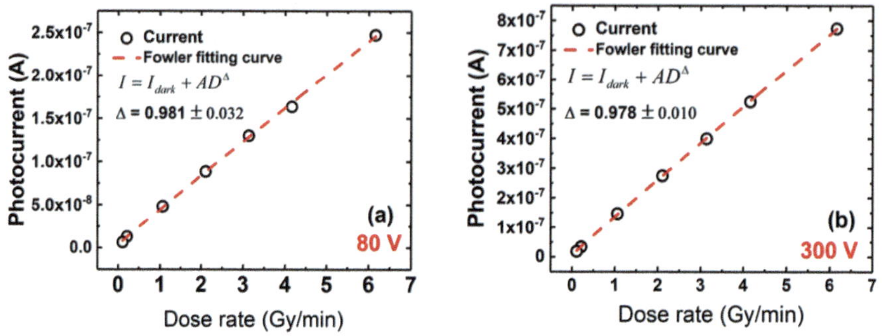


**Fig. 3.53** Linearity of the Fowler model fit for the polycrystalline detector. **a** 80 V; **b** 300 V

Figure 3.54 displays the device's sensitivity at bias levels of 80 V and 300 V. The sensitivity of the device is determined by the slope of the linear fit in the figure. The sensitivity of the device is 2.3608 μC/Gy at a bias of 80 V and 7.4832 μC/Gy at a bias of 300 V. The device has a sensitive volume of 4 × 4 × 0.5 mm$^3$. To get the characteristic sensitivity of the device, divide the sensitivity by the sensitive volume. At 80 V bias, the sensitivity is 0.2951 μC/Gy mm$^3$, and at 300 V bias, the sensitivity is 0.9354 μC/Gy mm$^3$.

Figure 3.55 depicts the device being exposed to X-ray radiation at a tube voltage of 120 kV and a tube current of 30 mA. The device current values were measured numerous times across several cycles (Cycle > 35), and the repeatability was found to be 0.677% when the device bias voltage was set to 80 V. The device exhibits a repeatability of 0.724% under a bias voltage of 300 V. It is evident that the variation in current responsiveness at 300 V is marginally greater than the current fluctuation at 80 V.

3. Analysis and Discussion

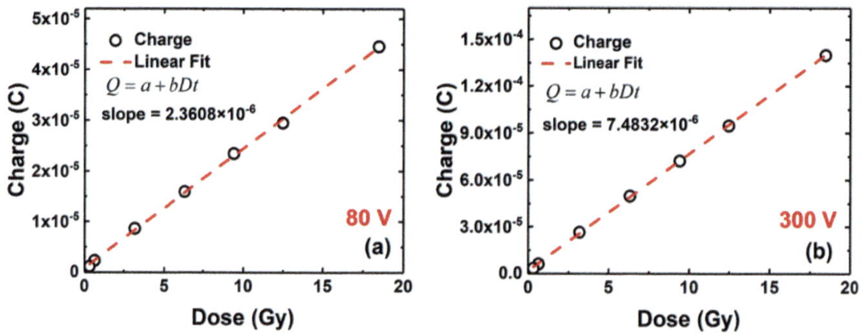


**Fig. 3.54** Sensitivity of the polycrystalline detector. **a** 80 V; **b** 300 V

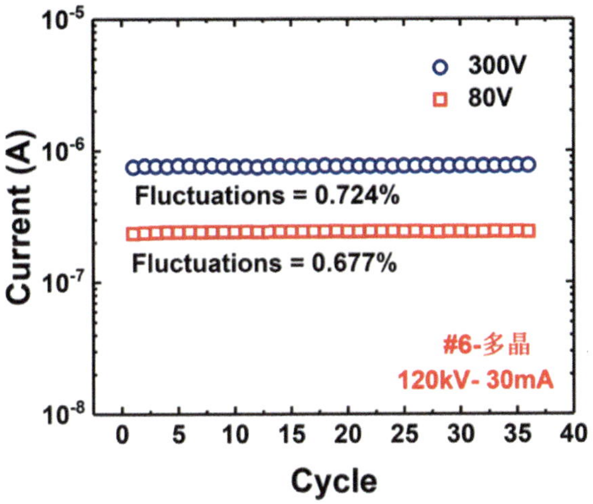


**Fig. 3.55** Repeatability of polycrystalline diamond core detectors

From the provided measurements, it is evident that the current generated in response to X-rays by a nuclear detector with a hydroxide terminal/gold electrode (H/O/Au) structure, made using "electron-grade" polycrystalline diamond, is considerably lower compared to that of a nuclear detector made with "electron-grade" single-crystal diamond using the same process. Based on the material analysis results presented in this chapter, it was found that both samples have a very low impurity content. No significant impurity peaks were observed in the PL spectroscopy. Additionally, the Raman peak positions and half-height widths of both samples are highly similar. Furthermore, the FTIR results indicate that both samples are classified as Type IIa diamonds. Hence, the presence of grain boundaries in a polycrystalline structure will greatly diminish the current amplification and characteristic responsiveness.

Table 3.12 presents a comparison between the performance of the current study and previously published X-ray detectors made of polycrystalline diamond with different structures. The table reveals that, apart from the decline in performance due to the quality of the material, the electrode contact also significantly impacts the detector's performance. The H/O/Au electrode configuration has exceptional ohmic contact qualities, which enable the device to attain a gain, SNR, characteristic sensitivity, and exponent of 1.31, $2 \times 10^3 \sim 5 \times 10^4$, 0.2951 μC/Gy $mm^3$, $0.981 \pm 0.032$; The device exhibits a gain, SNR, typical sensitivity, and exponent of 3.92, $2 \times 10^3 \sim 10^5$, 0.9354 μC/Gy $mm^3$, $0.978 \pm 0.010$ at a bias of 300 V.

### *3.5.4 Summary*

Current mode diamond nucleus detectors require high output current and sensitivity, excellent linearity and repeatability, as well as rapid response time. Based

**Table 3.12** Comparison of the performance of this work with other published polycrystalline diamond X-ray detectors

| Electrode type | Gain | SNR | Feature sensitivity | Δ |
|---|---|---|---|---|
| H/O/Au (80 V) | 1.31 | $2 \times 10^3 \sim 5 \times 10^4$ | 0.2951 μC/Gy mm$^3$ | 0.981 ± 0.032 |
| H/O/Au (300 V) | 3.92 | $2 \times 10^3 \sim 10^5$ | 0.9354 μC/Gy mm$^3$ | 0.978 ± 0.010 |
| DLC [39] | $2.47 \times 10^{-3}$ | 70–80 | 3.8 nC/Gy mm$^3$ | 0.73 |
| DLC/Ni [39] | $175 \times 10^{-3}$ | 3300 | 43.5 nC/Gy mm$^3$ | 0.86 |
| Pt [39] | $24.3 \times 10^{-3}$ | 1000 | 23.6 nC/Gy mm$^3$ | 0.85 |
| Ag [41] | – | 20–100 | 7–43 nC/Gy mm$^3$ | 0.95–1.00 |
| Cr/Au [48] | – | – | – | 0.88–0.90 |

on the findings of the previous study, the performance of the device is hindered by both the electrode contact and the material body features. Hence, this part delineates the present reaction of a hydroxide-terminated CVD single-crystal diamond core detector to continuous X-rays with dose rates varying from 0.108 to 6.157 Gy/min, utilizing a "electron-grade" diamond material. The findings indicate that carrier charge reinjection enhances the current gain and characteristic sensitivity of the detector's response to X-rays. This is attributed to the high-quality ohmic contact and the favorable properties of the bulk material. Additionally, the suppression of the trap effect at the interface leads to an Δ index close to 1. The device exhibited a gain, SNR, characteristic sensitivity, and Δ index of 151.83, $10^4$–$10^6$, 41.441 μC/Gy mm$^3$, 1.033 ± 0.014 at a bias of 200 V (0.66 V/μm). The existence of grain boundaries in hydroxide-terminated CVD polycrystalline diamond nucleation detectors leads to a response current that is considerably lower compared to single-crystal diamond nucleation detectors produced through the same method. The device has a gain, SNR, characteristic sensitivity, and exponent of 3.92, $2 \times 10^3$ to $10^5$, 0.9354 μC/Gy mm$^3$, 0.978 ± 0.010 at a bias of 300 V (0.60 V/μm). When compared to other single crystal or polycrystalline diamond core detectors operating in current mode, these performance indexes offer notable advantages.

## 3.6 Effect of Diamond Material Properties on the Performance of Nuclear Radiation Detectors

### *3.6.1 Introduction*

The performance of nuclear detectors is influenced by the qualities of the diamond body material, as revealed by the working principle of diamond nuclear detectors. Due to its excellent insulating qualities, diamond does not require the formation of p-n junctions or Schottky junctions in order to mitigate leakage currents. In general terms, a diamond nucleus detector with a vertical type MiM structure functions in a manner analogous to that of a solid ionization chamber. In the traditional diamond classification, the highest grade of diamond is known as Type IIa. These diamonds have an extremely low concentration of N impurities (< 1 ppm) and no detectable B impurities when tested using FTIR and secondary ion mass spectrometry (SIMS) up to their detection limit. Type IIa diamonds make up no more than 2% of all natural diamonds. The excessive price and lack of uniformity in natural diamonds contribute to the elevated expenses involved in manufacturing nuclear detectors, exacerbating the unpredictability of their performance. When comparing natural diamond to diamond produced by the CVD method, the latter offers the benefits of being able to be produced on a larger scale, at a lower cost, and with performance that can be controlled. Unintentionally doped CVD diamond is typically classified as Type IIa. However, the literature states that the CCE performance index of nuclear detectors made with unintentionally doped Type IIa CVD diamond, which is a crucial performance measure, ranges from a few percentage points to nearly 100%. Hence, investigating the impact of diamond material properties on the functionality of nuclear detectors holds immense importance in developing advanced diamond nuclear detectors with superior performance.

CVD diamonds are categorized into two groups: single crystals and polycrystals. Single crystals lack grain boundaries, resulting in enhanced carrier transport capabilities. However, achieving a crystal size larger than 12 mm is frequently challenging. On the contrary, polycrystalline diamonds are produced through a heterogeneous epitaxial process, resulting in the ability to achieve large sizes through CVD epitaxy. However, these diamonds have noticeable grain boundaries, which negatively impact their energy-spectral properties.

The Hecht equation reveals that the CCE is influenced by three factors:

(1) The product of electron/hole mobility and lifetime ($\mu\tau$).
(2) The thickness of the diamond body material ($d$).
(3) The strength of the electric field ($E$).

Thinning the thickness of the nuclear detector is a possible approach to achieve high CCE performance [49], as shown in Fig. 3.25. However, if the thickness, d, becomes too thin, it will not be able to capture all the energy of α-particles, making it unsuitable for detecting high-energy charged particles. Hence, a diamond nucleus detector with a thickness above 100 μm is more feasible. Furthermore, the standard mechanical polishing thinning procedure has a processing limit of around 100 μm. Therefore, the diamond core detectors under discussion here all have thicknesses exceeding 100 μm.

Conversely, augmenting the product can also improve the CCE performance of diamond core detectors. Based on the previous investigation, it is evident that the primary material defects in diamond are comprised of different impurities and dislocations. All of these material defects affect the $\mu\tau$ of nonequilibrium carriers in diamond. By cultivating high-grade CVD diamond, the presence of impurities and crystal flaws may be effectively reduced, hence improving the overall quality of the product. Advanced diamond growth approaches encompass ultra-high-purity growth and high-crystallization-quality growth, characterized by low dislocation density.

However, these two methodologies differ significantly in their approach to diamond formation [3]. When considering diamond materials, it is crucial to choose the appropriate growth process, crystal thickness, and material testing and screening methods to effectively build high-performance diamond core detectors. Hence, it is crucial to comprehend the degree to which impurities, dislocations, and thickness impact the performance of CCE.

## 3.6.2 Material Selection

The investigation of the impact of impurities, dislocations, and material thickness on the performance of CCE involves taking into account the following factors during material selection, as shown in Fig. 3.56.

Impurities: Diamond materials with elevated amounts of impurities were acquired by MPCVD growth. As mentioned earlier, the level of nitrogen and other impurities in diamond samples is influenced by the purity of $H_2$ used during the growth process. In this study, diamond samples with high impurity concentrations were developed using hydrogen with a purity of 6N (6 out of 9). After employing the previously mentioned PL spectra, we acquired samples with moderate levels of impurities. Diamond materials with minimal impurity levels were cultivated in a hydrogen environment with a purity level of 9N. Additionally, commercially available diamond single crystals of "electronic grade" were employed for comparative purposes.

Dislocations in homogeneous epitaxial CVD diamond primarily arise from dislocations in the HPHT substrate and the expansion of defects at the growth interface in the direction of growth. To mitigate this, we employed a low dislocation density HPHT substrate and employed the surface oxygen etching technique to cultivate CVD

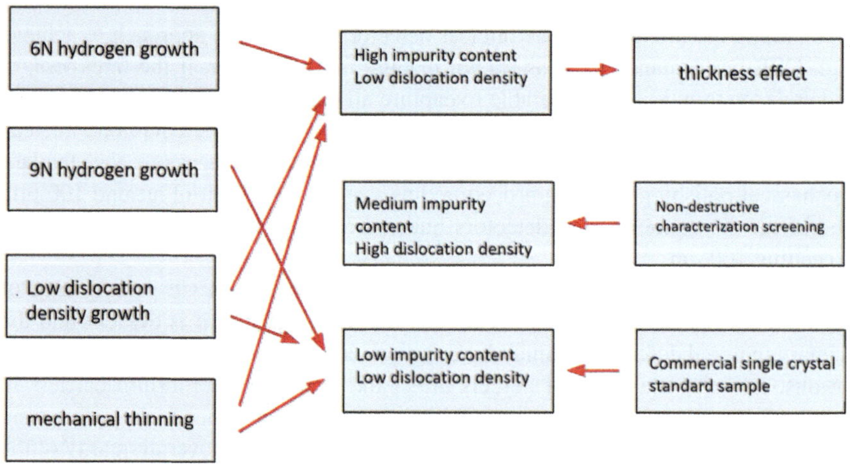


**Fig. 3.56** Selection of CVD single crystal diamond samples

single crystal diamond with a low dislocation density. Highly dislocated samples were acquired using XRD screening, as previously described.

The material thickness used for the investigation was achieved by cutting and polishing CVD single crystal diamond produced in 6N hydrogen to 200 μm and 500 μm. The purpose was to investigate how thinning the material affects the CCE properties.

### 3.6.3 Material Properties Characterization

In the case of single crystal diamond, six pieces of CVD single crystal diamond were produced and chosen. These six pieces were then divided into three groups. The first group consisted of two samples that were grown under 6N hydrogen to achieve a high impurity and low dislocation level. These samples were labeled as #7 and #8. The second group consists of two samples, which have a medium impurity content and high dislocation. These samples were selected using the non-destructive characterisation approach and are labeled as sample numbers #9 and #10. The third group consists of two samples belonging to the low impurity, low dislocation category. One of these samples, labeled as #11, is a high-quality single crystal diamond generated under 9N hydrogen conditions in this chapter. The other sample, labeled as #5, is a commercially available single crystal diamond of "electron grade" quality.

1. Surface appearance

To achieve uniformity in surface roughness, the surfaces of six samples were meticulously polished. The AFM testing results in Fig. 3.57 display the surface topography of the polished materials. All samples exhibit surface Rq values ranging from 1 to 2 nm within the 5 μm × 5 μm test region. The Rq values for samples #7, #8, #9, #10, #11, and #5 were 1.50 nm, 1.99 nm, 1.96 nm, 2.00 nm, 1.15 nm, and 1.09 nm, respectively. The white dots observed on samples #7 and #9 are remnants of polishing abrasives, whereas the remaining samples exhibit fewer traces of polishing abrasives, which can be eliminated through cleaning throughout the device's manufacturing process. Sample #11 exhibited the lowest surface undulation value, measuring 7.6 nm. In contrast, sample #9 had the highest surface undulation, measuring 15.1 nm. The image also demonstrates that all the samples have a distinct diagonal polishing pattern on their surfaces. The variation in surface roughness among the six samples is attributed to the tolerance of the polishing process. The average Rq value of the six samples is 1.615 nm, with a standard variation of 0.39 nm. Based on this, we conclude that the surface consistency of the six samples after polishing is satisfactory, and there are no noticeable changes in their morphology.

2. Sample type

The diamond samples of all sorts were filtering using FTIR, as illustrated in Fig. 3.58. The primary distinctive peaks manifest in the di-phonon region spanning from 1800 to 2700 $cm^{-1}$. These peaks are inherent and arise from carbon–carbon bonding. No

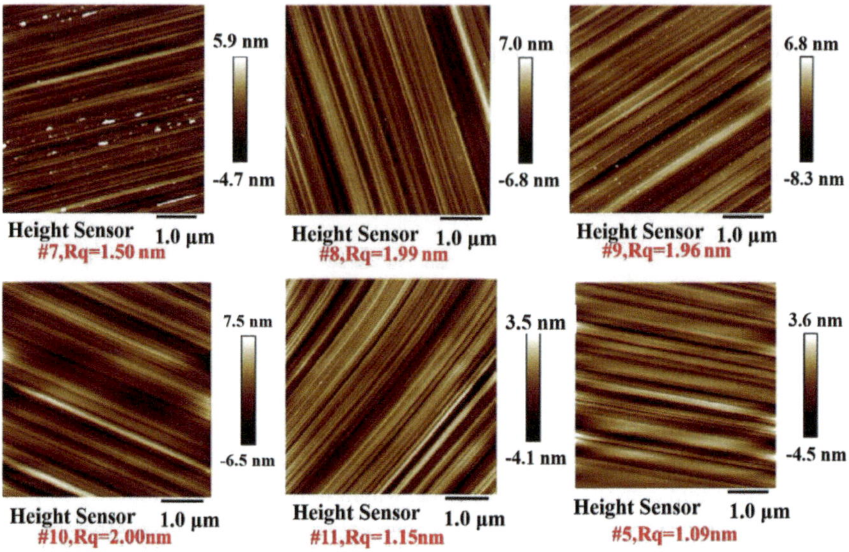


**Fig. 3.57** Sample surface after fine polishing

impurity peaks were found in the characteristic N region from 1000 to 1400 $cm^{-1}$, the characteristic H region from 2786 to 3394 $cm^{-1}$ and the characteristic B region from 2700 to 3000 $cm^{-1}$ in the figure; Based on the diamond classification guidelines, diamonds that do not show any N impurities when tested with FTIR are classified as Type II. Additionally, if these diamonds also lack B elements, they are further categorized as Type IIa. Therefore, despite the fact that the six CVD diamonds were produced using various methods and obtained from diverse origins, they all belong to the Type IIa diamond category.

3. Crystalline quality and dislocation density

Figure 3.59 shows the (004) faceted X-ray rocking curves of all diamond single crystal samples. The FWHM values of the rocking curves for samples #7, #8, #9, #10, #11, and #5 are 35.7, 45.3, 127.6, 102.8, 46.3, and 51.3 arc sec, respectively. Theoretically, the $CuK\alpha_1$ X-ray diffraction rocking curve value for diamond is 4.2 arcsec [50]. Given that the instrument's resolution is 10 arcsec, the broadening of the full width at half maximum (FWHM) of the sample's rocking curve is attributed to lattice defects, specifically the broadening caused by dislocations. The dislocation density (DD) can be calculated according to the following equation:

$$DD = \beta^2/9b^2 \tag{3.27}$$

Here, β is the FWHM of the sample rocking curve in radians; The length of the dislocations, denoted as b, is determined by the Burgers vector. In this case, the lattice constant of diamond, which is 0.357 nm, is used to estimate the dislocation densities

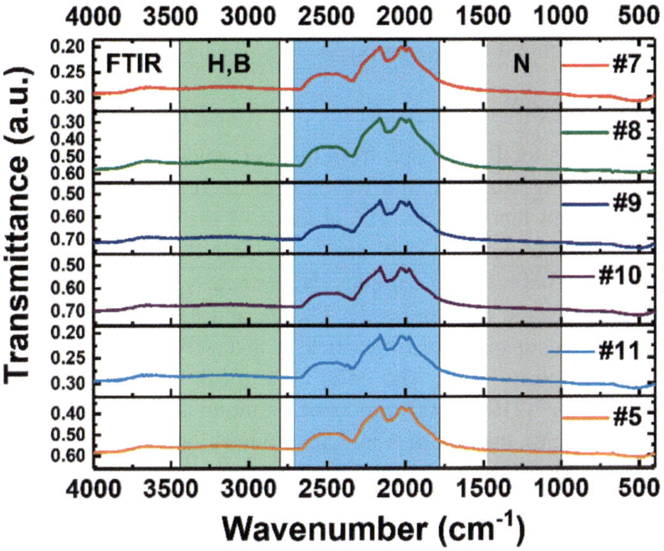


Fig. 3.58 Infrared spectrum of the sample at room temperature

Fig. 3.59 X-ray wiggle curve of diamond (004) surface

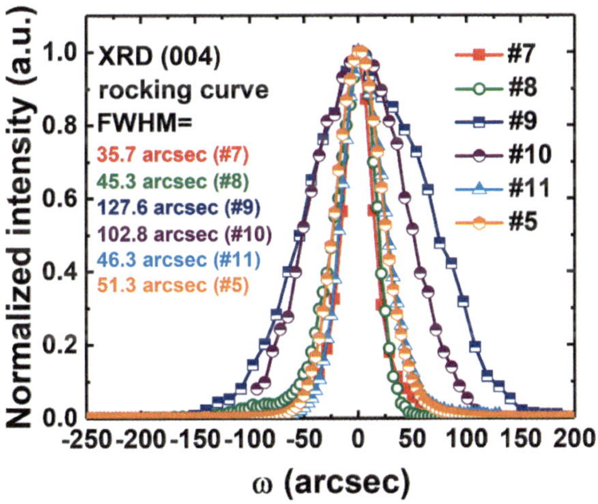


of all the samples. The results are presented in Table 3.13. It can be observed that the first group of low dislocation density samples, namely #7 and #8, grown in a 6N hydrogen environment, as well as the third group of samples, #11, grown in a 9N hydrogen environment, and sample #5, have dislocation densities on the order of magnitude of $10^6$ $cm^{-2}$. The dislocation densities in samples #9 and #10, which belong to the second group, are one order of magnitude greater than the dislocation densities in the other samples. They reach a value of $10^7$ $cm^{-2}$. The diffraction angles of all the samples are concentrated within the range of 119.6° to 120.0°, which is characteristic of diamond diffraction peaks.

4. Raman spectroscopy

Figure 3.60 displays the Raman spectrum test outcomes of the materials. The tests were conducted using a laser with a wavelength of 514 nm, and all samples were examined at ambient temperature. The absence of the diamond defect D peak at 1350 $cm^{-1}$ and the G peak at 1580 $cm^{-1}$, which reflect the graphite phase, can be noticed in Fig. 3.60a for all samples. Nevertheless, the presence of a peak at 2060 $cm^{-1}$ was detected in samples #7, #8, #9, and #10, indicating the existence of defects attributed to nitrogen impurities. Conversely, samples #11 and #5 did not exhibit any impurity peak. The 2460 $cm^{-1}$ peak is observed in the second-order Raman region (2100–2600 $cm^{-1}$) of diamond in samples #11 and #5. In other diamond samples, the luminescent background along the direction of the larger Raman frequency shift has combined with the second-order Raman peaks, making it difficult to distinguish.

Figure 3.60b displays the amplification of the primary Raman peaks for each sample, while the corresponding outcomes are presented in Table 3.14. The first

**Table 3.13** XRD measurement results of six samples

| | ω (arc sec) | 2θ (°) | Dislocation density $(cm)^{-2}$ |
|---|---|---|---|
| #7 | 35.7 | 119.6920 | $2.6 \times 10^6$ |
| #8 | 45.3 | 119.6417 | $4.2 \times 10^6$ |
| #9 | 127.6 | 119.9046 | $3.3 \times 10^7$ |
| #10 | 102.8 | 119.9093 | $2.2 \times 10^7$ |
| #11 | 46.3 | 119.9284 | $4.4 \times 10^6$ |
| #5 | 51.3 | 119.6983 | $5.4 \times 10^6$ |

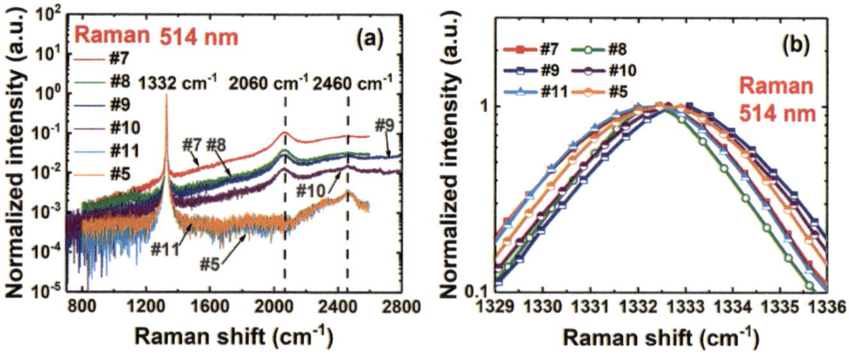


**Fig. 3.60** Comparison of Raman spectra of diamond single crystal samples at room temperature (514 nm laser)

order Raman peaks of all the samples are observed within the region of 1332–1333 $cm^{-1}$. The Raman peaks of the initial batch of samples, specifically #7 and #8, are positioned at a wavelength of 1332.4 $cm^{-1}$. Additionally, both the Raman half-height and width measure 3.6 $cm^{-1}$. The Raman peaks of samples #9 and #10 from the second batch are positioned at 1332.9 $cm^{-1}$ and 1332.7 $cm^{-1}$, respectively. Both samples have a Raman half-height and width of 3.9 $cm^{-1}$. The Raman peaks of samples #11 and #5 in the third set are positioned at 1332.2 $cm^{-1}$ and 1332.4 $cm^{-1}$, respectively. Additionally, both peaks have a Raman half-height of 3.7 $cm^{-1}$ and a breadth of 3.7 $cm^{-1}$. The Raman peaks seen in the aforementioned samples closely resemble those of a perfect diamond (1332.3 $cm^{-1}$). When considering the XRD results, it is observed that samples #9 and #10, which have a higher dislocation density, exhibit a greater deviation from the ideal diamond peak position in their Raman peaks. Conversely, samples with a lower dislocation density show a smaller deviation. This discrepancy is attributed to the fact that dislocations in diamond typically take the form of penetrating types that are generated along the growth direction, and these dislocations have a more significant impact on the lattice structure. The higher impurity concentrations in samples #7 and #8, as well as the lower impurity concentrations in samples #11 and #5, are relatively close to the theoretical value. This is because the impurities in the diamond primarily exist as point defects within the internal structure. Despite the relatively high impurity levels in samples #7 and #8, the FTIR results indicate that the impurity concentrations in these samples are still very low. As a result, they do not significantly affect the overall lattice structure of the diamond. The internal stress of diamond is calculated using Eq. (3.4). It is observed that samples #9 and #10, which have higher dislocation densities, exhibit internal compressive stresses of 0.31 GPa and 0.21 GPa, respectively. In contrast, the remaining samples, with lower dislocation densities, have a stress value of only 0.05 GPa, which is close to a state of no stress.

The Raman spectroscopy findings indicate that the single crystal diamond samples utilized for creating nuclear detectors had excellent crystal quality, with little impurities and flaws. This outcome aligns with the observation that all the CVD diamonds produced using FTIR are of Type IIa variety. Nevertheless, the samples with both high and medium impurity quantity still exhibited faint luminescence peaks from

**Table 3.14** Statistics of Raman test results for diamond samples (514 nm laser)

| | Raman peak level $(cm)^{-1}$ | Half height and width $(cm)^{-1}$ | Stress σ (GPa) |
|---|---|---|---|
| #7 | 1332.4 | 3.6 | – 0.05 |
| #8 | 1332.4 | 3.6 | – 0.05 |
| #9 | 1332.9 | 3.9 | – 0.31 |
| #10 | 1332.7 | 3.9 | – 0.21 |
| #11 | 1332.2 | 3.7 | + 0.05 |
| #5 | 1332.4 | 3.7 | – 0.05 |

impurities, which could not be discerned by FTIR spectroscopy. Hence, it is imperative to conduct a more in-depth analysis of the aforementioned phenomena by the utilization of PL spectroscopy.

5. PL spectrum

All diamond samples were tested for impurities by PL spectroscopy. Figure 3.61 displays the photoluminescence spectra obtained at room temperature using a 514 nm laser. To assist the comparison of the strength of the impurity peaks, the PL spectra of all the samples were normalized using the first order Raman peaks. The image reveals the presence of luminescence peaks at wavelengths of 575, 637, 662, 686, 696, 704, 737, and 742 nm in samples #7, #8, #9, and #10. However, similar luminescence peaks were absent in samples #11 and #5 from the low impurity group. Additionally, the figure also shows the presence of first-order Raman peaks. The wavelengths 575 nm, 637 nm, 737 nm, and 742 nm correspond to the $NV^0$, $NV^-$, $SiV^-$, and $GR^1$ centers, respectively. The underlying physical mechanisms responsible for the peaks at 662, 686, 696, and 704 nm remain unidentified at present. These peaks have been linked to luminescence caused by the replication of $NV^0$ and $NV^-$ zero phonon lines, or complex defect complexes related to nitrogen impurities [51, 52]. Despite being grown in a 6N hydrogen atmosphere, samples #7 and #8 exhibit slightly different impurity luminescence compared to samples #7 and #8. This difference is attributed to the presence of nitrogen impurities originating from residue in the growth gas and leakage from the growth chamber. The entry of these impurities into the diamonds is somewhat random, but their peaks are significantly higher than those observed in samples #9 and #10. The presence of silicon impurities in the diamond is a result of the plasma's etching action on the cavity. The process of linking these impurities into the diamond interior is also rather unpredictable. Samples #9 and #10 are categorized as medium impurity content samples due to their impurity luminescence region normalized maxima being below 0.2, whereas samples #7 and #8 had values above 0.5. The impurity luminescence peaks of samples #11 and #5, which have a low impurity content, are no longer visible in linear coordinates. Instead, there is only a weak luminescence background.

Additionally, Fig. 3.62 displays the PL spectra of samples #11 and #5 at various temperatures. Specifically, Fig. 3.62a presents the PL spectra in logarithmic coordinates when excited by a 514 nm laser at room temperature. It is evident from the figure that sample #11 exhibits a slightly higher luminescent background compared to sample #5, but no distinct impurity peaks were observed. At a temperature of 3 K, sample #11 exhibited $NV^0$ and $NV^-$ impurity peaks with low luminescence intensity. In contrast, sample #5 did not display any noticeable impurity peaks, as depicted in Fig. 3.62b. Additionally, it demonstrates that CVD diamond produced with 9N hydrogen still retains some weak impurities, though the impurity level has been significantly diminished in comparison to diamond generated with 6N hydrogen.

6. SIMS

Secondary Ion Mass Spectrometry (SIMS) is a highly precise analytical technique used for quantitative characterization. It has a remarkable sensitivity, with a detection

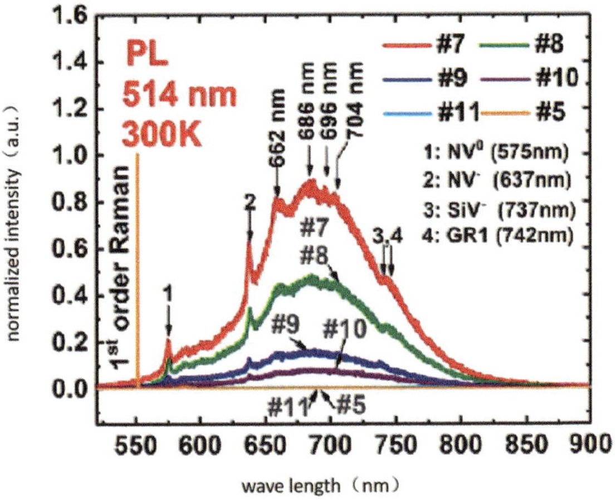


Fig. 3.61 PL spectrum of diamond sample at room temperature (514 nm laser)

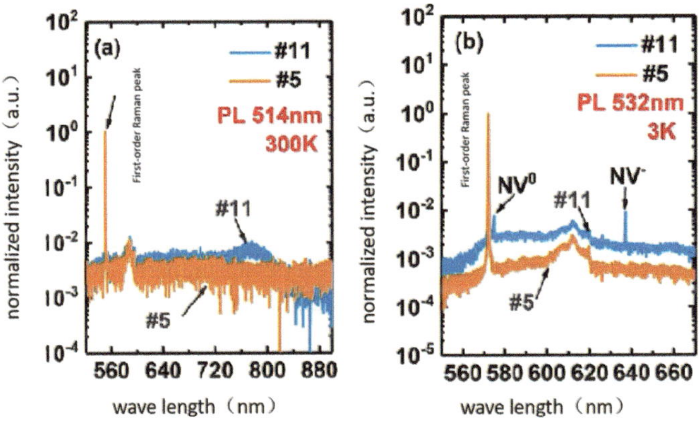


**Fig. 3.62** PL spectra of the third set of samples at different temperatures. **a** 300 K 514 nm laser; **b** 3 K 532 nm laser

limit of $10^{16}$ atom/cm$^3$ for nitrogen impurities and $10^{14}$ atom/cm$^3$ for boron impurities. Given that nitrogen and boron impurities are the primary sources of impurities in diamond, we conducted measurements using SIMS (Fig. 3.63) to determine the concentrations of these impurities in sample #7. Notably, sample #7 exhibited the highest impurity luminescence intensity among all the samples. Consequently, the measured nitrogen and boron impurity concentrations in sample #7 represent the upper limit of impurity concentrations for all the samples. The nitrogen (N) impurity concentration in sample #7 is $3 \times 10^{16}$ atom/cm$^3$, equivalent to approximately 170 ppb. On the other hand, the boron impurity concentration in sample #7 has surpassed the detection limit of SIMS, indicating that it does not exceed 1 ppb. Given that the N impurity concentration of sample #5 is below 5 ppb, we may conclude that

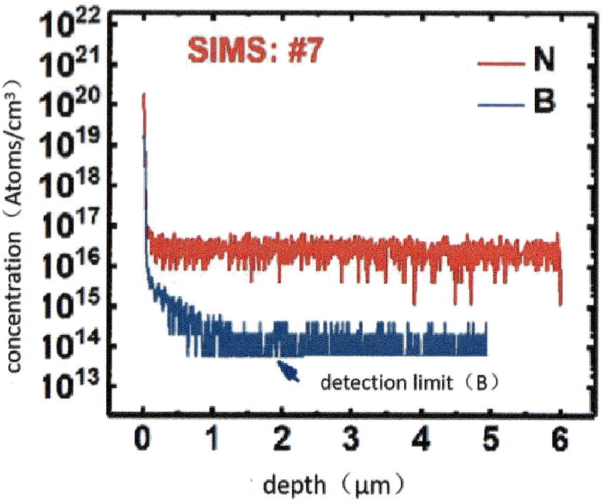


**Fig. 3.63** SIMS analysis results of sample #7

the nitrogen impurity concentration of all samples is within the range of 5–170 ppb. Additionally, the boron impurity concentration of all samples is below 1 ppb.

### *3.6.4 Device Preparation*

The device process follows the same steps as the device preparation process discussed earlier. The thickness of the detector is measured using a micrometer, and the size of the electrode is determined based on the dimensions of the specified mask plate. The particular measurements of the device are provided in Table 3.15.

### *3.6.5 Energy Spectrum Properties*

The energy spectral characteristics of the six CVD single-crystal diamond nucleus detector samples in this study were assessed using a pulsed height spectrometer, as explained in the preceding section. The radioactive source employed was a $^{241}$Am source, and the CCE properties were calibrated using a standard silicon detector. The CCE values were determined based on the highest point of the energy spectrum peaks.

Figure 3.64 displays the relationship between the bias electric field and CCE for six CVD single crystal diamond core detectors.

At a bias field of 1 V/μm, Sample #7 exhibits a CCE of 3.3%. However, as the bias field increases further, the CCE reaches a maximum value and does not increase any further, reaching saturation.

**Table 3.15** Prepared CVD diamond core detector samples dimensions

| | Sample number | Thickness (mm) | Electrode size ($mm^2$) | Remarks |
|---|---|---|---|---|
| Monocrystalline | #7 | 0.20 | 3 × 3 | 6N hydrogen growth, high impurity, low dislocation |
| | #8 | 0.50 | 3 × 3 | 6N hydrogen growth, high impurity, low dislocation |
| | #9 | 0.20 | 4 × 4 | Medium impurity content, high dislocation density |
| | #10 | 0.23 | 4 × 4 | Medium impurity content, high dislocation density |
| | #11 | 0.25 | 2 × 2 | 9N hydrogen growth, low impurity, low dislocation |
| | #5 | 0.30 | 4 × 4 | Electronic grade single crystal, low impurity, low dislocation |

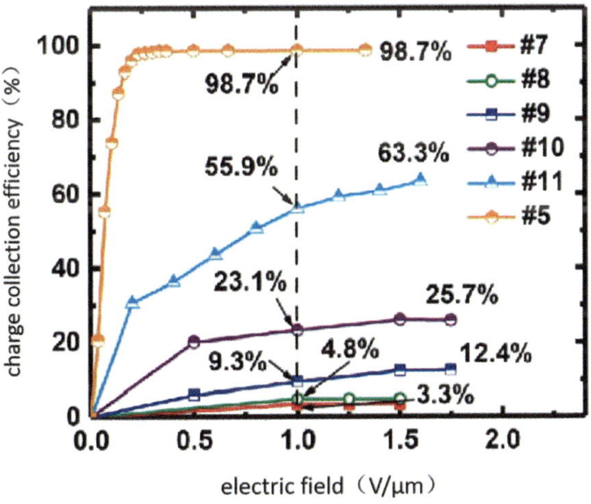


**Fig. 3.64** Detector bias electric field versus CCE

At a bias field of 1 V/μm, Sample #8 exhibits a CCE of 4.8%. However, as the bias field increases further, the CCE reaches a saturation point and no longer increases.

Sample #9 exhibits a CCE of 9.3% at an electric field strength of 1 V/μm. The CCE value shows a modest increment as the electric field is raised, reaching a saturated value of 12.4% when the electric field surpasses 1.5 V/μm.

Sample #10 has a CCE efficiency of 23.1% at an electric field strength of 1 V/μm. The CCE value shows a modest increment as the electric field is raised, reaching a saturated CCE of 25.7% when the electric field surpasses 1.5 V/μm.

Sample #11 exhibits a CCE of 55.9% at an electric field strength of 1 V/μm. As the electric field strength further increases, the CCE value shows a modest increment, reaching 63.3% when the electrical field strength reaches 1.6 V/μm.

The CCE of sample #5 is 98.7% at an electric field of 1 V/μm, and it becomes saturated as the electric field increases.

Based on the aforementioned findings, it is evident that the CCE characteristics of CVD single-crystal diamonds exhibit significant variations. To gain a deeper understanding of the energy spectral features of these diamond core detectors, we will systematically study the properties of these samples under an electric field of 1 V/μm.

The graph in Fig. 3.65 illustrates the energy spectrum characteristic curve of the detector at an electric field strength of 1 V/μm. The energy spectrum characteristic refers to the probabilistic statistical distribution of electron–hole pairs generated by the nuclear detector in response to α-particles. The center position of the peaks represents the region where the probability distribution of electron–hole pair collection is most concentrated. Therefore, the center position of the peaks corresponds to the average value of the CCE.

The maximum value of sample #7 corresponds to a CCE of 3.3%. The peak does not follow a Gaussian distribution, but rather exhibits semi-Gaussian characteristics. Once the peak surpasses its highest value, the distribution exhibits an exponential reduction. At this point, the maximum is capable of accumulating approximately 7% of the charge. However, it is important to note that the probability distribution at the maximum value is extremely low and does not accurately represent the overall distribution. Instead, it only reflects the discontinuous nature of charge collecting.

The peak center location of sample #8 corresponds to a CCE of 4.8%, and the peak exhibits a Gaussian distribution with a maximum CCE of approximately 12%.

The peak center location of sample #9 corresponds to a CCE of 9.3%. The peak exhibits a typical Gaussian distribution with a maximum CCE value of around 15%.

The peak center location of sample #10 corresponds to a CCE of 23.1%. The peak exhibits a typical Gaussian distribution, with the highest CCE value reaching almost 30%.

The sample #11 exhibits a peak center position that corresponds to a CCE of 55.9%. The peak is characterized by a standard Gaussian distribution, with a maximum value in close proximity to 70%.

The peak center position of sample #5 corresponds to a CCE of 98.7%. The peak exhibits a standard Gaussian distribution with a maximum value near 100%.

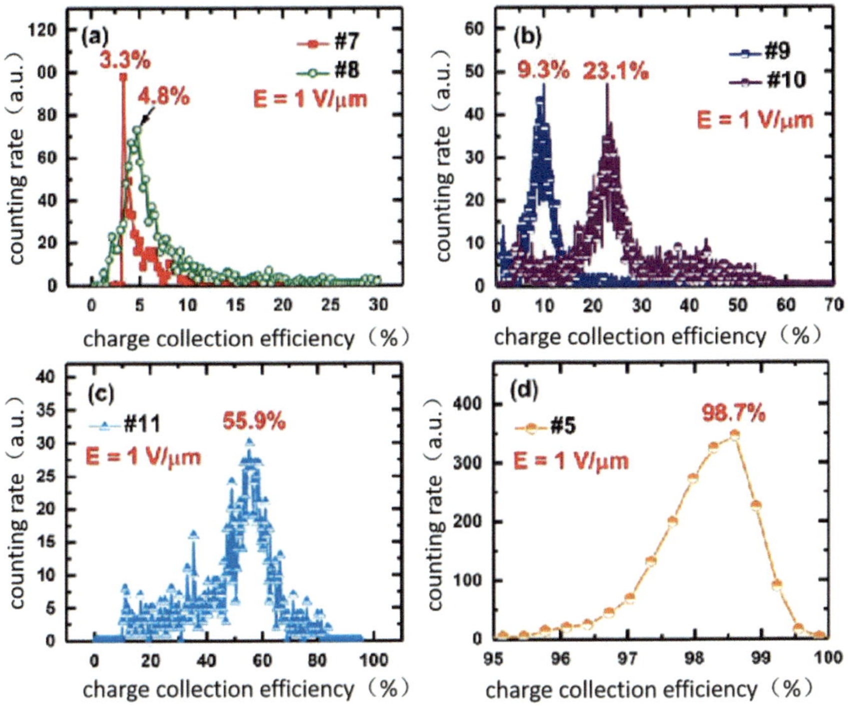


**Fig. 3.65** Energy spectrum characteristic curve at electric field strength of 1 V/μm

From the energy-spectral characterisation graphs provided, it is evident that there are significant variations in both the CCE values and the energy-spectral characterization curves among different samples.

Figure 3.66 displays the CCE characteristics versus energy resolution for the five samples, highlighting the concentrated nature of the statistical distribution of charge collection. However, sample #7 is excluded from the plot as it does not demonstrate Gaussian distribution characteristics, rendering its energy resolution statistically insignificant. The image illustrates a clear inverse relationship between the energy resolution of the five samples and their CCE. Specifically, samples #8, #9, #10, #11, and #5 exhibit energy resolutions of 68.3%, 52.28%, 30.8%, 18.9%, and 1.7%, respectively. Equation (3.16) reveals the relationship between the energy resolution and the values of $E_0$, Fano factor $F$, and average ionization energy $w$. In this study, the entire energy of the $\alpha$-particle was absorbed, and the variable w represents the mean ionization energy of diamond. Therefore, the value of $E_0$ is essentially dependent on the CCE value.

$$E_0 = \text{CCE} \cdot E_\alpha \tag{3.28}$$

Then Eq. (3.16) can be written as:

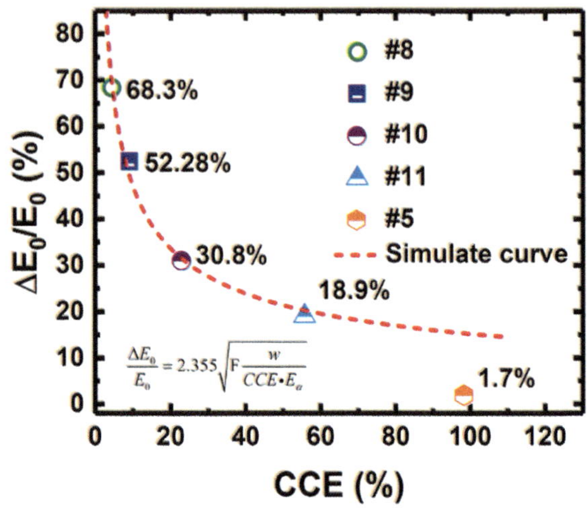


**Fig. 3.66** CCE characteristics versus energy resolution for all diamond core detectors

$$\frac{\Delta E_0}{E_0} = 2.355\sqrt{F\frac{w}{\mathrm{CCE}\cdot E_\alpha}} \tag{3.29}$$

Here, $E_\alpha$ represents the energy of the incident particle. Observation reveals that samples #8, #9, #10, and #11 adhere to the connection described in Eq. (3.29), whereas example #5 exhibits a minor deviation. This demonstrates that the CCE value significantly impacts the energy resolution. To achieve a lower energy resolution, enhancing the CCE value is a crucial determinant.

### 3.6.6 Analysis and Discussion

1. Dislocation density and the effect of impurities

Based on the aforementioned results, it is evident that samples #7 and #8, which exhibit a high impurity concentration, possess the lowest CCE values. Conversely, samples #9 and #10, characterized by a moderate impurity concentration, display slightly higher CCE values. Lastly, samples #11 and #5, which have a low impurity concentration, exhibit the highest CCE values. Nevertheless, the CCE values mentioned above are influenced by the combined impact of sample thickness d, impurity concentration, and dislocation density on the back of the product. However, it is not feasible to directly ascertain the extent to which these parameters affect the CCE performance. To exclude the influence of thickness ($d$), the $\mu\tau$ product of each sample was determined using the Hecht equation at an electric field strength of 1 V/μm. Considering the detection limit of SIMS, all samples except sample #7 have impurity concentrations that have reached the detection limit. Therefore, the relative quantitative values of the impurities were determined by integrating the impurity luminescence range from 560 to 850 nm. This integration was based on the PL spectra obtained under 514 nm laser excitation at room temperature. The dislocation density was characterized using Eq. (3.27) by using the square of the FWHM value of the XRD wiggle curve. The aforementioned results are displayed in Table 3.16.

The PL integral area values for the samples range from 0.93 to 104.52, differing by two orders of magnitude. These values align with the N impurity concentration data obtained from SIMS product information, which range from 5 to 170 ppb for sample #5. Figure 3.67a displays the relationship between the product and the PL integral area for the six samples. It is evident that the $\mu\tau$ product experiences a significant reduction as the PL integral area value increases. Specifically, when the PL integral area value increases by two orders of magnitude, the $\mu\tau$ product decreases by three to four orders of magnitude. In other words, it decreases from $1.14 \times 10^{-4}$ to $6.6 \times 10^{-8}$ $cm^2/V$. This suggests that impurities such as N, Si, and others in diamond, which contribute to deep trap energy levels with complex centers, have a substantial impact on the $\mu\tau$ product of carriers. The $\mu\tau$ product, as shown in Fig. 3.67a, takes into account the influence of dislocation density. When comparing the low dislocation density samples #5, #11, and #8 with sample #7, the $\mu\tau$ product for sample #5 remains the highest, despite having the highest dislocation density. This is evident when considering Fig. 3.67b in conjunction with Fig. 3.67a. The $\omega^2$ values of samples #9 and #10 exhibit a significant increase compared to the other samples. However, despite the fact that the PL integral values of samples #9 and #10 are significantly lower than that of sample #7, the $\mu\tau$ product of samples #9 and #10 with medium impurity concentrations remains an order of magnitude higher than that of sample #7 with high impurity concentration. The samples #8 and #9 exhibit an anomalous phenomenon, where the $\mu\tau$ product of sample #8 ($2.4 \times 10^{-7}$ $cm^2/V$) is slightly higher than that of sample #9 ($1.86 \times 10^{-7}$ $cm^2/V$). This difference can be attributed to the significantly lower dislocation density in sample #8 compared to sample #9. However, the increase in the PL integral value in sample #8 is insignificant compared to that of sample #9. At this point, the effect of dislocations on the $\mu\tau$ product becomes more prominent.

Figure 3.67b demonstrates that the $\mu\tau$ product of the sample does not exhibit a distinct and consistent distribution in relation to the $\omega^2$ values. Despite sample #7 having a high impurity concentration and low dislocation density, samples #9 and #10, which have a high dislocation density and medium impurity concentration, still yield a larger $\mu\tau$ product. This indicates that the impact of dislocation density on

**Table 3.16** Product of each sample derived from Hecht model $\mu\tau$

| | #7 | #8 | #9 | #10 | #11 | #5 |
|---|---|---|---|---|---|---|
| $\mu\tau$ ($cm^2/V$) | $6.6 \times 10^{-8}$ | $2.4 \times 10^{-7}$ | $1.86 \times 10^{-7}$ | $5.39 \times 10^{-7}$ | $1.92 \times 10^{-6}$ | $1.14 \times 10^{-4}$ |
| PL integral area (560–850 nm) | 104.52 | 59.12 | 20.56 | 10.59 | 1.41 | 0.93 |
| $\omega^2$ $(arcsec)^2$ | $(35.7)^2$ | $(45.3)^2$ | $(127.6)^2$ | $(102.8)^2$ | $(46.3)^2$ | $(51.3)^2$ |

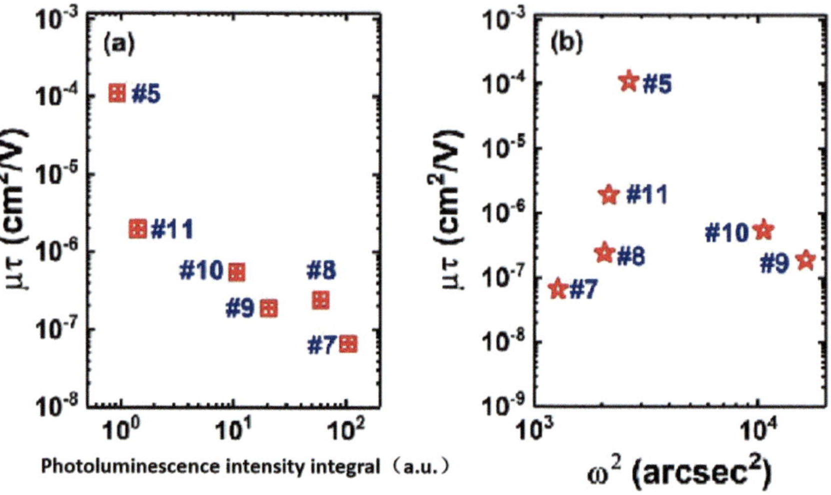


**Fig. 3.67** **a** $\mu\tau$ product versus PL spectrum integrated area; **b** $\mu\tau$ product versus XRD wiggle curve half-height width

the $\mu\tau$ product is not predictable within the tested range. Dislocations, like contaminants, can create traps and complex centers, leading to a decrease in the carrier $\mu\tau$ product. Schreck et al. investigated the effect of dislocation density on heterogeneous epitaxial single-crystal diamond, and after excluding the effect of impurities, they found that the carrier lifetime $\tau$ is proportional to 1/DD at dislocation densities from $10^7$ to $10^9$ cm$^{-2}$ and is about 10 to 0.1 ns, and for homogeneously epitaxialized single-crystal diamonds, the carrier lifetime is about 150 ns [53] at DD = $10^4$ cm$^{-2}$. Type IIa diamonds have mobilities $\mu$ ranging from 1000 to 4500 cm$^2$/V·s [54]. In the case of homogeneous epitaxial single-crystal diamond, the dislocation density is generally less than $10^7$ cm$^{-2}$. Consequently, it is feasible to get a sufficiently high $\mu\tau$ product, which can bring the CCE close to 100%, provided that the influence of impurities is disregarded. The dislocation densities of the current samples, as determined by Eq. (3.27), range from $10^6$ cm$^{-2}$ to $10^7$ cm$^{-2}$. It is seen that the relationship between the $\mu\tau$ product and 1/DD is sublinear, and as the dislocation density decreases further, the $\mu\tau$ product will eventually achieve saturation. At this stage, according to Fig. 3.67a, the impact of impurity concentration on the $\mu\tau$ product becomes increasingly significant. As the impurity concentration lowers, the $\mu\tau$ product also starts to grow in a superlinear manner.

From the above discussion, it is evident that the $\mu\tau$ product of CVD single crystal diamond is influenced by both impurities and dislocations. This joint influence is a crucial factor that affects the performance of the CCE of the nuclear detector. The impact of impurities on the $\mu\tau$ product is most significant when the density of dislocations falls within the range of $10^6$ to $10^7$ cm$^{-2}$. Specifically, when the nitrogen impurity reaches 170 ppb, the $\mu\tau$ product decreases to $6.6 \times 10^{-8}$ cm$^2$/V. Conversely, when the $\mu\tau$ product is less than 5 ppb, it can reach as high as 1.14 × $10^{-4}$ $cm^2/V$. Currently, the impact of dislocation density on the $\mu\tau$ product is of lesser significance.

The CVD single crystal diamond is often produced through uniform epitaxial growth, resulting in a dislocation density in the CVD layer that is generally comparable [3] to that of the HPHT Type Ib substrate utilized, ranging from $10^4$ to $10^6$ $cm^{-2}$. Based on the aforementioned findings, the dislocation densities of samples #7, #8, #11, and #5 are approximately $10^6$ $cm^{-2}$. The decrease in impurity content at such dislocation densities has led to a near-perfect CCE of 100%. Consequently, for uniform epitaxial-grown CVD diamond, it is not crucial to regulate the dislocation density of the CVD layer. While the leaking of the growth chamber and the presence of contaminants in the reaction gas can significantly impact the CCE values, it is important to prioritize the implementation of high-purity growth techniques.

2. The effect of thickness on the performance of CCE

The Hecht model can be used to determine the impact of thickness on CCE based on the calculated $\mu\tau$ product. Figure 3.68 displays the relationship between the thickness ($d$) and the CCE of the nuclear detector for six samples, with an electric field of 1 V/μm.

When the thickness is reduced from 500 to 200 μm:

For sample #7 with a $\mu\tau$ product of 6.6 × $10^{-8}$ $cm^2/V$, the CCE increases from only 1.3% to 3.3%;
For sample #8 with a $\mu\tau$ product of 2.4 × $10^{-7}$ $cm^2/V$, the CCE increased from 4.7 to 11.8%;
For sample #9 with a $\mu\tau$ product of 1.86 × $10^{-7}$ $cm^2/V$, the CCE increased from 3.7 to 9.3%;
For sample #10 with a $\mu\tau$ product of 5.39 × $10^{-7}$ $cm^2/V$, the CCE increased from 10.6 to 26.05%;

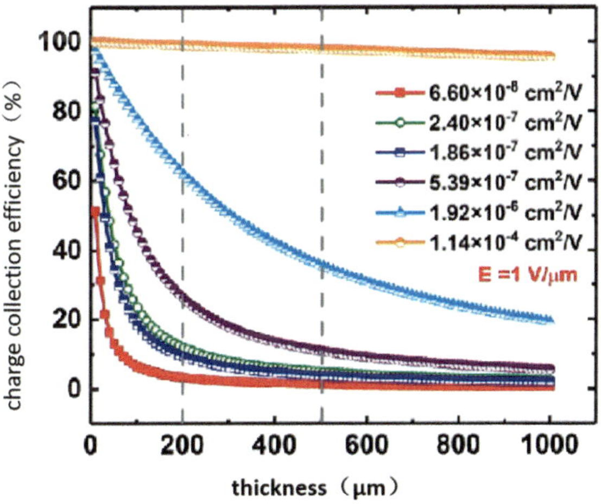


**Fig. 3.68** Effect of material thickness on CCE of diamond core detector under Hecht model (E = 1 V/μm)

For sample #11 with a $\mu\tau$ product of $1.92 \times 10^{-6}$ cm$^2$/V, the CCE increased from 35.3 to 61.8%;
For sample #5 with a $\mu\tau$ product of $1.14 \times 10^{-4}$ cm$^2$/V, the CCE increased from 97.7 to 99.08%;

Except for sample #5, reducing the thickness of the remaining samples to 50 μm does not result in a CCE value over 90% in theory. Additionally, thinning the samples sacrifices the ability to detect particles within a certain energy range, making the diamond nuclear detector less practical. Sample #5 exhibits negligible variations in CCE when the thickness, denoted as d, is altered within the range of 1 to 1000 μm. In the case of homogeneous epitaxially grown single-crystal diamond, reducing the thickness of the sample within the tested range of 200 to 500 μm does not lead to the development of a high-performance diamond nucleation detector with a CCE exceeding 90%, unless the impurity concentration and dislocation density of the diamond are also reduced.

### 3.6.7 Conclusion

For the fabrication of nuclear detectors, six accidentally doped Type IIa CVD monocrystalline diamond pieces were utilized. Six diamond nucleus detectors were examined for their CCE performance using alpha particle spectroscopic studies utilizing FTIR, XRD, Raman, PL, and SIMS techniques. The findings indicate that the primary factor constraining the CCE of diamond core detectors is the impurities present in the diamond, as discussed earlier. On the other hand, dislocations, which are common lattice defects, have a lesser impact when the density of dislocations is in the range of $10^6$–$10^7$ cm$^{-2}$. Merely reducing the thickness of diamond crystals within the specified range of 200 to 500 μm is not an effective method for achieving diamond core detectors with superior performance. The drop in the mobility lifetime product of the carriers, which occurs when the nitrogen impurity concentration grows from 5 to 170 ppb, is responsible for the reduction in the CCE from 98.7 to 3.3%. In order to achieve CCE close to 100% in homogeneous epitaxial CVD single-crystal diamonds, it is crucial to minimize the dislocation density to $10^6$ cm$^{-2}$. Therefore, implementing a high-purity growth approach is essential for enhancing the performance of these diamond detectors. The aforementioned material characterisation methods also serve as a potential screening process. To produce diamond core detectors with excellent performance, it is crucial to meticulously choose the diamond material before proceeding with electrode production. Moreover, the CCE characteristics of excellent diamonds (#11) produced with 9N hydrogen exhibited a noteworthy enhancement compared to the properties of regular diamonds grown with 6N hydrogen (#7 and #8). To cultivate CVD single-crystalline diamond materials suitable for high-performance diamond nucleation detector preparation, it is imperative to continue investing efforts in the growth process and make necessary modifications to the growth equipment.

# References


1. Hecht K (1932) Zum Mechanismus des lichtelektrischen Primärstromes in isolierenden Kristallen. Z Phys 77(3):235–245
2. Tallaire A, Achard J, Brinza O et al (2013) Growth strategy for controlling dislocation densities and crystal morphologies of single crystal diamond by using pyramidal-shape substrates. Diam Relat Mater 33:71–77
3. Tallaire A, Achard J, Silva F et al (2013) Growth of large size diamond single crystals by plasma assisted chemical vapour deposition: recent achievements and remaining challenges. C R Phys 14(2–3):169–184
4. Bergman L, Nemanich RJ (1995) Raman and photoluminescence analysis of stress state and impurity distribution in diamond thin films. J Appl Phys 78(11):6709–6719
5. Achard J, Tallaire A, Mille V et al (2014) Improvement of dislocation density in thick CVD single crystal diamond films by coupling $H_2/O_2$ plasma etching and chemo-mechanical or ICP treatment of HPHT substrates. Physica Status Solidi A 211(10):2264–2267
6. Teraji T (2006) Chemical vapor deposition of homoepitaxial diamond films. Physica Status Solidi A 203(13):3324–3357
7. Element six : https://www.e6.com/zh/
8. Breeding CM, Shigley JE (2009) The "type" classification system of diamonds and its importance in gemology. Gems Gemol 45(2):96–111
9. Wu G, Chen M, Liu J, et al (2014) Infrared raman spectroscopy analysis of CVD synthetic diamonds. Acta Mineralogica Sinica 34(03):411–415
10. Baron C, Ghodbane S, Deneuville A et al (2005) Detection of CHx bonds and sp2 phases in polycrystalline and ta-C:H films from Raman spectra excited at 325 nm. Diamond Rel Mater 14(3–7):949–953
11. Lee EH, Hembree DM, Rao GR et al (1993) Raman scattering from ion-implanted diamond, graphite, and polymers. Phys Rev B Condens Matter 48(21):15540–15551
12. Kanda H, Watanabe K (1999) Distribution of nickel related luminescence centers in HPHT diamond. Diam Relat Mater 8(8–9):1463–1469
13. Shimaoka T, Koizumi S, Tanaka MM (2018) Diamond photovoltaic radiation sensor using pn junction. Appl Phys Lett 113(9):093504
14. Galbiati A, Lynn S, Oliver K et al (2009) Performance of monocrystalline diamond radiation detectors fabricated using TiW, Cr/Au and a novel ohmic DLC/Pt/Au electrical contact. IEEE Trans Nucl Sci 56(4):1863–1874
15. Hoshino Y, Saito Y, Nakata J (2010) Interdiffusion analysis of Au/Ti and Au/Pt/Ti electrode structures grown on diamond (001) surface by Rutherford backscattering spectroscopy. Jpn J Appl Phys 49(10):101302
16. Abdel-Rahman MAE, Lohstroh A, Bryant P (2019) Alpha spectroscopy and X-ray induced photocurrent studies of a SC CVD diamond detector fabricated with PLD contacts. Rad Phys Chem 164:108357
17. Zhang Z, Huang J, Xi Y et al (2019) CVD diamond film detectors for α particles with a new electrode structure of reduced graphene oxide/Au. Mater Sci Semicond Process 91:260–266
18. Maier F, Ristein J, Ley L (2001) Electron affinity of plasma-hydrogenated and chemically oxidized diamond (100) surfaces. Phys Rev B 64(16):165411
19. Maier F, Riedel M, Ristein J et al (2001) Spectroscopic investigations of diamond/hydrogen/metal and diamond/metal interfaces. Diam Relat Mater 10(3):506–510
20. Kono S, Saitou T, Kawata H et al (2009) Characteristic energy band values and electron attenuation length of a chemical-vapor-deposition diamond (001)2×1 surface. Surf Sci 603(6):860–866
21. Pakes CI, Garrido JA, Kawarada H (2014) Diamond surface conductivity: properties, devices, and sensors. MRS Bull 39(6):542–548
22. Edmonds MT, Pakes CI, Mammadov S et al (2011) Work function, band bending, and electron affinity in surface conducting (100) diamond. Physica Status Solidi A 208(9):2062–2066

23. Kono S, Teraji T, Takeuchi D et al (2017) Direct determination of the barrier height of Au ohmic-contact on a hydrogen-terminated diamond (001) surface. Diam Relat Mater 73:182–189
24. Kono S, Sasaki T, Inaba M et al (2017) Sheet resistance underneath the Au ohmic-electrode on hydrogen-terminated surface-conductive diamond (001). Diam Relat Mater 80:93–98
25. Yang HS, Li Y, Norton DP et al (2005) Low-resistance ohmic contacts to p-ZnMgO grown by pulsed-laser deposition. Appl Phys Lett 86(19):192103
26. Kumar A, Kumar A, Topkar A et al (2017) Prototyping and performance study of a single crystal diamond detector for operation at high temperatures. Nucl Instrum Methods Phys Res, Sect A 858:12–17
27. Knoll GF (1989) Radiation detection and measurement, 2nd ed. United States: John Wiley and Sons Inc.,New York, NY
28. Shimaoka T, Kaneko JH, Sato Y et al (2016) Fano factor evaluation of diamond detectors for alpha particles. Physica Status Solidi A 213(10):2629–2633
29. Pan Z, Qin K, Zhou J et al (2015) The diamond radiation detector with an Ohmic contact using diamond-like carbon interlayer. Mater Res Innov 19(sup5):S5-828–S825-831
30. Bolshakov AP, Zyablyuk KN, Kolyubin VA et al (2017) Thin CVD diamond film detector for slow neutrons with buried graphitic electrode. Nucl Instrum Methods Phys Res, Sect A 871:142–147
31. Liu JW, Chang JF, Zhang JZ et al (2019) Design, fabrication and testing of CVD diamond detectors with high performance. AIP Adv 9(4):045205
32. Guo Y-Z, Liu J-L, Liu J-W et al (2020) Comparison of α particle detectors based on single-crystal diamond films grown in two types of gas atmospheres by microwave plasma-assisted chemical vapor deposition. Int J Miner Metall Mater 27(5):703–712
33. Tsubota M, Kaneko JH, Miyazaki D et al (2015) High-temperature characteristics of charge collection efficiency using single CVD diamond detectors. Nucl Instrum Methods Phys Res, Sect A 789:50–56
34. Fraimovitch D, Erez E, Ruzin A (2019) Printed silver contacts to CVD diamond detectors. Nucl Instrum Methods Phys Res, Sect A 927:349–352
35. Kaneko J, Fujita F, Konno Y et al (2012) Growth and evaluation of self-standing CVD diamond single crystals on off-axis (001) surface of HP/HT type IIa substrates. Diam Relat Mater 26:45–49
36. Claps G, Murtas F, Foggetta L et al (2018) Diamondpix: a CVD diamond detector with Timepix3 chip interface. IEEE Trans Nucl Sci 65(10):2743–2753
37. Ren Z, Yuan G, Zhang J et al (2018) Hydrogen-terminated polycrystalline diamond MOSFETs with $Al_2O_3$ passivation layers grown by atomic layer deposition at different temperatures. AIP Adv 8(6):065026
38. Imanishi S, Horikawa K, Oi N et al (2019) 3.8 W/mm RF power density for ALD $Al_2O_3$-based two-dimensional hole gas diamond MOSFET operating at saturation velocity. IEEE Electron Device Lett 40(2):279–282
39. Abdel-Rahman MAE, Lohstroh A, Jayawardena I et al (2012) The X-ray detection performance of polycrystalline CVD diamond with pulsed laser deposited carbon electrodes. Diam Relat Mater 22:70–76
40. Lansley SP, Betzel GT, Metcalfe P et al (2010) Comparison of natural and synthetic diamond X-ray detectors. Australas Phys Eng Sci Med 33(4):301–306
41. Fidanzio A, Azario L, Kalish R et al (2005) A preliminary dosimetric characterization of chemical vapor deposition diamond detector prototypes in photon and electron radiotherapy beams. Med Phys 32(2):389–395
42. Ade N, Nam TL (2015) The influence of defect levels on the dose rate dependence of synthetic diamond detectors of various types on exposures to high-energy radiotherapy beams. Radiat Phys Chem 108:65–73
43. Schirru F, Kupriyanov I, Marczewska B et al (2008) Radiation detector performances of nitrogen doped HPHT diamond films. Physica Status Solidi A 205(9):2216–2220
44. Schirru F, Kisielewicz K, Nowak T et al (2010) Single crystal diamond detector for radiotherapy. J Phys D Appl Phys 43(26):265101

45. Abdel-Rahman MAE, Lohstroh A, Sellin PJ (2011) The effect of annealing on the X-ray induced photocurrent characteristics of CVD diamond radiation detectors with different electrical contacts. Physica Status Solidi A 208(9):2079–2086
46. Kassel F, Guthoff M, Dabrowski A et al (2016) Severe signal loss in diamond beam loss monitors in high particle rate environments by charge trapping in radiation-induced defects. Physica Status Solidi A, 213(10):2641–2649
47. Lan L, Xiaoping O, Xinjian T et al (2012) Priming effect on a polycrystalline CVD diamond detector under 60Co $\gamma$-rays irradiation. Nucl Instrum Methods Phys Res, Sect A 672:29–32
48. De Angelis C, Casati M, Bruzzi M et al (2007) Present limitations of CVD diamond detectors for IMRT applications. Nucl Instrum Methods Phys Res, Sect A 583(1):195–203
49. Pomorski M, Caylar B, Bergonzo P (2013) Super-thin single crystal diamond membrane radiation detectors. Appl Phys Lett 103(11):112106
50. Sumiya H (2014) HPHT synthesis of large, high-quality, single crystal diamonds. Elsevier Ltd., pp 195–215
51. Nasieka I, Strelchuk V, Boyko M et al (2015) Raman and photoluminescence characterization of diamond films for radiation detectors. Sens Actuators, A 223:18–23
52. Li HD, Zou GT, Wang QL et al (2008) High-rate growth and nitrogen distribution in homoepitaxial chemical vapour deposited single-crystal diamond. Chinese Phys Lett 25(5):1803–1806
53. Schreck M, Scajev P, Trager M et al (2020) Charge carrier trapping by dislocations in single crystal diamond. J Appl Phys 127(12):125102
54. Lohstroh A, Sellin PJ, Wang SG et al (2007) Effect of dislocations on charge carrier mobility–lifetime product in synthetic single crystal diamond. Appl Phys Lett 90(10):102111

# Chapter 4
# Preparation of SiC Nuclear Radiation Detectors

The quality of SiC crystals used as the radiation detection medium in SiC nuclear radiation detectors significantly affects the detectors' detection capability and susceptibility to radiation damage failure. Furthermore, the performance of SiC nuclear radiation detectors is eventually influenced by several crucial procedures, such as the study and adjustment of the material's characterization results and electrical properties. Hence, this chapter will provide a comprehensive account of the meticulous procedures involved in the production and characterisation of the detectors.

## 4.1 Several Basic Structures of SiC Nuclear Radiation Detectors

### *4.1.1 PN Junction Type Detectors*

PN junction type detectors are radiation detectors that function by utilizing the reverse features of the PN junction. A metallic substance is applied onto the surfaces of P-type and N-type SiC materials to create an ohmic contact electrode, as seen in Fig. 2.3. The depletion layer of the detector is a region that is sensitive to detection. Its thickness increases as the reverse voltage increases. In this condition, the movable carriers in the depletion layer region are mostly depleted due to the high electric field strength and resistivity. The P-type region is often used as the particle incident surface of the detector due to two main factors: ① Given the same excess carrier concentration, the diffusion length and mobility of electrons surpass those of holes. To optimize the sensitivity and response speed of the detector, it is advantageous to utilize the P-type region as the surface for particle incidence. This is because the electric field in the depletion region will attract the holes created by irradiation towards the P-type region, minimizing their drift distance; ② The presence of a P-type semiconductor region as



Y. Zhang et al., *Wide Band-Gap Semiconductor Nuclear Radiation Detector*,
Wide Bandgap Semiconductors, https://doi.org/10.1007/978-981-96-0340-4_4

a dead layer enhances the efficiency of collecting electron–hole pairs generated by particle excitation within the dead layer. This is due to the superior diffusion length and mobility of electrons compared to holes. Consequently, the P-type semiconductor region as a dead layer helps to minimize the diffusion distance of the irradiated holes. The primary drawbacks of the PN junction kind of detector are as follows: ① The capacitance of the PN junction is significant, resulting in a sluggish response; ② The thin depletion layer of the PN junction exhibits limited sensitivity when nuclear radiation particles deeply penetrate.

### 4.1.2 PiN Junction Detector

In order to overcome the limitations of PN junction detectors in nuclear radiation detection, a modified device structure has been developed. This structure involves incorporating a thicker layer of intrinsic semiconductor material (known as the i-type layer) between the P-type and N-type semiconductor materials. This modification results in the creation of a specialized PiN diode. Under high reverse bias voltages, the depletion layer expands into the i-region, resulting in a wider region of inherent electric field. As a result, PiN junction detectors exhibit greater detection sensitivity than PN junction detectors, and the wider depletion region decreases the junction capacitance and enhances its reaction speed. Conversely, the increased thickness of the intrinsic semiconductor layer (i-layer) in the PiN junction detector will lengthen the time it takes for electron–hole pairs to transition, thereby restricting the response speed. However, due to the PiN junction's high breakdown field and low leakage current properties, the reverse bias operating voltage can be suitably raised to maintain its response speed. The PiN junction is characterized by two fundamental structures: planar and mesa structure, as seen in Fig. 4.1. Although planar structure diodes can be readily created using typical planar methods, mesa structures necessitate specific countertop fabrication techniques like as etching or grooving. The advantages of mesa structure are: ① The planar structure offers the benefit of enhancing the electric field concentration between the curved section of the planar junction edge and the surface of the device, hence improving the breakdown characteristics of the device. ② It offers the benefit of minimizing edge inductance and capacitance, hence enhancing the device's response speed.

### 4.1.3 Schottky Junction Detectors

Schottky junction detectors are also known as surface barrier detectors. A specific metal is deposited on the surface of an N-type low-doped SiC epitaxial layer to generate a Schottky contact electrode, and another metal is deposited on the surface of an N-type severely doped substrate to form an ohmic contact electrode. When a reverse voltage is applied, a depletion layer is created within the 4H–SiC material

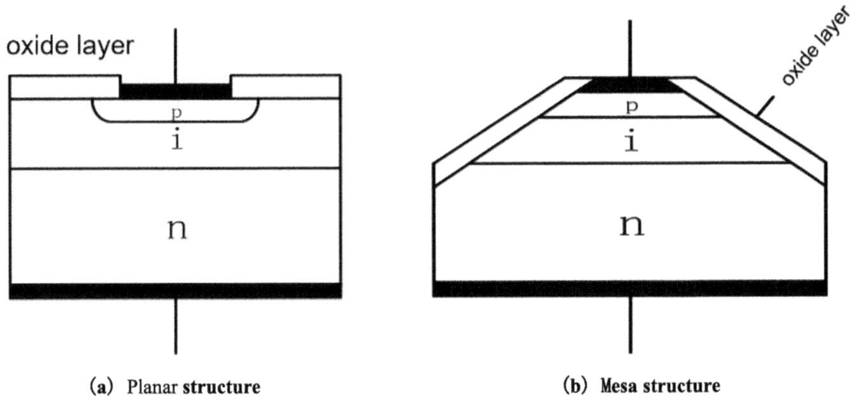


**Fig. 4.1** Schematic diagram of a simple PiN junction diode

due to the Schottky barrier effect at the metal–semiconductor contact. This depletion layer acts as a region in the detector that is sensitive to probes. The thickness of this layer increases as the reverse voltage increases, until the entire epitaxial layer is depleted and reaches its maximum value. The advantages of Schottky junction detectors are: ① The manufacturing process is relatively simple; ② The surface of dead layer is characterized by its minimal thickness, lacking a P-type region and consisting solely of the Schottky metal layer. ③ The PN junction and PiN junction type detectors exhibit both barrier capacitance and diffusion capacitance, while the Schottky junction detectors only exhibit barrier capacitance, resulting in a smaller detector capacitance. Thus, Schottky junction type detectors exhibit the benefits of little inherent noise, excellent energy resolution, strong energy linearity, and rapid response. The disadvantage is: ① The depletion layer region of the Schottky junction is characterized by a minimal thickness. ② Under reverse high voltage, the Schottky junction exhibits a relatively high leakage current and a low reverse breakdown voltage. ③ The Schottky barrier is not ideal for operating circumstances with high temperatures.

### *4.1.4 Neutron Detectors*

Due to their lack of charge and inability to directly induce ionization or excitation in the detecting medium, the detection of neutrons is typically carried out in two separate stages [1]. During the first phase, the neutron penetrates the detecting medium, SiC, and engages in a nuclear interaction with C or Si atoms, resulting in the creation of a charged secondary particle. The threshold nuclear reactions that commonly occur with Si and C atoms in the SiC material as high-energy neutrons of 2.5 and 14 MeV pass through can be represented by the following equations.

$$\begin{cases} {}^{28}\mathrm{Si} + \mathrm{n} \rightarrow {}^{28}\mathrm{Si} + \mathrm{n}' \\ {}^{12}\mathrm{C} + \mathrm{n} \rightarrow {}^{12}\mathrm{C} + \mathrm{n}' \\ {}^{28}\mathrm{Si} + \mathrm{n} \rightarrow {}^{28}\mathrm{Al} + \mathrm{p} \\ {}^{12}\mathrm{C} + \mathrm{n} \rightarrow {}^{12}\mathrm{B} + \mathrm{p} \\ {}^{28}\mathrm{Si} + \mathrm{n} \rightarrow {}^{25}\mathrm{Mg} + {}^{4}\mathrm{He} \\ {}^{12}\mathrm{C} + \mathrm{n} \rightarrow {}^{9}\mathrm{Be} + {}^{4}\mathrm{He} \\ {}^{12}\mathrm{C} + \mathrm{n} \rightarrow \mathrm{n}' + 3{}^{4}\mathrm{He} \end{cases} \tag{4.1}$$

The reaction generates secondary particles, which are very energetic and ionized. These particles include Si, C, P, He, Al, B, Mg, Be, and others. They carry some of the kinetic energy information from the incident neutrons. During the second phase, these highly energized ionized particles penetrate the detector's sensitive area, resulting in the creation of electron–hole pairs (the quantity of which is directly related to the energy transferred by the neutron in the sensitive area). The movement of these electrons and holes, under the influence of a reverse bias voltage, induces a modification in the charge on the two electrodes. This alteration is subsequently gathered as electrical pulses, forming the detector's response signal and enabling the detection of neutrons. Hence, PN junction detectors, PiN junction detectors, and Schottky junction detectors are all equally appropriate for neutron detection. Nevertheless, the neutron -C and neutron -Si scattering cross sections are minimal, leading to a diminished cross section for neutron interaction with SiC crystals. Consequently, the detection efficiency is greatly reduced when SiC crystals are employed as a direct medium for neutron detection, particularly for thermal neutrons with energies ranging from 0.025 to 100 eV. SiC neutron detectors typically incorporate a neutron conversion layer with a high neutron interaction cross-section, which is positioned alongside a PN (PiN) junction-type or surface barrier-type diode. This distinguishes them from SiC nuclear radiation detectors used for detecting other types of nuclear radiation particles. To enhance the detection sensitivity and efficiency of thermal neutrons by the nuclear reaction method, a conversion layer of 6LiF can be utilized. This is applicable for thermal neutrons with an energy range of 0.025–100 eV. To detect fast neutrons with energies more than 1 MeV, a conversion film made of polyethylene (PE) is commonly applied to the surface of the detector. The recoil proton method, which involves detecting nuclear recoil, is then employed. Figure 4.2 [2] provides a concise representation of the fundamental structural diagram of the neutron conversion layer detector and the corresponding particular reactions. During the nuclear reaction, the neutrons undergo an interaction with 6LiF, resulting in the production of primarily two secondary charged particles, α and 3H. Neutrons undergo nuclear recoil and interact with PE to primarily generate recoil protons. The article includes Figs. 4.3 and 4.4, which depict the schematic structure of Schottky junction type and PiN junction type detectors utilized for thermal neutron detection, respectively.

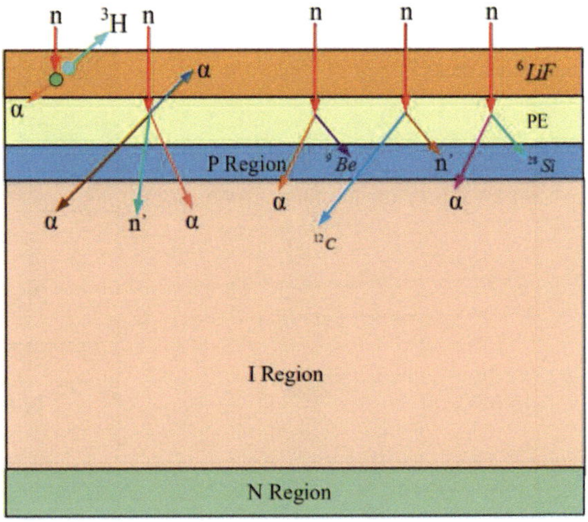


**Fig. 4.2** Schematic diagram of the neutron incidence reaction process [3]

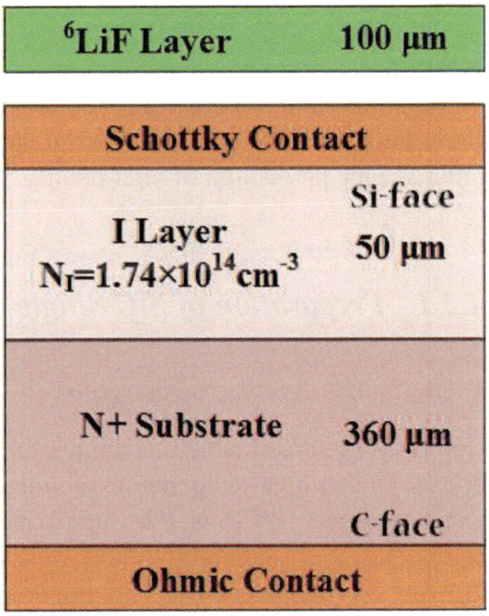


**Fig. 4.3** Schematic structure of Schottky junction SiC thermal neutron detector [4]

## 4.2 Detector-Grade SiC Epitaxial Material Preparation and Analysis

High-quality, thick SiC single-crystal layers with low doping concentrations and defect densities are necessary for detector-grade SiC materials. The SiC bulk materials do not meet all the requirements for directly making detectors due to their inadequate quality and surface qualities. Therefore, detector-grade SiC single-crystal materials are produced via epitaxial growth processes. However, despite the high level

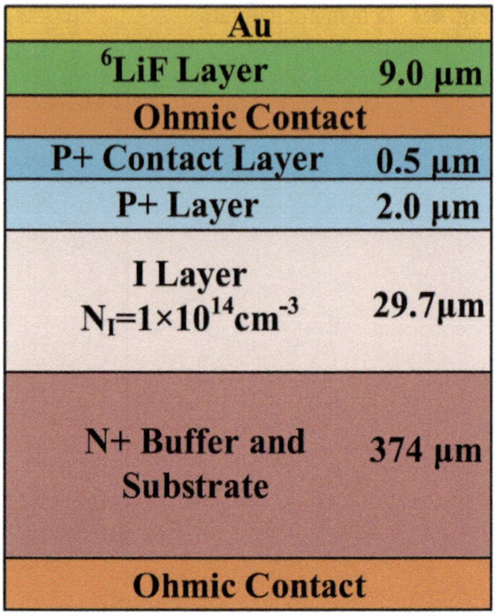


**Fig. 4.4** Schematic structure of PiN junction SiC thermal neutron detector [5]

of advancement in epitaxial growth technology, the parameters of the epitaxial layer may deteriorate if the quality of the substrate material or the method itself is poor. Both the growth of SiC bulk material and epitaxial growth processes significantly influence the production of high-quality SiC epitaxial materials for detectors.

### *4.2.1 Preparation of SiC Single Crystal Substrates*

1. SiC single crystal growth method

Currently, the predominant technique for producing SiC single crystals is the seed crystal-guided high-temperature sublimation method, also referred to as physical vapor transport (PVT) or the improved Lely method [6]. This method, depicted in Fig. 4.5, is considered the most well-established and widely used approach for growing SiC single crystals. The fundamental principle is as follows: SiC powder, serving as the growth source, is positioned at the bottom of a graphite crucible with a higher temperature in a quasi-closed growth system. Through the application of voltage in an inert gas environment at a pressure ranging from 0 to 120 mbar, the SiC powder is heated to a high temperature of approximately 2100–2500 °C using high-frequency induction heating. As a result, the SiC powder undergoes sublimation and decomposition, transforming into gas-phase components such as Si, $SiC_2$, and $Si_2C$. These gas-phase components are then transported to the seed crystal, which is at a lower temperature, driven by the temperature gradient. Consequently, the gas-phase components on the surface of the seed crystal become supersaturated and subsequently undergo recrystallization. The gas-phase constituents are carried to the seed crystals positioned at the upper part of the crucible at lower temperatures,

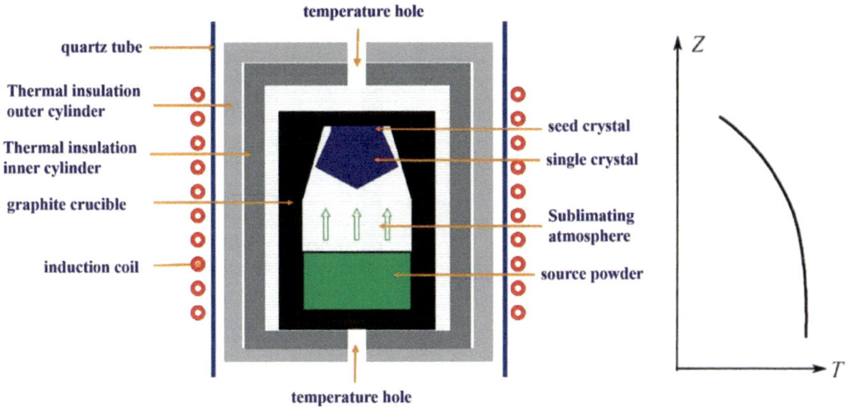


**Fig. 4.5** Schematic diagram of SiC single crystal growth system and temperature distribution along the crucible direction [7]

propelled by the temperature gradient. This results in the gas-phase constituents, which are in a supersaturated state, recrystallizing on the surface of the seed crystals. Consequently, SiC crystals are able to grow [8].

2. Defects in SiC substrates

SiC single crystals grown through the vapor-phase method cannot achieve dislocation-free crystals using the necking and expanding technique, unlike Si single crystals grown through the melt method. Consequently, the majority of dislocation defects in SiC crystals originate from the inherited extension of dislocations in the seed crystals, as depicted in Fig. 4.6a [9]. In this figure, TSD represents helical dislocations, TED represents edge dislocations, and BPD represents basal dislocations. Furthermore, the crystals experience slip paradigmatic deformation due to thermal stresses, resulting in the creation of many dislocation defects within the elastic strain field. This serves to alleviate the elastic stress field, as depicted in Fig. 4.6b [10]. The flaws present in SiC single crystals also affect the epitaxial growth layer utilized for device fabrication, causing a deterioration in the parameters of the epitaxial layer. This is illustrated in Fig. 4.7, which demonstrates the expansion of defects in the epitaxial layer and the typical morphology of these faults [12]. As SiC single crystal growth technology continues to advance, the crystal's integrity has significantly improved. This is mostly shown in a significant decrease in the density of microtubules, the eradication of small-angle crystal borders, and a reduction in the number of enclosure. Nevertheless, the crystal contains a limited quantity of microtubules and a significant concentration of dislocations of various types, including edge-type, screw-type, and base plane dislocations. These imperfections continue to affect the efficiency and dependability of SiC power electronic devices to some extent. The microtubules not only expand along the crystal growth direction [0001], but also propagate into the epitaxial layer of SiC. The presence of the microtubules defect in SiC crystals poses a significant risk to the functionality and longevity of SiC

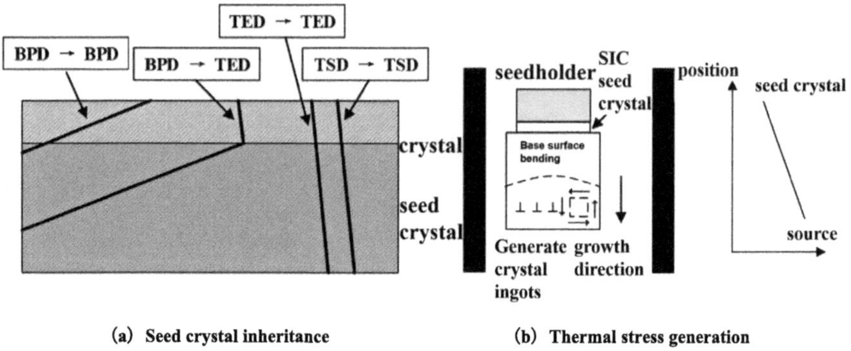


**Fig. 4.6** Main sources of defects in SiC substrates [8]

devices, particularly when working under conditions of high voltage or high current. Consequently, it has earned the reputation of being a highly detrimental defect, often referred to as the "killer" defect. As the size of single crystals grows larger, it becomes increasingly challenging to precisely control each process parameter. Consequently, the density of microtubules shows a significant upward trend. However, thanks to continuous enhancements in the growth process, the microtubule density of SiC single crystals has consistently decreased in recent years [8]. For instance, in 2007, Cree introduced a 4-in. ($\Phi = 100$ mm) N-type 4H–SiC substrate with zero microtubes (Micropipe), which served as a fundamental step towards the utilization of SiC. In 2010, a 6-in. ($\Phi = 150$ mm) N-type 4H–SiC substrate was released with a microtubule density below 10/cm$^2$; In 2012, we successfully brought 6-in. 4H–SiC epitaxial wafers with low microtubule density and a thickness of 100 μm into the commercial market. Regarding dislocation flaws, the company recorded a decrease in screw-type dislocation density to 447 dislocations/cm$^2$ on its 4-in. substrate and 56 dislocations/cm$^2$ on the base plane in 2016. The 6-in. substrate exhibits a screw-type dislocation density of 230 dislocations/cm$^2$ and a base plane dislocation density of 112 dislocations/cm$^2$, which is the highest level observed in the industry [13]. On a domestic level, significant advancements have been made in controlling defects in SiC single-crystal substrate microtubules. The fundamental principles of microtubule growth, extension, and elimination have been thoroughly understood, allowing for precise control of microtubule density on 6-in. substrates at a rate of 1 microtubule/cm$^2$.

### *4.2.2 SiC Thick Epitaxial Layer Growth Process*

Presently, the quality of SiC bulk material and its surface is inadequate for direct device production. However, the performance of semiconductor nuclear radiation

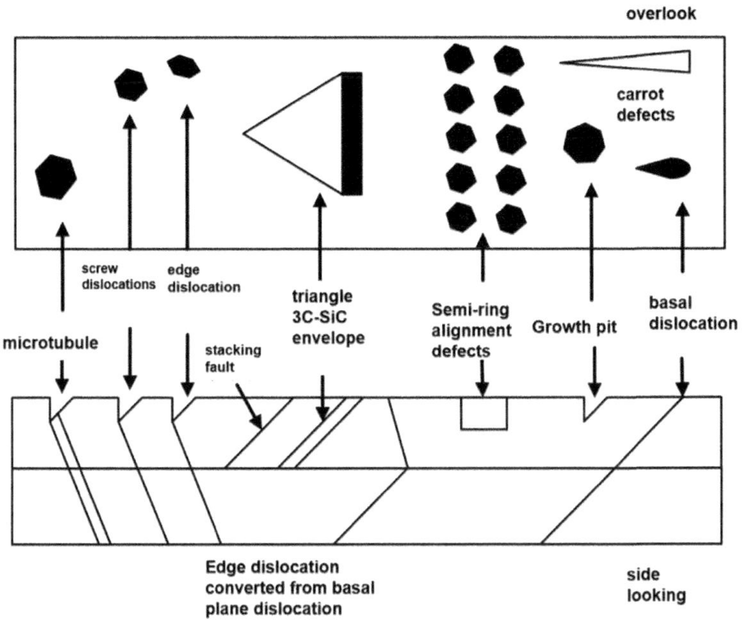


**Fig. 4.7** Defect morphology and structure of epitaxial layer [11]

detectors relies on high-quality SiC material due to its operating mechanism. Consequently, achieving high-quality epitaxial growth of SiC is a crucial aspect in the fabrication of SiC nuclear radiation detectors.

Epitaxial growth, also known as epitaxy [14], is the controlled and orderly development of atoms or molecules of a substance on the surface of a meticulously prepared crystal, typically referred to as a substrate, under specific conditions. It is a single crystal layer that is continuous, smooth and matches to the lattice structure of the substrate. The epitaxial layer refers to this single crystal layer. Epitaxial growth refers to the process of growing an epitaxial layer. Epitaxial growth is classified as homogeneous and heterogeneous epitaxy based on the substrate materials used. Heterogeneous epitaxy allows for the integration of different material substrates. For example, Si heteroepitaxial structures are highly efficient in completely utilising both materials, manufacturing them at a low cost, and integrating them with Si microelectronic monoliths. Nevertheless, the incongruity of lattice characteristics and the disparity in thermal expansion coefficients between SiC and Si will generate lattice defects and additional stresses that will degrade the quality of the SiC epitaxial layer. In addition, at present, SiC heteroepitaxy practically only provides 3C-SiC, which is widely employed in the field of electrical devices and microcomputer systems, whereas for the field of wide-band nuclear radiation detectors, 4H–SiC is the most appropriate material among all SiC homoepitaxy polymorphs at present. Therefore, high-quality epitaxial materials for SiC nuclear radiation detectors are generated by uniform epitaxial growth of 4H–SiC on 4H–SiC substrates.

SiC epitaxial growth is primarily carried out using many methods, as seen in Fig. 4.8: chemical vapour deposition (CVD), sublimation or physical vapour transfer (PVT), molecular beam epitaxy (MBE), and liquid phase epitaxy (LPE). Table 4.1 [15] presents a comprehensive analysis of the pros and cons of the SiC epitaxy methods mentioned above. As can be observed from the chart, each approach has its own advantages and disadvantages, e.g., in terms of crystallization quality, the epitaxial layer obtained by MBE is of the best quality, followed by CVD, LPE and other ways. MBE and LPE are the slowest growth rates in terms of growth rate and are appropriate for research rather than large-scale production. CVD is presently the standard process technology for epitaxial development of compound semiconductors.

1. SiC CVD homogeneous epitaxial growth

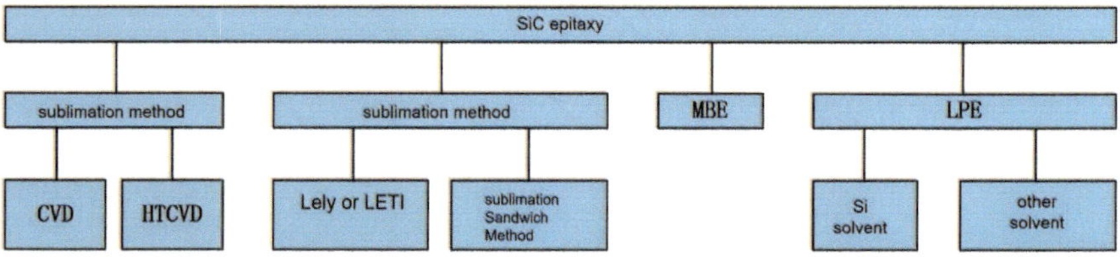


**Fig. 4.8** SiC epitaxial growth method [15]

**Table 4.1** Comparison of the advantages and disadvantages of several commonly used SiC epitaxy methods

| Epitaxial growth method | Advantages | Disadvantages |
|---|---|---|
| Sublimation method (PVT) | Simple growth technique, high temperature and fast growth rate | Uneven evaporation, difficult to obtain high quality SiC epitaxial material, mainly used for SiC substrate material growth |
| Liquid phase epitaxy (LPE) | Low cost, high efficiency of microtubule defect closure and high growth rate | Need to accurately control the thermal equilibrium conditions, doping concentration is difficult to control, and the surface morphology is rough |
| Molecular beam epitaxy (MBE) | The low growth temperature allows the growth of different SiC crystalline forms, which facilitates the growth of ultra-fine structures and heterogeneous structures | Low growth rate, not suitable for the preparation of epitaxial materials required for power devices |
| Chemical vapor deposition (CVD) | Precise control of epitaxial material thickness and doping concentration, good surface morphology and suitable growth rate | Highly pure growth source required |

**Fig. 4.9** Main links of gas phase growth

In order to create nuclear radiation detectors made of 4H–SiC that have minimal leakage at high reverse bias voltages, it is necessary to grow thick epitaxial layers of high quality. The thickness of these layers should be determined based on the desired reverse bias operating voltage and the range of the nuclear radiation particles that will be detected in the epitaxial layer of the SiC material. Additionally, these epitaxial layers should have a doping concentration of less than $(5 \times 10^{14}–1 \times 10^{15})$ $cm^{-3}$.

Chemical Vapor Deposition (CVD) is a sophisticated technique used to create complex epitaxial structures. It is an evolution of the Vapor Phase Epitaxy (VPE) process. The epitaxial growth of SiC thin layers using this method is similar to gas-phase epitaxy of Si. Both methods involve using $H_2$ or Ar as the carrying gas to introduce $SiH_4$, $SiHCl_3$, or $SiCl_4$ as the growth source into the reaction chamber. The main difference is that in SiC growth, a carbide such as $CH_4$, $C_2H_6$, $C_3H_8$, or $CCl_4$ needs to be added to the growth source gas. Conversely, the utilization of organic silicides for SiC thin-layer epitaxy eliminates the need for additional carbon-containing gases. The chemical vapor phase growth process has four primary stages, as illustrated in Fig. 4.9.

The main advantages of CVD over sublimation method, liquid phase epitaxial method and molecular beam epitaxy are:

(1) During CVD growth, the mass flow controllers accurately regulate the flow of growth source gases and doping source gases. This ensures exact control over the components, thickness, and doping concentration of the epitaxial material.
(2) Epitaxial materials produced by CVD technique exhibit excellent homogeneity and a wide range of doping capabilities, making them well-suited for the fabrication of epitaxial materials needed for diverse devices.
(3) The equipment construction is uncomplicated, user-friendly, and suitable for large-scale industrial production, making it more practically valuable compared to molecular beam epitaxy. Currently, CVD is the prevailing technique for SiC epitaxial growth. It is primarily categorized into atmospheric-pressure chemical vapor deposition (APCVD), low-pressure chemical vapor deposition (LPCVD), plasma-enhanced chemical vapor deposition (PECVD), and high-temperature chemical vapor deposition (HTCVD). There are three different methods of chemical vapor deposition: low-pressure chemical vapor deposition (LPCVD), plasma-enhanced chemical vapor deposition (PECVD), and high-temperature chemical vapor deposition (HTCVD). Each method has its own set of advantages and limitations.

The APCVD (Atmospheric Pressure Chemical Vapor Deposition) reactor is relatively simple to design and exhibits a high reaction rate. However, it is susceptible to gas-phase reactions and the films produced have inadequate step coverage. Additionally, due to operating in the "substance transport limitation" region, the reaction stream must be uniformly distributed across the entire substrate.

LPCVD (Low Pressure Chemical Vapor Deposition) offers excellent step coverage uniformity and precise control over the composition and structure of the epitaxial layer. Reducing the pressure significantly decreases the particle pollution caused by the formation of gas particles, but it also leads to greater temperatures during the process. LPCVD operates within the zone where surface reactions dominate, therefore the reactor design must possess excellent control over the deposition temperature.

PECVD, or plasma-enhanced chemical vapor deposition, involves the activation of reactive gas molecules to facilitate their chemical reaction and deposition into a film at low temperatures, including room temperature. However, PECVD has a low yield and its weak adhesion can lead to particle contamination.

HTCVD [16], or high temperature chemical vapor deposition, is a process that can achieve temperatures as high as 2300 °C. This allows for the production of high quality SiC epitaxial layers at fast growth rates. The HTCVD technique holds great significance in the development of SiC, particularly for the cultivation of high purity SiC semi-insulating materials. This method offers precise manipulation of the gas phase of the crystals being grown, allowing for effective management of inherent point defects and the attainment of highly pure crystals, in contrast to the evaporation growth technique. Nevertheless, the expense associated with epitaxial growth via the HTCVD technique is exorbitant. Consequently, for the production of high-quality 4H–SiC epitaxial layers suitable for detectors, the more cost-effective regular epitaxial CVD procedure is typically employed instead of HTCVD. Furthermore, the LPCVD technique, commonly employed for SiC development, exhibits a rapid growth rate and effectively addresses the challenge of producing high-quality thick-film epitaxial materials within a reasonable timeframe by minimizing gas-phase nucleation under low pressure conditions. The need for superior thick film epitaxial growth for nuclear radiation detectors.

2. 4H–SiC LPCVD Thick Epitaxial Growth Process Design

The usual approach for developing SiC devices is the chemical vapor deposition (CVD) of hexagonal SiC polymorphs in the off-axis SiC {0001} direction. Monosilane ($SiH_4$) and propane ($C_3H_8$) or ethylene ($C_2H_4$) are commonly used as precursor gases. The carrier gas is usually hydrogen ($H_2$) and sometimes argon (Ar) is added. Typical growth temperatures are typically 1500–1650 °C, corresponding to growth rates of 3–15 μm/h. The CVD growth process for SiC usually includes in-situ etching and main epitaxial growth. In situ etching is performed with pure $H_2$, $HCl/H_2$, hydrocarbons/$H_2$, or $SiH_4/H_2$ at elevated temperatures, typically the same temperatures used for primary epitaxial growth. The purpose of in-situ etching is to eliminate subsurface damage and obtain regular step structures. The main epitaxial growth of N-type or P-type SiC (or multilayers thereof) will be performed immediately after etching.

The primary process parameters of CVD include the gas flow distribution, temperature distribution, pressure, shape of the substrate, and positioning of the substrate. The uniformity of the flow field and temperature distribution within the CVD reactor significantly affects the surface quality and ultimate morphology of the CVD deposition products. Jia et al. [17] from Xidian University developed a Grove model to study the formation of 4H–SiC homogeneous epitaxy. They analyzed the chemical reaction and growth factors involved in this process. The reaction occurs by combining silane ($SiH_4$) and propane ($C_3H_8$) as reactant gases, while hydrogen ($H_2$) acts as the carrier gas, resulting in the production of SiC. The model's conclusions pertain to the growth chamber of an EPIGRESS VP508GFR epitaxial furnace, which utilizes a horizontal hot-wall CVD technique. The overall equation for the reaction is [18, 19]:

$$C_3H_8 + 3SiH_4 \rightarrow 3SiC + 10H_2 \tag{4.2}$$

The following are reasonable assumptions for modeling: (i) The gas passing through the growth cavity has a constant density. (ii) The gas undergoes complete reaction inside the growth cavity. (iii) All gases are considered ideal gases, following the ideal gas equation of state. (iv) The heat produced by the chemical reaction is disregarded, and the temperature inside the growth cavity remains constant.

Epitaxial development can be categorized into three primary processes: diffusion, adsorption/desorption, and chemical reaction. The pace of epitaxial growth is determined by the slowest of these three processes, as it is directly influenced by the rates of all three. In normal/decompression epitaxy, the adsorption/desorption process occurs at a quicker rate than diffusion and chemical reactions. As a result, the growth rate is predominantly determined by the diffusion and chemical processes.

The process of growth is illustrated by reaction Eq. (4.2) in the field of chemistry. In order to achieve consistent epitaxial thickness, the substrate undergoes rotational motion during the growth process, as depicted in Fig. 4.10. The coordinate system is defined by setting the y-axis as the left boundary of the tray, the origin as the middle of the boundary, and the x-axis as the straight line over the center of the substrate connecting to the origin. The substrate is represented by the solid circular section, with ω denoting the angular velocity of substrate rotation.

The gas flow rate $u$ at any point on the substrate surface is:

$$u = \sqrt{(\omega|r + a - x| \sin\theta + A)^2 + \omega|r + a - x| \cos\theta^2} \tag{4.3}$$

where A is the main gas flow rate, obtained from the ideal gas equation of state:

$$A = \frac{P_1 V_1}{P_2 S} \tag{4.4}$$

where $P_1$ is the standard atmospheric pressure, $P_2$ is the growth air pressure, $V_1$ is the carrier gas flow rate (L/min), and S is the cross-sectional area of the growth cavity. The distance of any point $x$ on the substrate from the head of the base is:

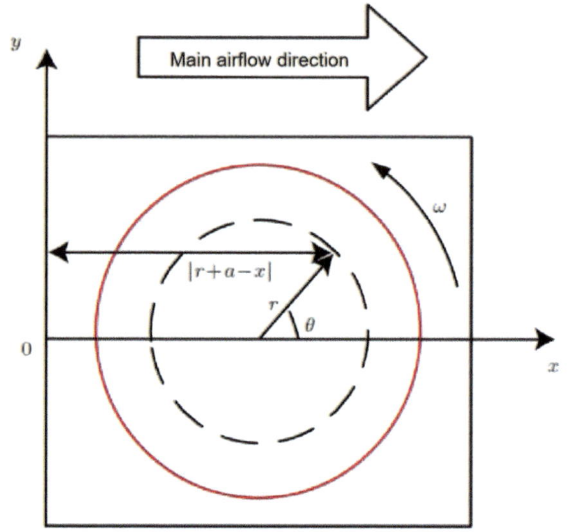


**Fig. 4.10** Top view of the substrate tray structure [17]

$$L = r + a + |r + a - x| \cos\theta \tag{4.5}$$

Its gas phase transfer coefficient is:

$$h_G = D_{G_0}\left(\frac{T}{T_0}\right)^{\delta}\left(\frac{\rho u}{\mu L}\right)^{1/2} \tag{4.6}$$

$\mu$ is the gas viscosity coefficient, $u$ is the gas flow rate, $\rho$ is the gas density, $T_0$ is the room temperature expressed in terms of thermodynamic temperature, $T$ is the experimental temperature expressed in terms of thermodynamic temperature, $D_{G_0}$ is the diffusion coefficient of the gas at room temperature, which is usually in the range of 0.1–1 $cm^2/s$, and $\delta$ is a constant with the value of 1.75–2. Due to the continuous rotation of the substrate during the growth process, the growth environment at any location on the substrate, except the center of the circle, undergoes periodic changes. Hence, it is necessary to ascertain the instantaneous rate of a specific location on the substrate and subsequently integrate it to calculate the cumulative growth thickness following one complete rotation of said location. This value can then be transformed into the growth thickness per hour, also known as the growth rate per hour. Therefore, based on the Grove model and accounting for the continuous rotation of the substrate during the development process, the growth rate $v$ can be expressed as:

$$v = \frac{21600}{2\pi}\int_0^{2\pi}\frac{k_s h_G}{k_s + h_G}\frac{N_T}{N_{SiC}} Y \, d\theta \tag{4.7}$$

$$k_s = k_{s0}\exp\left(-\frac{E_0}{kT}\right) \tag{4.8}$$

where $k_s$ is is the growth rate of the surface chemical reaction, $k_{s0}$ is an experimental constant, $E_0$ is the reactant activation energy, hG is the gas-phase mass-transfer coefficient, NT is the number of molecules per unit volume of the gas mixture, NSiC is the molecular density of the SiC crystals, and Y is the percentage of reactant concentration.

(a) Key process parameters for 4H–SiC LPCVD thick epitaxial growth

According to the above Grove model 4H–SiC epitaxial growth rate model based on Grove model, and combined with the experimental results can be a series of analysis and verification.

(1) The distribution of the growth rate in the direction of the carrier gas flow

Equation (4.3) reveals a gradual drop in the relative velocity u from the periphery to the center of the substrate circle. Equations (4.6) and (4.7) demonstrate that a reduction in relative velocity results in a steady decline in the growth rate. Furthermore, due to the continuous rotation of the tray during the growth process, the epitaxial layer on the substrate exhibits identical periodic changes in growth conditions from the center to the circumference. Consequently, the epitaxial layer on the substrate forms a circular pattern with the same growth rate and thickness. This results in a distribution of epitaxial layer thickness that is thicker at the edges and thinner at the center, resembling a "bowl-shaped" structure. This observation is in accordance with the experimental findings depicted in Fig. 4.11. *R* denotes the distance between a point on the substrate and the center of the substrate circle, whereas *d* represents the thickness of the epitaxial layer. The solid line represents the observed thickness distribution. However, inside a certain area spanning from the center to the periphery, the measured thickness exceeds the predicted value. The model assumes a homogeneous temperature field, but in reality, the high temperature zone will shift downstream due to the influence of the main airflow. This causes an increase in the growth rate at that point. This region can be used to determine the precise location of the highest temperature point in the growth cavity. It also allows for a comprehensive consideration of the influence of the main airflow on the high temperature point and thickness distribution. This, in turn, helps improve the uniformity of the epitaxial layer's thickness.

(2) Effect of carrier gas flow rate on epitaxy rate and thickness uniformity

According to Eqs. (4.3) and (4.4) the relative velocity u keeps increasing with the increase of the carrier gas flow rate, which in turn increases the epitaxial rate. When u increases to such an extent that the gas phase transfer coefficient $h_G \gg k_s$, Eq. (4.7) can be simplified to:

$$v = \frac{3600\omega}{2\pi} \int_0^{2\pi} k_s \frac{N_T}{N_{SiC}} Y \mathrm{d}\theta \tag{4.9}$$

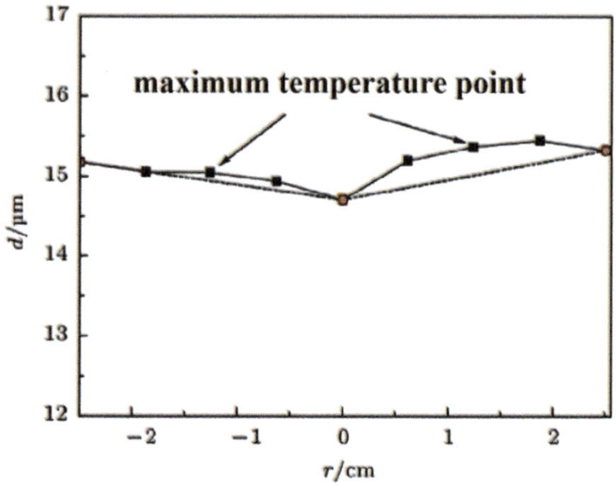


Fig. 4.11 Thickness distribution on epitaxial layer

At present, the epitaxy reaches the reaction control zone, as shown by Eq. (4.9). During this stage, the epitaxial rate is independent of the relative velocity $u$. Increasing the flow rate of the carrier gas will not result in an increase in the epitaxial rate. Increasing the flow of the carrier gas not only impacts the relative velocity ($u$), but also reduces the concentration of the reactant ($Y$). According to formula (4.9), if the carrier gas flow rate continues to increase, the rate of epitaxial growth will steadily decrease. The experiment involved a gradual increase in the flow rate of the carrier gas from 40 to 120 L/min. During this procedure, as the flow rate of the carrier gas surpassed 60 L/min, the rate of epitaxial growth steadily decreased. These findings are illustrated in Fig. 4.12.

At a constant temperature, the thickness distribution on the substrate is mostly influenced by the gas phase transfer coefficient $h_G$. Therefore, it can be inferred that

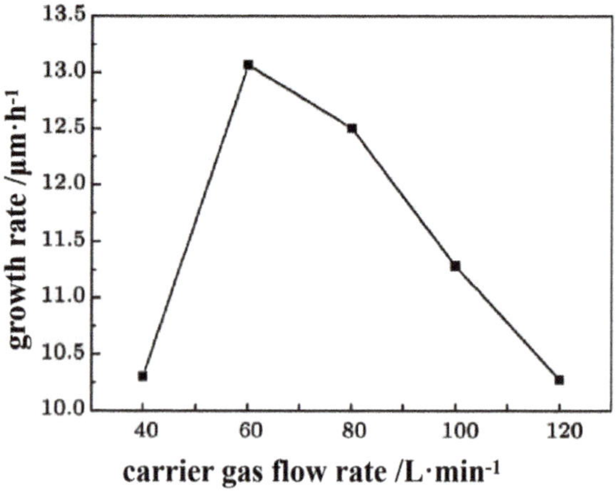


Fig. 4.12 Epitaxial rate variation with carrier gas flow rate [17]

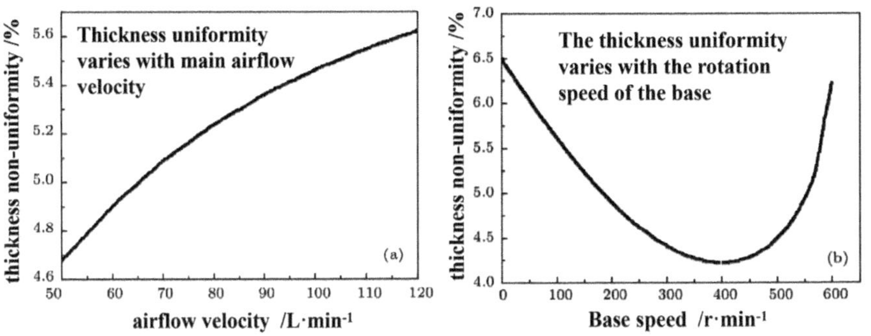


**Fig. 4.13** **a** Effect of main gas flow rate on epitaxial uniformity; **b** Effect of substrate speed $\omega$ on epitaxial uniformity [17]

the uniformity of $h_G$ corresponds to the uniformity of the epitaxial layer's thickness to some degree. Figure 4.13 displays the outcomes of the theoretical computations. Figure 4.13a demonstrates that the unevenness in thickness of the epitaxial layer worsens progressively as the main gas flow rate increases. The experimental findings demonstrate a steady deterioration in the thickness non-uniformity of the epitaxial layer, increasing from 2.7 to 3.1%, while the main gas flow rate is incrementally raised from 75 to 85 L/min. The observed high temperature point, depicted in Fig. 4.11, is responsible for the achieved homogeneity in the thickness of the epitaxial layer. According to Fig. 4.13b, the homogeneity of the substrate improves when the rotating speed remains below 400 rpm and the substrate speed increases. The empirical findings surpass the theoretical assertion that the rotational velocity of the substrate deviates significantly from 400 rpm. The portion of the graph beyond 400 rpm indicates that while the gas levitation approach can enhance the uniformity of the epitaxial layer, excessive substrate rotational speed in relation to the main gas flow rate leads to a steady decline in epitaxial uniformity.

(3) Effect of temperature on epitaxy rate

Equations (4.6) and (4.7) demonstrate that the epitaxial rate is positively correlated with temperature. With an increase in temperature, there is a simultaneous increase in both the rate of chemical reactions and the coefficient of gas phase transfer. Consequently, this leads to an increase in the rate of epitaxial growth. Nevertheless, the impact of temperature on $k_s$ is far greater than that of $h_G$. Consequently, when the temperature rises excessively, $k_s$ will increase significantly more than $h_G$ in order to ensure that the growth falls within the quality control range. Currently, Eq. (4.7) can be condensed as:

$$\mathrm{v} = \frac{3600\omega}{2\pi}\int_0^{2\pi} h_G \frac{N_T}{N_{SiC}} Y \mathrm{d}\theta \tag{4.10}$$

Equations (4.7) and (4.10) demonstrate that further temperature rise at this juncture will not yield a substantial enhancement in the epitaxial rate. This has been empirically demonstrated. Nevertheless, during the experiment, while keeping other parameters constant, elevating the temperature from 1580 to 1650 °C resulted in a reduction in the growth rate from 10.0214 μm/h to 9.5262 μm/h. Disregarding any experimental error, the elevated temperature prompts the chemical raw materials to undergo a reaction in the air, generating solid particles (which may not necessarily be SiC). Consequently, the quantity of particles available for epitaxy growth decreases, leading to a reduction in the reaction rate. The epitaxial growth process is hindered by a decrease in the quantity of particles employed, resulting in a reduction of the reaction rate. If the solid particles produced during the gas-phase reaction come into contact with the epitaxial layer, they might have a negative impact on its quality. Therefore, it is crucial to carefully regulate the temperature during the epitaxial process to prevent the occurrence of the reaction in the nucleation zone of the gas phase.

(4) The influence of other factors

Several empirical research have demonstrated that the LPCVD preparation of the 4H–SiC epitaxial layer is influenced by factors such as growth gas pressure and the ratio of reaction gas flow rate, which impact both the quality and growth rate of the layer.

① Reaction pressure

At pressures below 100 Torr, the growth pressure has no impact on the growth rate, which remains stable (as depicted in Fig. 4.14). However, when the growth pressure fluctuates within the range of 100–450 Torr, the growth rate decreases as the pressure increases [20]. The observed variations in the aforementioned experimental phenomena mostly stem from the increase in epitaxial growth pressure, which leads to a greater thickness of the boundary layer. Consequently, the passage of reactants through the boundary layer to reach the growth surface becomes more challenging. Consequently, the rate of epitaxial growth diminishes gradually as the pressure increases. Simultaneously, the growing pressure will also exert a discernible influence on both the surface shape and the defects of the epitaxial layer.

② Reaction gas flow ratio

In the $SiH_4$–$C_3H_8$–$H_2$ system that is frequently utilized, the ratio of gas flow rates for the reaction can also be represented as the ratio of carbon to silicon (C/Si) versus the ratio of $SiH_4/H_2$ flow rates. To enhance the growth rate, one can reduce the hydrogen flow rate, so extending the duration that the reaction gas remains in the reaction chamber. By manipulating the C/Si ratio, the growth rate remains consistent both in the direction of gas flow and perpendicular to it. The primary purpose of varying the C/Si ratio is to influence the "competition principle" whereby nitrogen can replace carbon in the SiC lattice. Increasing the C/Si ratio reduces the likelihood of nitrogen dominating the lattice, thereby decreasing the concentration of nitrogen in the epitaxial layer. Thus, it is feasible to decrease the amount of inadvertently introduced nitrogen in the epitaxial layer and achieve a superior, unintentionally doped epitaxial layer with a minimal concentration of N-type carriers.

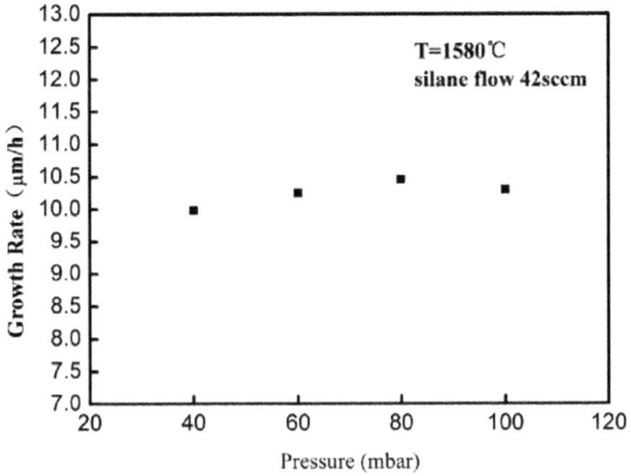


**Fig. 4.14** Effect of growth pressure on growth rate under low pressure conditions [15]

(b) 4H–SiC CVD doping control [21]

The idea of the site-competition effect, as identified by Larkin et al., plays a crucial role in attaining comprehensive doping control in SiC CVD [22, 23]. The nitrogen doping efficiency experiences a substantial rise in Si-rich conditions (low C/Si ratio) and a decrease in C-rich conditions (high C/Si ratio). The phenomena can be elucidated by the rivalry between nitrogen and carbon atoms on the growing surface, resulting in the substitution of carbon lattice positions in SiC with nitrogen atoms. Nitrogen doping is facilitated by a growth surface with low carbon atom coverage, whereas high carbon atom coverage hinders nitrogen atom doping. Conversely, when aluminum and boron replace the Si lattice sites, there is an inverse relationship: the doping of aluminum and boron decreases in the presence of an excess of Si and increases in the presence of an excess of carbon.

(1) Unintentional adulteration

By optimizing the method and purifying the source material, it is possible to attain a high level of purity in undoped (or inadvertently doped) SiC epitaxial layers. Nitrogen is the primary source of impurity in unintentional doping. To achieve high purity epitaxy, one can either increase the C/Si ratio or decrease the growing pressure. Figure 4.15 depicts the relationship between the doping concentration and the C/Si ratio for accidentally doped 4H–SiC{0001} epitaxial layers that were produced using hot-wall chemical vapor deposition (CVD). The size densities of both C- and Si-side epitaxials are around $5 \times 10^{15}$ cm$^3$ at a C/Si ratio of 0.5. By raising the C/Si ratio, epitaxial growth on the (0001) Si facets enables a significant decrease in the sizer density, reaching values as low as $5 \times 10^{12}$ cm$^3$ when the C/Si ratio is 2. When the C/Si ratio is increased in unwanted epitaxial layers, the conductivity type switches from N-type to P-type. P-type materials are acquired by reducing nitrogen doping and increasing aluminum or boron doping, in accordance with the site-competition theory. Nevertheless, the correlation between the doping concentration and the C/Si ratio is far less pronounced for the (0001)C facets. In this specific scenario, the minimum density of dopants is around 8 × $10^{14}$ $cm^3$. Figure 4.16 displays the relationship between the concentration of doping and the growth pressure for accidentally doped and nitrogen-doped 4H–SiC{0001} epitaxial layers that were formed using hot-wall CVD. The reduction of growth pressure severely hampers nitrogen doping. This can be partially explained by the rise in the real ratio of carbon to silicon on the surface where growth occurs at low pressure. This increase is mostly caused by the intensified release of silicon atoms. Furthermore, under conditions of reduced pressure, the release of nitrogen atoms during the movement of the surface is intensified.

(2) N-type doping

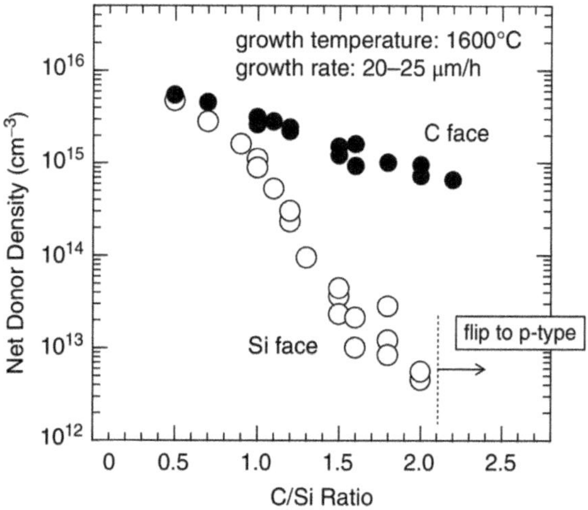


**Fig. 4.15** Hot-wall CVD growth of unintentionally doped 4H–SiC{0001} epitaxial layer doping concentration as a function of C/Si ratio [21]

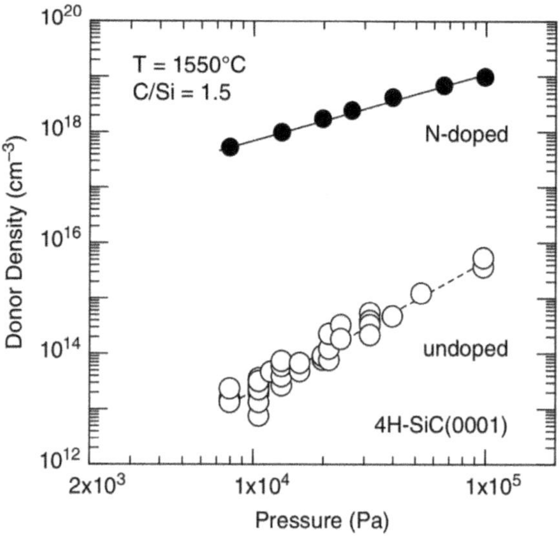


**Fig. 4.16** Hot-wall CVD growth of unintentionally doped and nitrogen-doped 4H–SiC{0001} epitaxial layers doping concentration versus growth pressure [21]

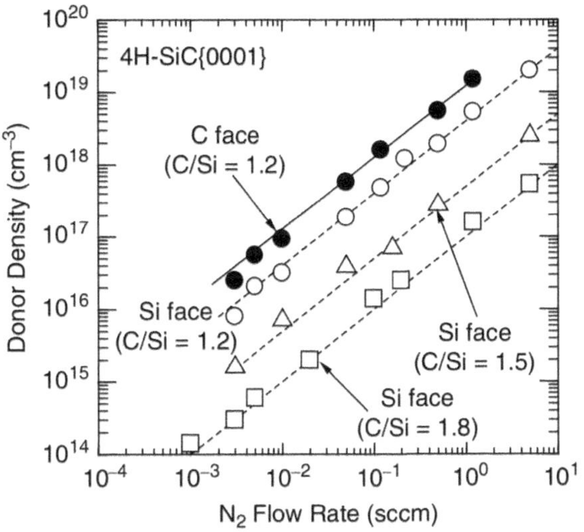


**Fig. 4.17** Hot-wall CVD growth of unintentionally doped and nitrogen-doped 4H–SiC{0001} epitaxial layer doping concentration vs. growth pressure [21]

Nitrogen doping can be readily accomplished by injecting nitrogen gas ($N_2$) during the CVD growth process. The graph in Fig. 4.17 illustrates the relationship between sizer density and $N_2$ flux in hot-wall CVD 4H–SiC(0001) at a temperature of 1550 °C. The sizing density may be derived from the capacitance–voltage (C-V) characteristic, and it is directly proportional to the N2 flow rate for CVD epitaxial growth on both (0001) Si and (0001) C surfaces. The $N_2$ flow rate and the C/Si ratio are crucial factors for achieving precise control over the nitrogen doping concentration (ranging from $1 \times 10^{14}$ to $2 \times 10^{19}$ $cm^3$) when the growth temperature and pressure are kept constant. At higher growth temperatures, the process of cold-wall CVD for SiC inhibits nitrogen doping on both the (0001) and (000) surfaces [24]. Nevertheless, in hot-wall CVD, the nitrogen doping on the (0001) surface intensifies while it diminishes on the (000) surface with an increase in the growth temperature [25].

(3) P-type doping

Introducing modest quantities of trimethylaluminum (TMA: $Al(CH_3)_3$) is an efficient method for in situ P-type doping in SiC CVD [26]. The graph in Fig. 4.18 illustrates the relationship between subject density and TMA flux in hot-wall CVD 4H–SiC{0001} at a temperature of 1550 °C. The density of aluminum atoms determined by secondary ion mass spectrometry (SIMS) is very comparable with the host-accepted density obtained from the capacitance–voltage (C-V) characteristic measurements.

Thus, by regulating the C/Si ratio in the CVD process, the doping concentration range can be substantially expanded. Furthermore, regulating the C/Si ratio proves to be a highly efficient method for achieving a distinct boundary between N-type and P-type epitaxial layers [27]. By correlating the variation in the C/Si ratio with the transition of dopants (such as from nitrogen to aluminum), it is possible to achieve a sudden alteration in the concentration of dopant atoms along the depth profile.

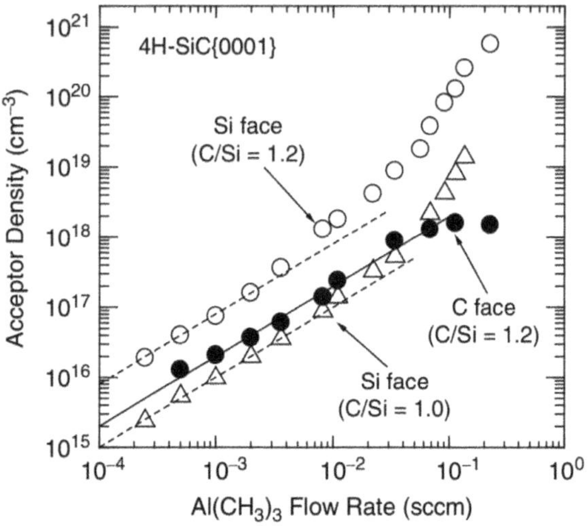


**Fig. 4.18** Hot-wall CVD growth of 4H–SiC{0001} epitaxial layer subjected to principal density versus TMA flux at 1550 °C [21]

## 4.3 Characterization and Analysis of SiC Epitaxial Layer

In order to create effective 4H–SiC nuclear radiation detectors, it is crucial to use high-quality thick-film homogeneous epitaxial materials that have large diameters, low doping concentrations, and low defect densities. These materials ensure that the detectors have low leakage currents and strong detection output signals. Additionally, they allow for the creation of a thick depletion region and a high-intensity electric field at high reverse voltages. Therefore, accurately characterizing the physical properties, doping concentrations, and defects of SiC epitaxial layers is a vital step in the development of 4H–SiC nuclear radiation detectors. The characterization of the epitaxial layer and the optimization of the CVD process are mutually beneficial in producing superior epitaxial materials for the fabrication of high-performance 4H–SiC nuclear radiation detectors.

### *4.3.1 SiC Epitaxial Layer Characterization Techniques*

The majority of widely used methods for characterizing materials can be used to SiC, with the exception of certain features that necessitate particular consideration. The following is a concise summary of the key distinctions in SiC characterization in comparison to conventional semiconductor materials:

(1) Exemplary arrangement (comprising substrate): It is often vital to characterize the qualities or faults of the SiC epitaxial layer as they can directly impact the device's performance. The influence of the substrate cannot be disregarded when the SiC epitaxial layer has a relatively small thickness of 5–10 μm. Hence, it is crucial to ascertain the precise thickness measured by each characterization

method. Even if the epitaxial layer has a considerable thickness (30–50 μm), it is necessary to consider the influence of the substrate.

(2) Wide bandgap: Conventional systems used to analyze the properties of Si and GaAs need to be adjusted when optical stimulation is needed for the analysis. To analyze 4H–SiC at room temperature, an excitation source with UV light at wavelengths below 370 nm, bigger than the bandgap width, is required. Due to SiC's nature as an indirect bandgap semiconductor, its optical absorption coefficient remains low even when the photon energy significantly exceeds the forbidden bandwidth. Furthermore, in order to observe the interband states in the characterisation process, it is necessary to apply thermal excitation at extremely high temperatures (450 °C) due to the fact that the band gap of 4H–SiC is roughly three times larger than that of Si.

1. Photoluminescence (Photoluminescence)

Excitation of a suitable light source can generate surplus carriers that display radiative composite features, often known as photoluminescence (PL). The excitation source typically has a photon energy that exceeds the prohibited bandwidth. When examining a material that is a flawless single crystal with extremely high purity, only peaks corresponding to band-to-band or free excitons are detected. However, the existence of flaws and/or impurities in the material leads to the occurrence of several composite routes, which in turn causes distinct emission peaks in the PL spectrum. Therefore, PL is a potent method for evaluating the presence of flaws and the level of purity in semiconductor materials. Photoluminescence spectroscopy is commonly conducted at low temperatures, such as 2, 4.2, or 77 K, in order to reduce the widening of spectral peaks. The temperature dependency of photoluminescence spectroscopy assists in comprehending the many processes involved in the composite. Figure 4.19 depicts a standard intricate trajectory followed by surplus carriers in SiC. Within SiC, the presence of impurities or point defects can take the place of specific uneven places, resulting in the formation of various energy levels. This occurrence contributes to the increased complexity of the photoluminescence (PL) spectra of SiC compared to other semiconductors. Typically, the photoluminescence spectrum of SiC comprises peaks corresponding to free excitons, excitons bound to neutral doped impurities, complex peaks arising from donor–acceptor pairs, and peaks resulting from different impurities. The PL energy spectrum of a 50 μm thick SiC epitaxial layer, grown via chemical vapor deposition (CVD) in a chlorine-based environment at a temperature of 1600 °C, is depicted in Fig. 4.20. The inset displays the energy spectra produced by excitons bound to nitrogen (such as the $Q_0$ and $P_{76}$ lines), excitons that are not bound (the I76 line), and excitons bound to aluminum (the Al-BE) in the near-forbidden-band region. The inherent flaws in the DI line and Ti line are extremely faint. The Al-BE, $Q_0$, and $P_{76}$ peaks correspond to exciton peaks that are bound by aluminum and nitrogen. These peaks arise from incomplete removal of trimethylaluminum (TMA) during growth in the CVD system. On the other hand, the $I_{76}$ peak represents a free exciton peak.

2. X-ray diffraction (XRD)

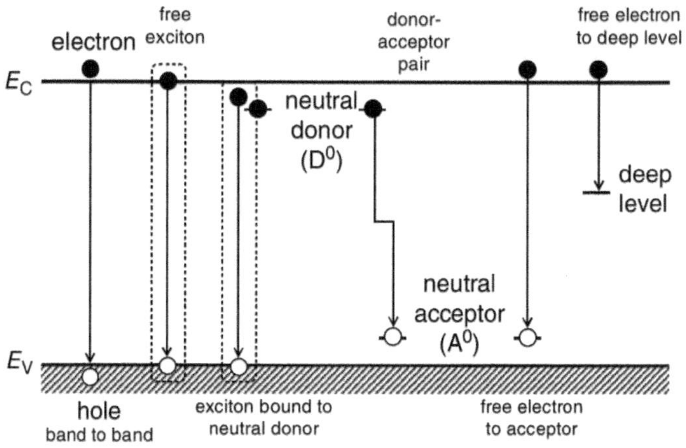


**Fig. 4.19** Schematic diagram of typical complex paths of excess carriers in SiC [21]

**Fig. 4.20** PL energy spectrum of 50 μm thickness epitaxial layer obtained by growth at 1600 °C [28]

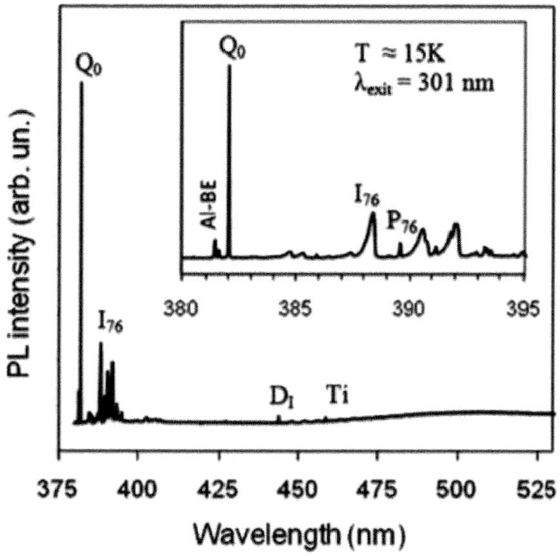


X-ray diffraction (XRD) analysis is a technique used to determine the spatial arrangement of atoms within a substance by measuring the diffraction pattern of X-rays that are scattered by the crystal structure. When X-rays with a specific wavelength are directed at a crystal, the X-rays are scattered by the regularly arranged atoms or ions in the crystal. This scattering causes the X-rays to become stronger in certain directions, resulting in a distinct diffraction pattern that corresponds to the crystal's structure. X-rays that have undergone diffraction adhere to the W. L. Bragg equation:

$$2d \sin\theta = n\lambda \tag{4.11}$$

The variables in the equation are as follows: θ represents the diffraction angle, $\lambda$ represents the wavelength of the X-rays, $n$ represents an integer, and $d$ represents the distance between crystal faces. The wavelength $\lambda$ of X-rays can be utilized to calculate the known diffraction angle, which in turn allows for the determination of the surface spacing. This surface spacing refers to the regular arrangement of atoms or ions within the crystalline form. By analyzing the diffracted X-ray intensity and surface spacing using the established table, it is possible to detect the crystalline structure of the specimen, which is referred to as qualitative analysis. Figure 4.21 displays the schematic schematics of various XRD testing techniques. (a) Linkage scanning involves collecting reflections formed by a matrix of dots parallel to the surface, known as out-of-plane scanning. (b) The q-2qB-B diffraction geometry is used for individual scanning to measure the oscillation curve in a plane perpendicular to the surface. (c) In-plane scanning involves fixing the grazing angle at 2a and scanning in a plane nearly parallel to the surface, collecting reflections formed by a matrix of dots that are nearly perpendicular to the surface. Images created by the surface.

Figure 4.22 depicts the ω-2θ continuous scanning curve test, where the scanning angle ranges from 5 to 85°, performed on the 4H–SiC epitaxial wafer (0004) face. The prominent peak observed at 35.67° corresponds to the diffraction peak of the 4H–SiC (0004) face. No significant diffraction peaks are observed at other angles, indicating that the 4H–SiC epitaxial layer possesses a well-formed crystalline structure without any impurity crystal images. Figure 4.23 displays the wiggle curves of CVD epitaxial

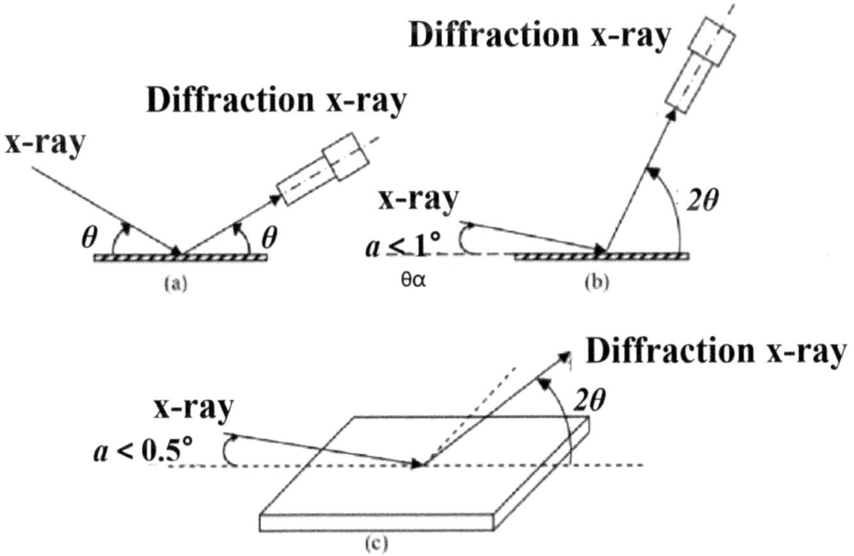


**Fig. 4.21** XRD schematic [29]

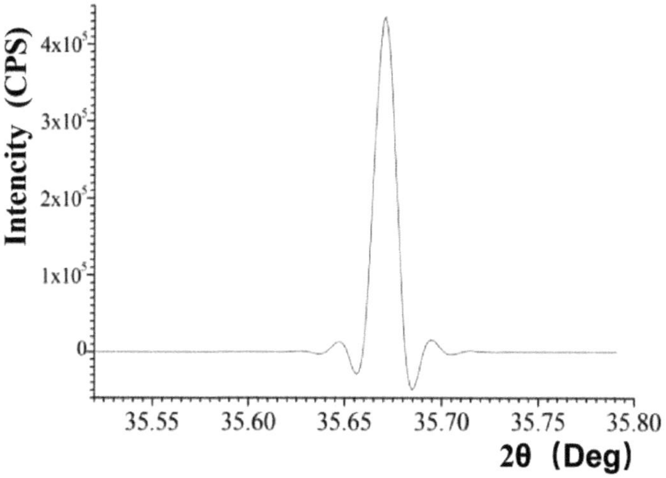


**Fig. 4.22** *ω-2θ* continuous scan curve [29]

4H–SiC samples at various growth pressures, as observed using high resolution X-ray diffraction (HRXRD). The epitaxial sample (0004) exhibits asymmetric peak shapes in its rocking curves under various pressures. According to the literature, this phenomenon is attributed to the dispersion effect, which causes the separation of diffraction peaks when both $K\alpha_1$ and $K\beta_1$ are involved in diffraction simultaneously. The HRXRD curves were subjected to Gaussian fitting to determine the full width at half maximum (FWHM) of the diffraction peaks. The FWHM values for the samples grown at pressures of 40, 60, and 100 mbar were around 23–25 arcsec, while the FWHM for the samples grown at 80 mbar was approximately 50 arcsec. The HRXRD diffraction peak width at half maximum is primarily influenced by the density of structural flaws in the crystal and the level of stress. Therefore, the test results obtained under various CVD growth circumstances serve as an indicator of the quality of epitaxial material crystallization.

3. Raman scattering (Raman scattering)

Light scattering that occurs when it interacts with an object can be categorized into two types: inelastic scattering and elastic scattering. Inelastic scattering encompasses Raman scattering and Brillouin scattering, as depicted in Fig. 4.24. Elastic scattering is characterized by a consistent photon energy and frequency, while inelastic scattering involves a modification in photon energy and a potential alteration in frequency, referred to as a frequency shift. Raman scattering has a significant frequency shift that differs among various films. This characteristic enables us to utilize Raman scattering as a means of distinguishing between different film materials. The Raman shift $\upsilon$ represents the disparity in frequency between the incoming light and the dispersed light. It only depends on the arrangement of the scattering molecule and is determined by the alteration in the molecule's vibrational energy level. Raman scattering

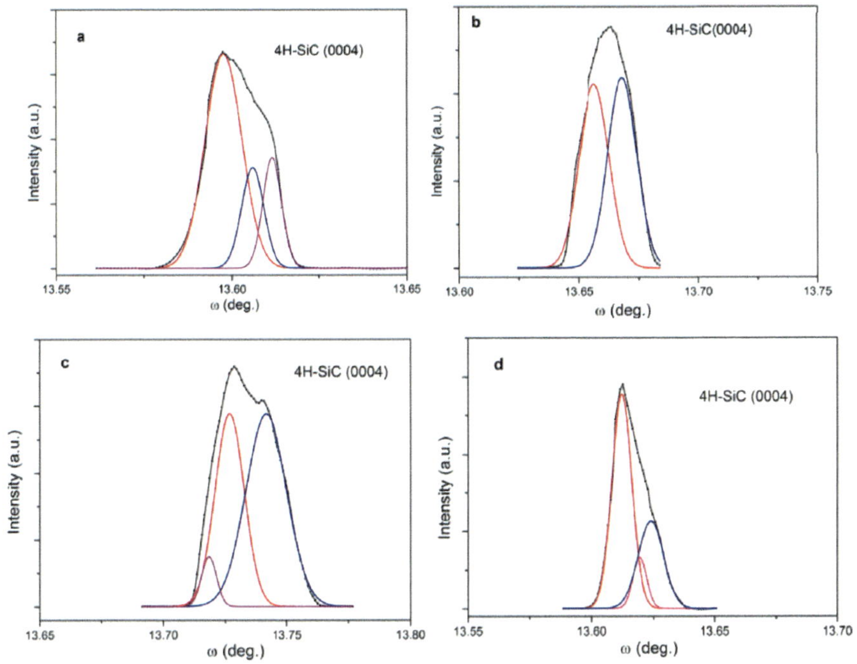


**Fig. 4.23** XRD swing curves of epitaxial samples prepared at different pressures. **a** 100 mbar; **b** 80 mbar; **c** 60 mbar; **d** 40 mbar [15]

tests rely on the distinctive nature of Raman shifts associated with various chemical bonds or groups. Raman spectroscopy enables researchers to ascertain the structure and symmetry of semiconductor materials, electron concentration, and other pertinent information by analyzing factors such as scattering frequency, line width, and intensity. Raman spectroscopy is being extensively employed for characterizing SiC crystals in terms of their crystalline form [30, 31], stacking fault [32], stress [33], and other material properties. Various SiC crystal types exhibit distinct frequency shift features in their Raman spectra, as seen by Table 4.2, which presents the typical Raman spectral data for frequently employed SiC crystal types.

Raman scattering is a non-invasive testing method that can definitively distinguish between different SiC polytypes. The Raman test results of four randomly selected spots on the 4H–SiC epitaxial substrate are displayed in Fig. 4.25. By comparing the Raman spectra with the standard 4H–SiC, it is evident that the material composition of all four sites exclusively consists of 4H–SiC. The Raman scattering spectra exhibit high sensitivity to the arrangement of stacking faults and deformation. The presence of stacking faults results in the emergence of stacked peaks and distortion of Raman peaks. Additionally, localized stresses cause the displacement of Raman peaks from their initial positions. Consequently, these phenomena enable the accurate measurement of stresses. The displacement of Raman peaks from their original positions due to localized stresses enables accurate quantitative assessment of the stresses.

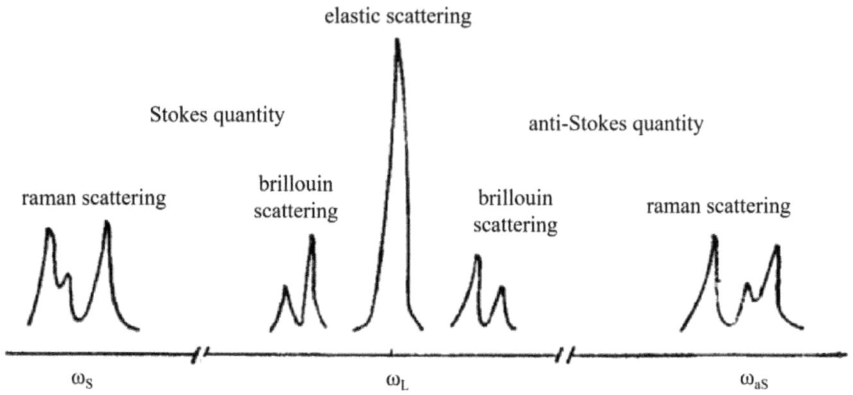


**Fig. 4.24** Scattered light spectrogram distribution [34]

**Table 4.2** Standard Raman spectral data of commonly used SiC crystal types

| Crystalline | Crystalline system | Raman spectrum peak $(cm)^{-1}$ |
|---|---|---|
| 3C–SiC | Cube | 796, 972 |
| 4H–SiC | Hexagon | 196, 204, 266, 610, 776, 796, 964 |
| 6H–SiC | Hexagon | 145, 150, 236, 241, 266, 504, 514, 767, 789, 797, 889, 965 |

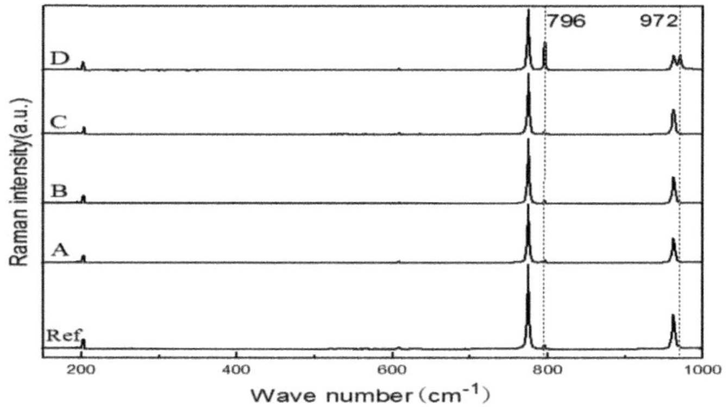


**Fig. 4.25** Raman pattern of 4H–SiC epitaxial layer [34]

4. Epitaxial layer thickness and doping concentration detection

(a) Infrared Fourier Transform Spectroscopy (FTIR)

Infrared Fourier Transform refers to the technique of using Fourier Transform to analyze infrared radiation. A spectrometer operates by exploiting the phenomenon of infrared light interference. The light emitted by the light source passes through the interferometer and forms interference light. This interference light then passes

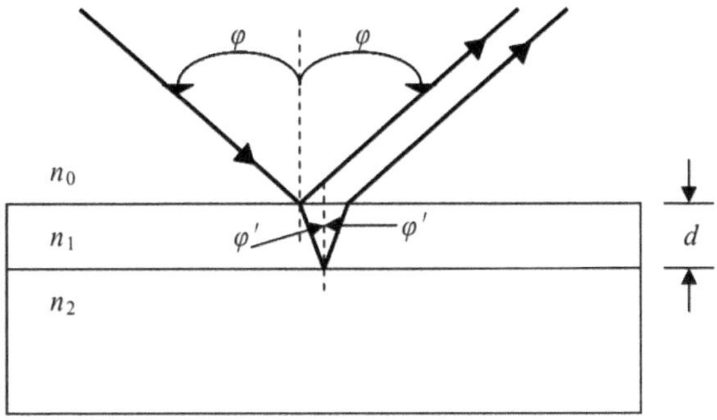


**Fig. 4.26** FTIR thickness test schematic

through the sample, and detectors are used to measure the resulting interference spectrum. This spectrum contains information about the sample. By applying a Fourier transform to the detected spectra, we can obtain the absorption rate or transmittance as a function of the infrared absorption wavelength. The spectrogram allows for the calculation of the thickness of the epitaxial film, as demonstrated in Fig. 4.26. The refractive indices of air, epitaxial layer, and substrate are denoted as $n_0$, $n_1$, and $n_2$, respectively. The thickness of the epitaxial layer is represented by $d$, while $\varphi$ represents the irradiation deflection angle. The refractive angle $\varphi'$ is given by:

$$\varphi' = \sin^{-1}\left(\frac{n_0 \sin \varphi}{n_1}\right) \tag{4.12}$$

The thickness of the epitaxial layer is calculated as:

$$d = \frac{i\lambda_0\lambda_i}{2n_1(\lambda_i - \lambda_0)\cos \varphi'} = \frac{i}{22n_1(1/\lambda_0 - 1/\lambda_i)\cos \varphi'} \tag{4.13}$$

Let $i$ represent the number of complete cycles between $\lambda_0$ and $\lambda_i$. $\lambda_0$ and $\lambda_i$ refer to the wavelength peaks or troughs that comprise $i$ cycles. The measure $2n_1\cos\varphi'$ is a constant that is determined by the experimental arrangement and the refractive coefficient of the film. It applies to two neighboring peaks when $i = 1$, one peak and one adjoining valley when $i = 1/2$, and two surrounding valleys when $i = 1$.

(b) Mercury probe C-V test

The schematic diagram of the mercury probe C-V test is depicted in Fig. 4.27. The utilization of the mercury probe C-V approach enables precise measurement of carrier concentration in the epitaxial material, providing valuable insights into the doping of the material under various process circumstances. When mercury (Hg) comes into contact with the 4H–SiC epitaxial layer to create a Schottky junction, the Schottky barrier is formed on the side of the epitaxial layer. The longitudinal distribution of the doping concentration in the SiC epitaxial layer can be determined by analyzing the variation of the barrier capacitance with the reverse bias voltage. The capacitance C of the Schottky diode can be mathematically represented as:

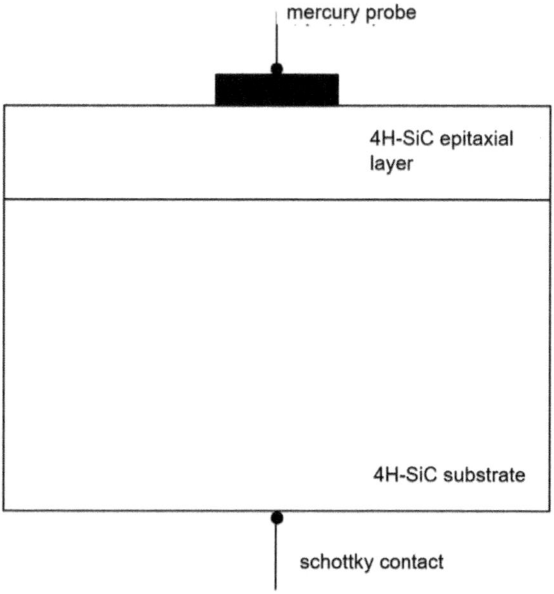


**Fig. 4.27** C-V test schematic

$$C = A\left[\frac{\varepsilon_s\varepsilon_0 N(X)}{2(V_{bi} + V)}\right]^{1/2} \tag{4.14}$$

In the context of the C-V tester, $A$ refers to the junction area, which corresponds to the area of the mercury probe. $\varepsilon_s$ represents the relative permittivity of the semiconductor, while $\varepsilon_0$ represents the permittivity of vacuum. $N(X)$ denotes the doping concentration at a specific depth $X$. $V_{bi}$ represents the built-in potential difference, and $V$ represents the reverse bias voltage. From this, it can be inferred:

$$\frac{1}{C^2} = \frac{2V_{bi}}{A^2 q\varepsilon_s\varepsilon_0 N(X)} + \frac{2V}{A^2 q\varepsilon_s\varepsilon_0 N(X)} \tag{4.15}$$

Thus the doping concentration of the epitaxial layer as a function of depth is:

$$N(X) = \frac{2}{A^2 q\varepsilon_s\varepsilon_0} \times \left(\frac{\mathrm{d}\frac{1}{C^2}}{\mathrm{d}V}\right)^{-1} \tag{4.16}$$

5. Microscopic characterization methods [35]

The surface morphology and quality of the epitaxial layer are crucial factors that impact the 4H–SiC device. The state of the epitaxial surface and the roughness of the contact surface during device fabrication directly influence the device's performance and quality. Defects and excessive roughness on the surface after epitaxial growth can result in uneven or non-functional conditions during processes such as metallization, lithography, and oxidation contact. These issues not only degrade performance but also disrupt the normal operation of the device process [36–38]. Deterioration not only impairs the functioning of the gadget, but also disrupts its regular operation [36–38]. Consequently, the utilization of surface microanalysis techniques such as optical microscopy (OP), scanning electron microscopy (SEM), and atomic force microscopy (AFM) is necessary to direct the enhancement of the epitaxial process, thereby improving the surface shape and quality of the epitaxial layer. Among these instruments, the high-resolution optical microscope allows for direct observation of the quality and structure of the 4H–SiC epitaxial layer. The SEM, on the other hand, is a non-destructive testing method that can be used for elemental analysis through X-ray energy dispersive spectroscopy. Lastly, the AFM provides a high level of precision in calculating the roughness of the epitaxial layer.

(a) Optical microscope (OP)

The optical microscope utilizes optical principles to magnify and visualize minuscule things that are indistinguishable to the naked eye. By using an optical microscope, one can directly observe the morphology and flaws of the 4H–SiC epitaxial layer. Prior to conducting optical microscope analysis of the surface flaws on the 4H–SiC epitaxial layer, it is important to perform a chemical etching process on it.

Chemical etching methods are frequently employed to generate etch pits caused by dislocations. SiC is highly unreactive, nevertheless, it can be corroded by molten KOH, NaOH, or $Na_2O_2$ at temperatures ranging from 450 to 600 °C. The SiC surface undergoes oxidation when exposed to the melt. During these melts, the SiC surface undergoes oxidation, and the resulting oxide is subsequently eliminated by the melt. Etch pits are formed at the intersection of dislocations or stacked laminations with the surface, where the high strain causes varied etching speeds, typically quicker. The image in Fig. 4.28 displays the top layer of a 4H–SiC(0001) crystal that has undergone etching in molten KOH at a temperature of 500 °C for a duration of 10 min. A substantial hexagonal void is formed at the site of the microtubule abnormalities, however, the microtubule abnormalities can be entirely eradicated at the present stage of the procedure. Figure 4.28 displays three prevalent variations of etch pits: The site consists of three distinct features: a sizable hexagonal pit, a smaller hexagonal pit, and an oval or shell-shaped pit. It was discovered that these etch pits corresponded to a screw-shaped dislocation (TSD), a blade-shaped dislocation (TED), and a basal plane dislocation (BPD), respectively [39, 40].

(b) Scanning electron microscope (SEM)

A scanning electron microscope uses a highly focused, high-energy electron beam to bombard the surface of a sample. This causes the incident electrons to excite the region and interact with the sample, generating various characteristic signals. These signals are then observed and analyzed to study the sample's surface or fracture morphology. The scanning electron microscope offers several advantages. Firstly, it provides great magnification, with the ability to reach a maximum magnification of 200,000 times and the option to alter it continually. Additionally, it offers a broad

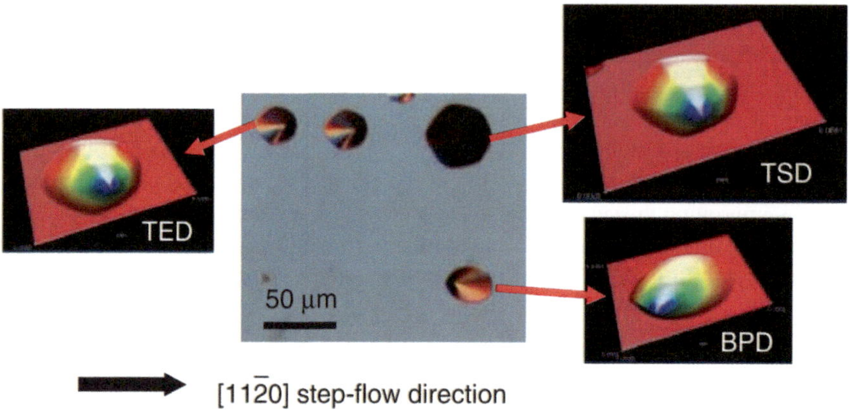


**Fig. 4.28** Off-axis 4H–SiC(0001) surface after etching in molten KOH at 500 °C for 10 min [21]

depth of field and field of view, resulting in three-dimensional imaging. Lastly, it allows for the observation of small structures on uneven surfaces. Simultaneously, sample preparation for SEM testing is uncomplicated. Currently, numerous scanning electron microscopy (SEM) devices can be integrated with other components like energy spectroscopy (EDS), enabling extensive use in the domains of materials, metallurgy, and biology. For instance, Fig. 4.29 displays an optical microscope image and a scanning electron microscopy (SEM) image of a triangular surface flaw.

(c) Atomic Force Microscopy (AFM)

Atomic Force Microscopy (AFM) is an analytical device utilized for studying the surface structure of solid materials. It is a potent technique frequently employed for surface research of semiconductor materials. AFM serves the purpose of examining

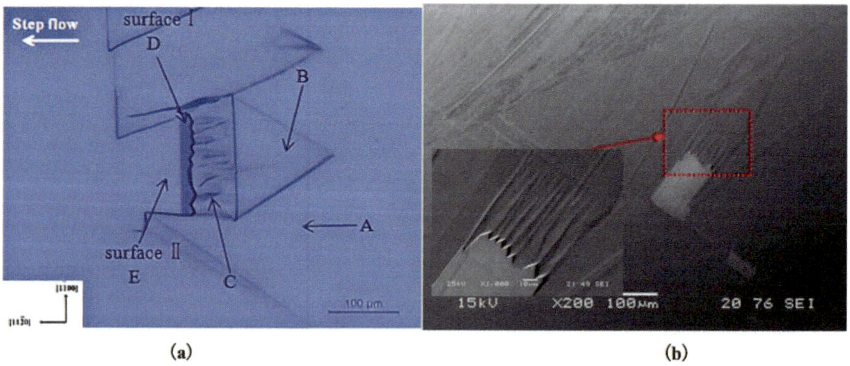


**Fig. 4.29** **a** Optical microscope image of the jagged triangular defect. **b** SEM image of the jagged triangular defect (inset is a magnified image in the dashed box area) [15]

the atomic-level structure of the sample surface, as well as evaluating the mechanical characteristics of the material surface for analysis. AFM acquires data on the morphology of the sample surface by quantifying the interaction force between the atoms located at the tip of the probe and the atoms present on the surface of the sample. The three often employed modes of AFM are the contact, non-contact, and tapping modes. Surface topography is determined in the contact mode by measuring the repulsive forces between atoms. Unlike the touch mode, the non-contact and tap modes utilize attractive forces to measure the surface topography. When comparing the two types, the contact type has the maximum resolution, whilst the non-contact type demonstrates a comparatively lower resolution. AFM testing offers advantages over scanning electron microscopy, as it does not necessitate the sample to possess electrical conductivity and provides a superior resolution of up to 0.15 nm in the transverse direction and 0.05 nm in the longitudinal direction. Figure 4.30 displays an AFM image of the epitaxial layer's surface, which is useful for assessing the surface roughness. The test region has a distinct step-like structure, with an average surface roughness (Ra) of approximately 0.378 nm and a root mean square roughness (RMS) of approximately 0.478 nm.

(d) Transmission electron microscopy (TEM)

A transmission electron microscope (TEM) is a device that uses an electron gun to emit a highly focused beam of electrons with a very short wavelength. The electrons are accelerated by a high voltage and then directed onto a thin sample after passing through an electromagnetic lens. When the electrons interact with the atoms in the sample, they can scatter, diffract, or pass through the sample directly (if the sample is less than 100 Å thick). Transmission electron microscopy employs three distinct imaging concepts based on the outcome of the collision between incident electrons and atoms. (1) Absorption picture—This refers to the formation of a phase in a sample when electrons collide with atoms, mostly due to the scattering effect caused by the high density and crystal quality of the sample. The magnitude of the scattering angle is correlated with the density, thickness, and internal composition of the sample, resulting in a distinct image that differs from the bright and dark image. (2) Diffraction image—The electron beam is diffracted by the sample, resulting in varying abilities of different crystal structures to diffract the electrons. This leads to distinct differences in the resulting imaging following diffraction. (3) Phase picture—Electrons are able to go through the sample when its thickness is reduced to 100 Å or below. In this scenario, the alteration in wave amplitude is insignificant, and the imaging is derived from the alteration in phase. Transmission electron microscopes possess exceptional resolution, often ranging from 0.1 to 0.2 nm, and offer substantial magnification, ranging from tens of thousands to millions of times. These instruments are primarily employed for the examination of extremely minute objects.

6. Hall test

Hall effect measurements provide information about the free carrier density (either electron density $n$ or hole density $p$) and mobility of the material. On the other hand,

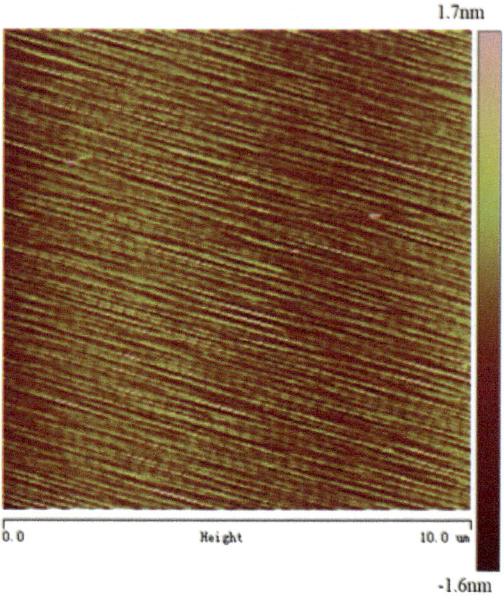


(a) Plan view

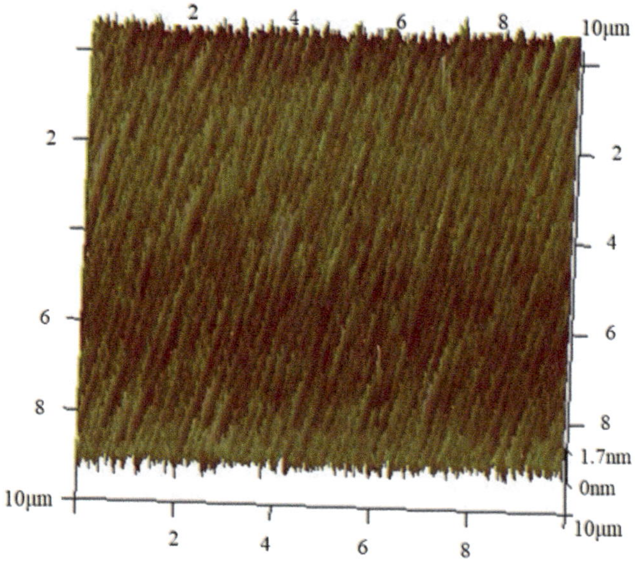


(b) 3D view

**Fig. 4.30** Sample 5 × 5 μm area AFM photo slice

the mercury probe C-V test mentioned earlier determines the net doping density, which is represented by $N_D$-$N_A$ for N-type materials and $N_A$-$N_D$ for P-type materials. Here, $N_A$ and $N_D$ refer to the donor and acceptor densities, respectively. As a result of the comparatively high ionization energy of acceptor ions in SiC, the density values obtained from P-type SiC Hall effect and C-V measurements at room temperature ($p < N_A$-$N_D$) exhibit notable discrepancies. Modulating the measurement temperature leads to a sole alteration in the hole density, while the overall doping density remains nearly constant. Figure 4.31 displays a schematic of the sample structure employed for Hall effect measurements. The SiC epitaxial layer under examination must be electrically insulated from the substrate. An approach involves utilizing a semi-insulating substrate to provide galvanic separation, as depicted in Fig. 4.31a. An alternative approach involves utilizing a pn junction for the purpose of isolation, as depicted in Fig. 4.31b. In the first scenario, it is crucial to compare the resistance of the measured epitaxial layer with the resistance of the substrate at the measured temperature to verify that the leakage current through the substrate is insignificant. In the second scenario, interlayers of varying conductivity kinds must be highly doped and have a significant thickness to prevent their infiltration. Additionally, it is imperative to ensure that the leakage current at the connection remains sufficiently low throughout a substantial surface area. The electrode structure necessary for precise readings is depicted in Fig. 4.31c. When conducting low-temperature measurements, it is necessary to create a highly doped area beneath the ohmic contacts to prevent the formation of non-ohmic connections.

7. Carrier life test

Carrier lifetime is a crucial physical characteristic that governs the efficiency of bipolar devices. The primary methods used to measure carrier lifetime are time-resolved photoluminescence (TRPL), photoconductive decay (PCD), and reverse recovery (RR). The decay curve of the measured signal *I(t)* can be mathematically represented as:

$$I(t) = I_0 \exp\left(-\frac{t}{\tau_{\text{decay}}}\right) \tag{4.17}$$

The initial signal strength is represented by $I_{0,}$ while the decay time is denoted as $\tau_{\text{decay}}$. It is important to note that $\tau_{\text{decay}}$ does not always match the carrier lifetime (τ). The relationship between decay time and carrier lifetime for each measurement method is provided in Table 4.3, where the measured $\tau_{\text{decay}}$ values correspond to the entries in the table. Additionally, it is worth mentioning that the decay curves frequently deviate from a straightforward exponential pattern, indicating that $\tau_{\text{decay}}$ is not constant.

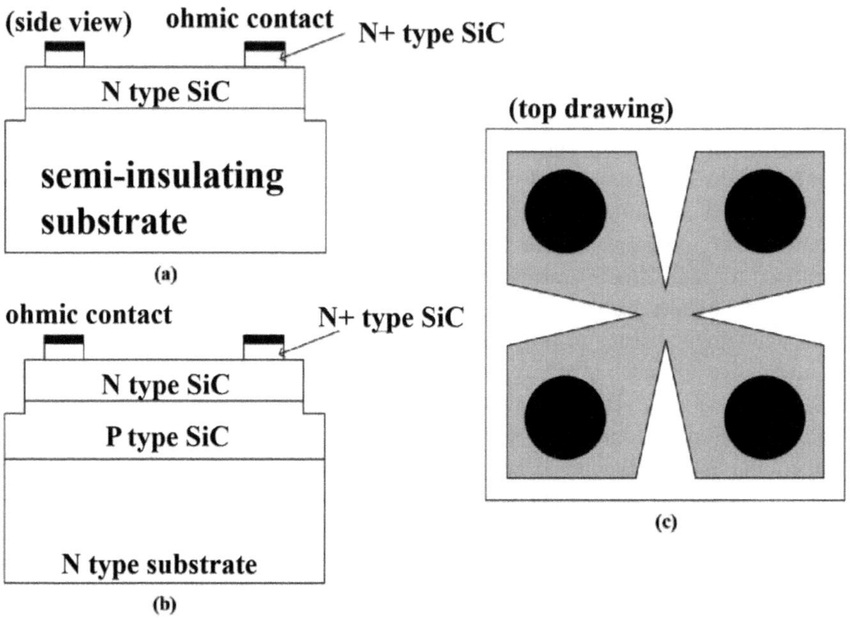


**Fig. 4.31** Schematic of the sample structure suitable for Hall effect measurements. Electrical isolation is achieved by using either **a** a semi-insulating substrate or **b** a pn junction [21]

**Table 4.3** Carrier lifetime ($\tau$) and decay time ($\tau_{\text{decay}}$) are related to each other [21]

| Injection method | TRPL | PCD | RR |
|---|---|---|---|
| High-level injection | $\tau/2$ | $\tau$ | $\tau$ (limited by surface compound) |
| Low-level injection | $\tau$ | $\frac{\mu_n d(\Delta n)}{dt} + \frac{\mu_p d(\Delta p)}{dt} \approx \tau$ (when $\Delta p = \Delta n$ time) | $\tau$ (limited by surface compound) |

## 4.3.2 Defects in SiC Epitaxial Layers

Defects in semiconductors can be categorized into: point defects and extended defects [41]:

(1) Point defects are localized at specific positions within the lattice structure, affecting only a small number of neighboring atoms and lacking any spatial extension. Point defects are universally present in semiconductor materials, albeit in small quantities, and are primarily created as a result of vacancies, gaps, and replacements. Point defects are defined as aggregations of a small number of defects that cause disturbances at or near lattice points. Examples of such defects include double vacancies and vacancy-donor complexes. Point defects inside a

semiconductor introduce additional electronic energy levels within the bandgap. These defects can function as “traps”, “recombination centers”, or “generation centers”, and have the potential to greatly modify the properties of the semiconductor and affect the performance of devices utilizing it. Point faults can selectively benefit specific devices and be deliberately introduced. Point flaws in switching devices can serve as composite centers, introducing energy levels that effectively eliminate minority carriers and enhance the device’s switching speed, leading to improved efficiency [42, 43]. Nevertheless, point faults typically have a negative impact on the performance of devices. Point defects can generate energy states that function as intricate centers for electron–hole pairs, leading to a decline in the efficiency of radiation detectors and photovoltaic solar cells.

(2) Extended defects, such as grain boundaries, dislocations, and stacked layer errors, will expand in all directions, originating from both the substrate and the process of epitaxial growth. The electrically active areas of most SiC devices are exclusively found within the epitaxial layer that is grown on top of the crystalline substrate. The electrical properties of these devices are significantly influenced by the caliber and surface texture of the semiconductor. Hence, the presence of flaws inside the epitaxial layer holds significant importance for any semiconductor radiation detection system. The primary flaws in the SiC epitaxial layer that impact the functionality of electronic devices are screw dislocations (TSD), edge dislocations (TED), basal plane dislocations (BPD), microscopic growth pits, as well as triangles, “carrots,” and “comets.” Defects in comets. A substantial reduction in defects is observed in the epitaxial layer compared to the body crystal, as most of the flaws formed in the substrate body crystals do not propagate into the epitaxial layer.

1. Surface morphological defects

The differential interference contrast optics (NDIC) and atomic force microscopy (AFM) were utilized to monitor the epitaxial surface morphology, as depicted in Fig. 4.32. The presence of step-bunching aggregation clusters on the epitaxial surface is evident. Veneroni et al. demonstrated that during the epitaxial growth process with a low C/Si ratio (e.g., C/Si = 1), the decomposition of $SiH_4$ generates a significant quantity of gaseous Si. This Si then clusters together to form nuclei, which subsequently deposit on the surface of the epitaxial layer. Over time, these nuclei arrange themselves into step-grooves that align with a specific crystallographic direction, resulting in a phenomenon known as step-bunching. This structure is a component of the SiC crystals found on the surface of the non-stoichiometric portion of the Si monomer composition, as well as the unstable $Si_2C$ and $SiC_2$ compounds. Presently, the production of large epitaxial layers necessary for SiC devices is often carried out using the traditional Si–C–H-based step-flow growth method on substrate materials with a 4° off-axis. Step-bunching frequently occurs during this growth process, where the epitaxial surface develops small clusters of steps. This phenomenon is highly responsive to minor changes in epitaxial growth conditions, such as the C/

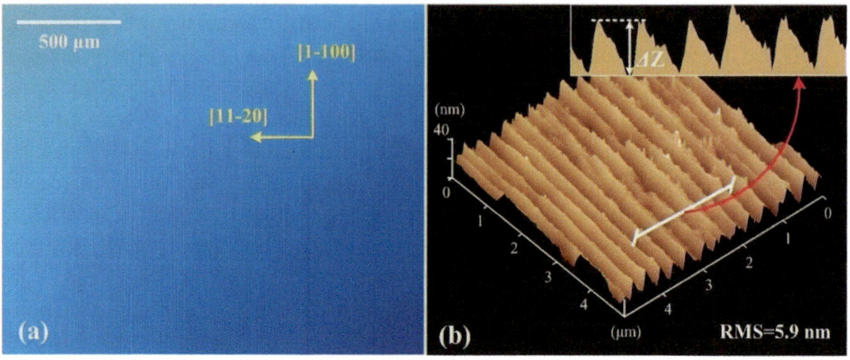


**Fig. 4.32** Original epitaxial surface morphology. **a** NDIC image; **b** AFM scan $(5 \times 5\ \mu m)^2$ [44]

Si ratio and growth temperature, as well as the preparation of the substrate material before the epitaxial growth. Their uniform distribution and alignment along the step-flow direction significantly impacts the surface flatness.

SiC epitaxial layers produced on off-axis {0001} substrates commonly display various surface defects, in addition to step-bunching. Figure 4.33 displays common surface imperfections seen in 4H- and 6H–SiC {0001} uniform epitaxial layers, including (a) "carrot" defects and shallow pits, (b) triangle defects, and (c) particle settling. While the precise mechanism behind the production of these flaws remains incompletely understood, they typically arise from technical issues such as inadequate elimination of polishing damage or suboptimal growth procedures. Down-Fall, a surface defect with lethal consequences, is believed to be formed when droplets fall on the surface either before or during epitaxial growth. The decrease in 4H–SiC breakdown voltage caused by Down-Fall is significantly greater (up to 90%) compared to the decrease caused by "carrot" defects and triangular defects [45]. The average density of these visible flaws in SiC uniform epitaxial layers is around 0.02–2 $cm^{-2}$. To prevent the creation of shallow pits, it is necessary to minimize the development of silicon droplets while etching in real-time. SiC devices have a considerable increase in leakage current and a notable fall in breakdown voltage (about 20–40% [46]) when they contain carrot (or comet) flaws or triangular defects. On the other hand, the impact of shallow pits on the devices is minimal [36]. The formation of leakage channels is a consequence of the stacking layer faults associated with the carrot, comet, and triangle defects.

2. Spiral dislocations (TSD) and microtubules (Micropipes)

Screw dislocations (TSD) have the ability to traverse the entire crystal along the crystallographic c-axis. Screw-type dislocations exhibit a tendency to halt exclusively at the surface of the crystal. These dislocations are found in all wafers obtained from the crystal growth process. They can also spread across epitaxial layers formed using CVD techniques. Furthermore, it is possible for new screw-type dislocations to arise during the process of epitaxial development. The measurement of helical dislocations

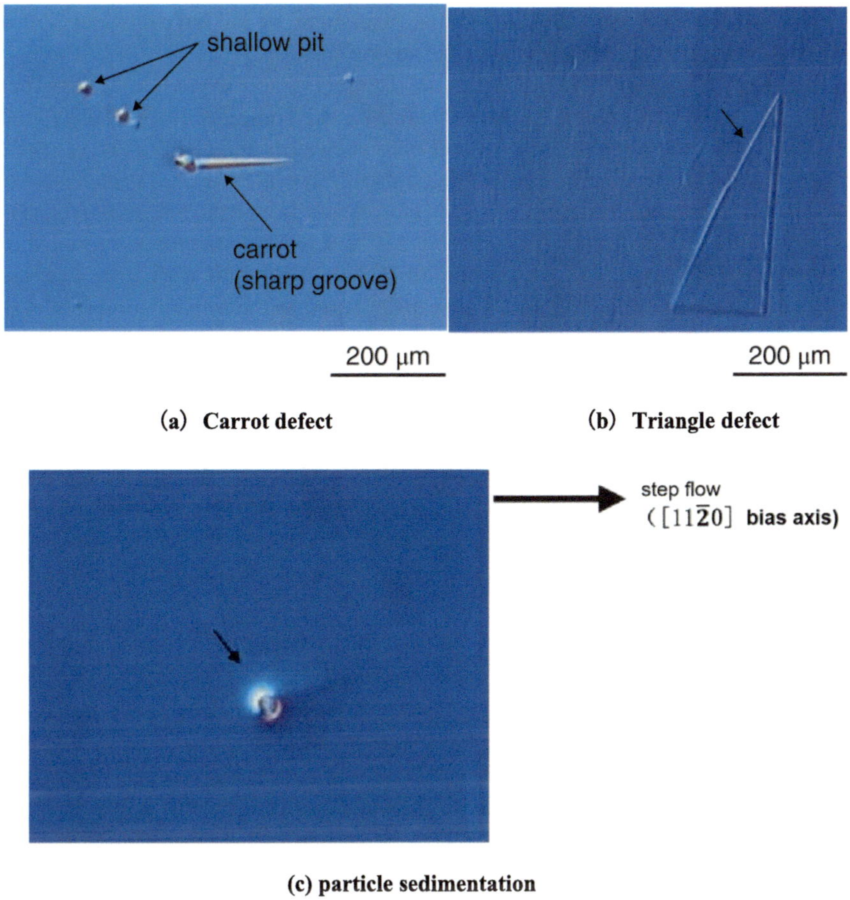


**Fig. 4.33** Typical surface defects observed in 4H- and 6 h sic{0001} coepitaxial layers. **a** "Carrot" defects and shallow pits; **b** triangular defects; **c** sedimentation [21]

is often determined by the magnitude of the Burgers vector (b). In the case of pure screw dislocations, the Burgers vector is aligned with the crystallographic c-axis, and the length of the Burgers vector is connected to the height of the step formed by the screw dislocation.

Microtubules refer to hollow nuclei that are created by screw-type dislocations possessing significant Burgers vectors. Microtubules are cylindrical structures that have a hollow interior and can penetrate SiC single crystals. Their sizes can vary from tens of nanometers to tens of micrometers in radius. The presence of these microtubules significantly impairs the performance of SiC-based power devices and radiation detectors. Microtubule flaws greater than 1 $mm^2$ on substrates in epitaxially grown p–n junction devices can cause the reverse bias point to fail prematurely, prior to avalanche. Suppliers have successfully accomplished a significant decrease in microtubule density (from $10^4$ $cm^{-2}$ to less than 1 $cm^{-2}$) and have now developed epitaxial layers that are completely free of microtubules [21].

Screw-type dislocations in SiC with small Burgers vectors have closed-core structures, commonly known as basic screw dislocations. These dislocations are found in densities of thousands/$cm^2$ in 4H- and 6H–SiC wafers and epitaxial layers [47]. While closed-core dislocations are less damaging to devices compared to microtubes, they nonetheless adversely affect device performance. The gentle breakdown of 4H–SiC p–n junction diodes at voltages below 250 V was attributed to the presence of closed-core dislocations [48]. According to Wahab et al., their work demonstrated that increasing the number of closed-core dislocations in the active region results in a reduction in breakdown voltage [49].

3. Basal Dislocation (BPD)

Basal dislocations (BPDs) are a specific form of dislocations that exhibit a high density within the epitaxial layer. In the context of SiC bipolar devices, the presence of BPDs (Boron-Precipitate Defects) has a detrimental effect. These defects cause a localized reduction in the lifespan of charge carriers, resulting in an increase in on-resistance. Additionally, BPDs can cause an increase in the amount of leakage current.

4. Edge Dislocation (TED)

The Burgers vector of the edge dislocations (TEDs) is orthogonal to the crystal c-axis. The TEDs primarily propagate from the substrate. BPDs migrate from the off-axis 4H–SiC substrate to the uniform epitaxial layer where they undergo conversion into TEDs. The transformation from BPDs to TEDs takes place as a result of mirror forces acting within the epitaxial layer.

5. Stacking Faults (SF)

Stacking Faults (SF) are planar flaws mostly located within the primary slip plane of SiC {0001}.Stacking faults (SFs) arise from the misalignment of the Si–C diatomic layer with the ideal stacking sequence along the crystallographic c-axis. Schottky diodes experience a decrease in barrier height and breakdown voltage when subjected to the influence of SFs. Charge accumulation in the stacking layer defects of SiC PiN diodes can generate electrostatic potentials, resulting in an increase in the forward voltage drop of the diode [50].

The SEM image in Fig. 4.34 depicts the Si side of the etched SiC substrate. The SEM image displays hexagonal holes that are correlated with three distinct forms of dislocations. The closed-core screw-type dislocations are characterized by hexagonal pits with little black spots at their centers. Hexagonal pits with spacious central cavities are classified as open-core screw-type dislocations (microtubules). Edge-type dislocations are characterized by hexagonal holes that lack a central point. The brightness of the images of hexagonal pits associated with edge-type dislocations is lower compared to the brightness of the images of hexagonal pits associated with open-core or closed-core dislocations. The SEM picture in Fig. 4.34b displays basal plane dislocations as distinct elongated lines, characterized by a sequence of asymmetric pits that are interconnected. Figure 4.35 displays an enlarged view of the dislocations that were observed on the surface of the substrate. The SEM picture of the etched SiC epitaxial layer is depicted in Fig. 4.35a. Figure 4.36b displays an enlarged SEM image of the epitaxial layer, revealing the presence of closed-core dislocations and edge-type dislocations. The density of dislocations in the epitaxial layer is significantly lower than that of the substrate. Additionally, the SEM picture of the etched epitaxial layer does not show any microtubules or BPDs.

This section provides a concise overview of the primary methodologies employed for characterizing the physical attributes and identifying defects in SiC. Additionally, it includes illustrative data samples to enhance the reader's comprehension. Aside from the aforementioned commonly employed characterization techniques in SiC materials, there exist several other techniques, including Secondary Ion Mass Spectrometry (SIMS), X-ray Photoelectron Spectroscopy (XPS), Acoustic Electron Spectroscopy (AES), and others. These techniques offer diverse perspectives for characterizing SiC materials and can be chosen based on specific requirements to achieve comprehensive material characterization.

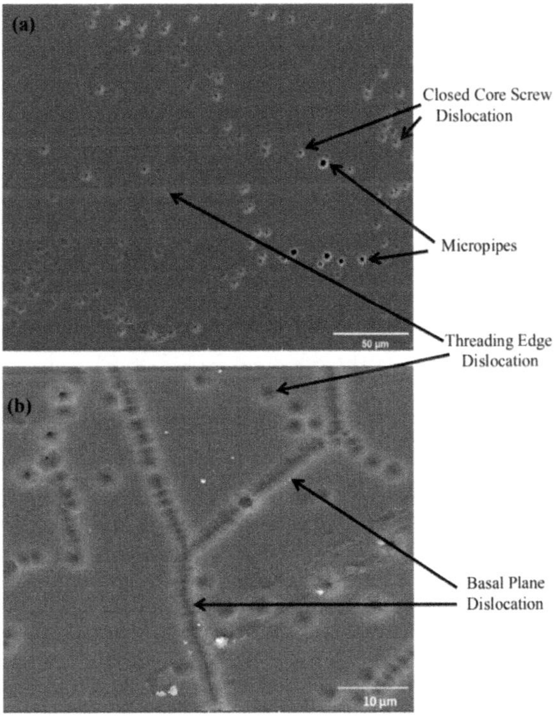


**Fig. 4.34** SEM images of molten KOH etched body SiC. **a** region with TSDs and TEDs; **b** region with BPDs and TEDs [41]

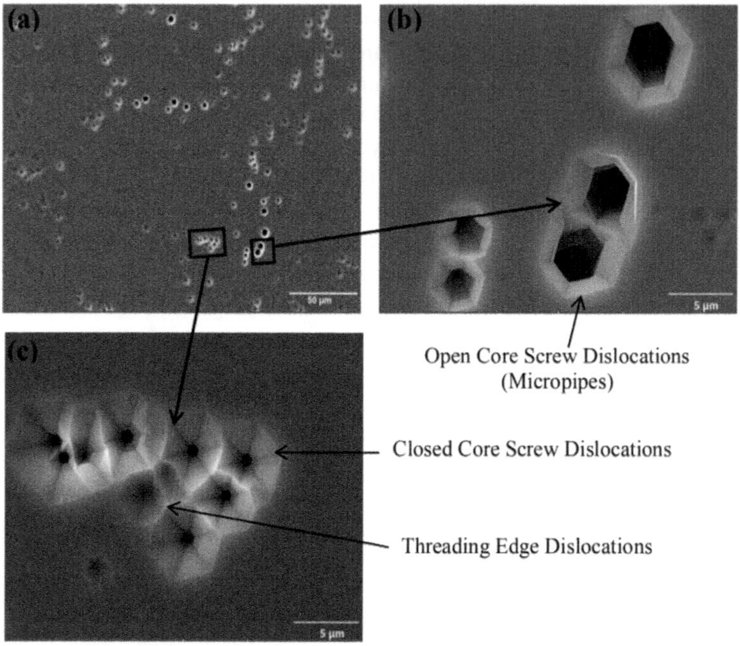


**Fig. 4.35** SEM images of fused KOH etched body SiC. **a** region with TSDs and TEDs; **b** magnified view of microtubules; **c** magnified view of closed core with threaded edge dislocations [41]

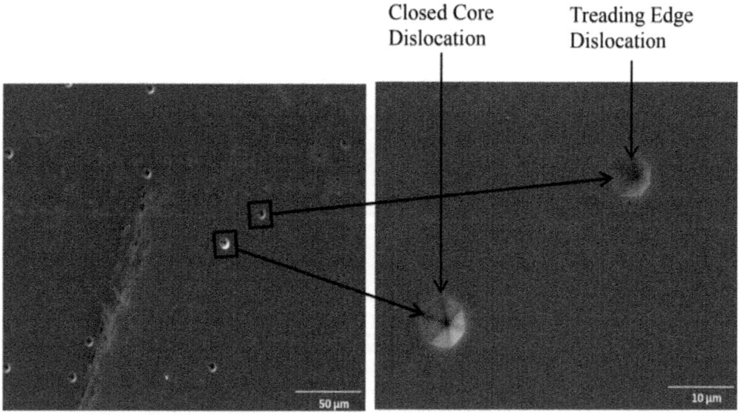


**Fig. 4.36** **a** SEM image of molten KOH etched SiC epitaxial layer; **b** enlarged view of closed core with threaded edge dislocation [41]

# 4.4 Preparation Process Method and Electrical Characteristics of SiC Nuclear Radiation Detectors

## *4.4.1 Detector Key Preparation Process and Basic Flow*

Following SiC epitaxial growth, a series of essential procedures must be undertaken to finalize the fabrication of SiC nuclear radiation detector devices. These procedures encompass metallization, ion injection doping, etching, oxidation, and passivation. Figure 4.37 depicts the fundamental configuration of a high-voltage PiN diode suitable for nuclear radiation detection. The terminal region is formed through an ion injection doping procedure, the passivation layer is formed through an oxidation and passivation procedure, the isolation table is formed through an etching procedure, and the ohmic contacts are formed through a metallization procedure. The aforementioned stages are fundamental and indispensable in the preparation of a nuclear radiation detector.

1. Metal contact [52, 53]

Metal–semiconductor (MS) contacts can be classified into two categories: rectified, also known as Schottky contacts, and non-rectified, also known as ohmic contacts. An ohmic contact is a type of contact that allows current to flow in both directions. The current–voltage relationship is linear, meaning that the current increases or decreases proportionally with the applied voltage, regardless of its polarity. Additionally, the resistance at the contact is very small relative to the resistance of the device it is connected to. The Schottky contact is characterized by its unidirectional conductivity, meaning it allows the flow of electric current only when a voltage is applied in a specific direction. In a typical 4H–SiC nuclear radiation detector with two structures, the PiN-structured detector utilizes an ohmic contact at both the cathode and anode to connect the device with the external environment. This is done to separate and collect electron–hole pairs that are generated by the pn junction and are affected by the

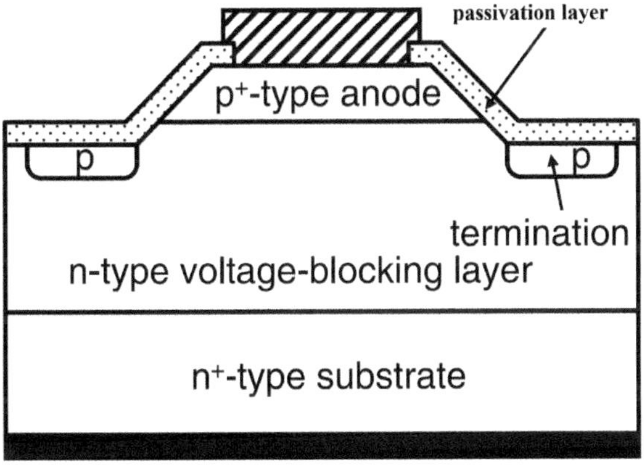


**Fig. 4.37** Basic structure of a high-voltage PiN diode for nuclear radiation detection [51]

space charge region. On the other hand, the Schottky-structured detector employs a Schottky contact at the anode and allows the passage of irradiated electron–hole pairs through the space charge region created by the Schottky junction. Schottky-structured detectors utilize Schottky contacts at the anode to isolate and gather the spatial charge regions produced by the Schottky junction. This allows for the separation and collection of irradiated electron–hole pairs. The cathode, on the other hand, employs an ohmic contact to establish a connection between the device and the external environment.

Due to the presence of the Schottky barrier at the interface between metal and semiconductor, it is typical for metal–semiconductor connections to exhibit rectifying behavior rather than non-rectifying behavior. In nuclear radiation detector applications, it is important to have a Schottky contact with a high Schottky barrier height (SBH) that is uniform and defect-free at the SiC/metal interface. This is because any variation or defects in the Schottky contact can decrease the SBH in specific areas of the gold-semiconductor contact, leading to an increase in the leakage current at high reverse bias. An ohmic contact can be viewed as the extreme scenario of a Schottky contact, featuring an enhanced Schottky barrier. If the size of the ohmic contact is significantly smaller than the contact resistance, it is possible to enable the 4H–SiC device to transmit through a high-current circuit, resulting in reduced power loss during device operation. Consequently, the prepared 4H–SiC nuclear radiation detectors will be better suited for extreme environments characterized by high-frequency, high-voltage, high-temperature, and high-power conditions. In summary, the metal contact of the 4H–SiC nuclear radiation detector must possess low specific contact resistance, high SBH (Schottky barrier height), and high stability in order to guarantee the device's performance and reliability. Therefore, the fabrication process of the metal contact holds a crucial role in the overall device manufacturing process.

(a) Schottky contact

Figure 4.38 illustrates the N-type Schottky barrier heights of four frequently employed metals utilized in the fabrication of 4H–SiC Schottky connections. It is evident that, for a specific metal, the barrier heights exhibit a minor increase at the $(000\bar{1})$C face, a slight decrease at the (0001) Si face, and an intermediate value at the $(11\bar{2}0)$ face. The variation could be attributed to the existence of polarity-dependent dipoles at the contact or surface states with distinct distributions. The slopes depicted in Fig. 4.38 range from 0.8 to 0.9, suggesting the absence of "pinning" of Fermi energy levels at the metal/SiC interface and indicating proximity to the Schottky–Mott limit. Figure 4.39 displays the magnitudes of the energy barriers for P-type Schottky connections formed with various metals. The plot exhibits a slope of around—0.8, whereas the combined Schottky contact barrier heights for the metal on both N-type and P-type SiC are in close proximity to the SiC bandwidth, denoted as $E_g$. It is crucial to achieve barrier heights that closely resemble the ideal Schottky metal contacts described in the literature. This requires using high-quality SiC epitaxial materials and optimized processes. The properties of the Schottky contacts are influenced by the surface conditions during the preparation process and

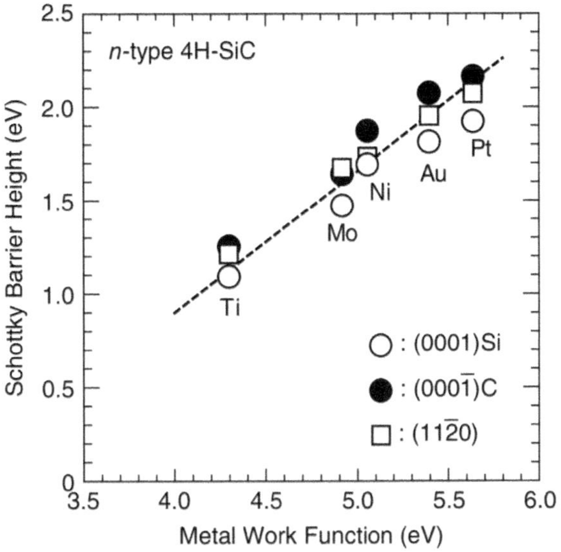


**Fig. 4.38** N-type 4H–SiC Schottky contact barrier height versus metal figure of merit for different metals [21]

the annealing conditions after the metal deposition. Preparation and annealing conditions following metal deposition. The process of creating Schottky connections on semiconductors involves three primary stages: surface cleansing, metal deposition, and annealing. The objective of surface cleaning is to facilitate a strong connection between the deposited metal and SiC. Previous research [54] has demonstrated that a treatment procedure involving sacrificial oxidation, 5% HF corrosion, and immersion in boiling water for 10 min can effectively decrease the density of surface states, resulting in a more desirable Schottky contact. Additionally, subjecting the metal/SiC Schottky interface to high-temperature annealing triggers a chemical reaction that modifies the energy-band structure and overall barrier height of the contact.

(b) Ohmic contact

Fabricating SiC devices with low ohmic contact specific contact resistance is more challenging compared to other semiconductor devices. The ohmic contacts of N-type and P-type SiC typically exhibit specific contact resistance values ranging from $10^{-5}$ to $10^{-6}$ $\Omega cm^2$ and $10^{-4}$ to $10^{-5}$ $\Omega cm^2$, respectively. The specific contact resistances are significantly influenced by factors such as the carrier concentration on the wafer surface, the choice of metal, the pre-treatment of the wafer surface, and the conditions of metallization heat annealing. To provide high-quality ohmic connections, commonly employed methods include ion implantation to enhance the carrier concentration on the surface of the wafer, performing high-temperature alloying annealing, and utilizing alloys or other compounds. The SiC material has distinct Si and C surfaces, and the disparity in their crystalline properties significantly impacts the electrical characteristics of the ohmic contact. Typically, the ohmic contact is predominantly constructed on the Si surfaces. The ohmic contact of N-type SiC is

**Fig. 4.39** P-type 4H–SiC Schottky contact barrier height versus metal figure of merit for different metals [21]

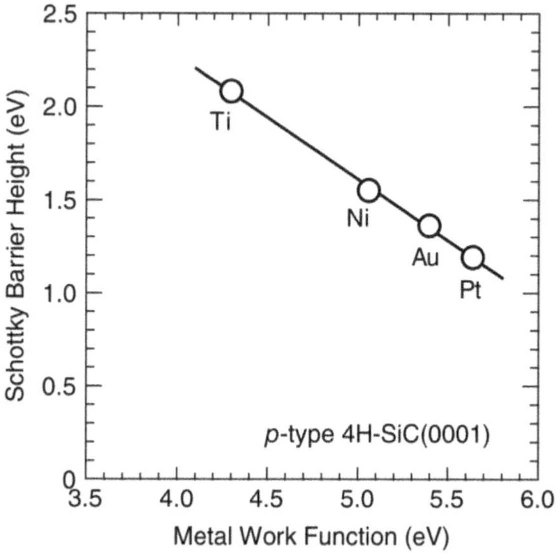


primarily achieved through the annealing of Ni metal at a resistance of 950 Ωcm$^2$. The primary method for creating ohmic connections on N-type SiC involves subjecting a nickel-based metal to metallization annealing at temperatures ranging from 950 to 1050 °C. The elevated barrier height of the contacts in P-type SiC materials poses challenges in achieving low specific contact resistance. This is often addressed by subjecting Al-Ti metals and Si compounds to metallization annealing at temperatures ranging from 900 to 1180 °C [53]. An ideal ohmic contact should possess the following primary attributes: The desired characteristics include: (i) the ability to be renewed with low specific contact resistance; (ii) a contact-semiconductor interface that is uniform and homogeneous, with planar homogeneity at the interfacial phase; (iii) a smooth surface morphology; (iv) thermal stability at high temperatures; (v) resistance to oxidation; (vi) good adhesion; (vii) ease of implementation of the process; and (viii) good mechanical properties.

The process of creating ohmic connections on semiconductors consists of three primary stages: surface cleaning, metal deposition, and annealing. Among these methods, the annealing treatment is commonly applied in the form of rapid thermal annealing (RTA). Due to its elevated temperature requirement, often ranging from 800 to 1200 °C. The heating and cooling pace of a typical annealing furnace is quite slow, resulting in a prolonged processing time. The RTA system has the ability to rapidly increase or decrease the wafer temperature. Typically, it takes less than 10 s for the wafer to reach the desired annealing temperature, which is approximately 1000 °C. The annealing process lasts around 10 s, after which the wafer undergoes rapid cooling by the cessation of heating lamps and the introduction of nitrogen cooling gas. In nuclear radiation detector applications, ensuring the thermal stability of the ohmic contacts is crucial for effective detection in challenging conditions.

Enhancing the thermal stability of the ohmic contacts can be achieved by employing high temperature thermal annealing at suitable temperatures and utilizing multilayer metals.

2. Ion injection

Selective doping in Si and most III-V semiconductors can be achieved through the use of diffusion and ion implantation techniques. Diffusion is a thermal process that enables impurity atoms to permeate the interior of a semiconductor. On the other hand, ion implantation involves ionizing and accelerating an impurity atom to a specific energy level, allowing it to pierce the semiconductor surface and enter the body. Typically, diffusion does less harm to the crystal structure than ion implantation. In ion implantation, the substrate material is bombarded with high-energy ions, resulting in the creation of various types of defects in the material. Greater diffusion constants for the dopant increase the likelihood of the formation of deep junctions during the diffusion process. Conversely, ion implantation has the ability to create doping profiles that are both more adaptable and precise.

Thermal diffusion methods for selective doping of materials have limited utility in SiC devices, unlike in the Si process. This is due to the strong chemical bonding in SiC, which causes the diffusion constants of dopant impurities to stay very low at temperatures above 1600 °C. The need for excessively high temperatures to achieve acceptable diffusion constants renders the diffusion processing of SiC unfeasible. This is due to the fact that such a high-temperature procedure results in a significant concentration of deep energy level defects and basal dislocation slippage. Furthermore, there is currently no mask material available that is adequate for the diffusion process at such elevated temperatures. Hence, the process of selectively introducing impurities into SiC is often accomplished using ion implantation. Currently, aluminum (Al) and boron (B) are primarily utilized for P-type doping, whereas nitrogen (N) and phosphorus (P) are predominantly employed for N-type doping. Due to its higher density, SiC allows for a smaller injection depth of ions compared to Si when subjected to the same energy injection. For instance, aluminum ions with an energy of 700 keV in 4H–SiC have a peak depth of approximately 0.75 μm. In some cases, injection energies may need to reach the MeV level in order to achieve a deeper injection region. Ion implantation is effective in creating a controllable doping region within a semiconductor material. However, only a minute portion of the injected doping ends up at the alternative lattice sites, and only this portion of the doping at the lattice sites becomes electrically activated. Furthermore, the injection of high-energy ions results in detrimental effects both on the surface and within the material in the region of injection. To address the issue of ion activation and reduce injection-related harm, it is necessary to inject SiC at a high temperature background ranging from 500 to 1000 °C. Additionally, the injected SiC must undergo annealing at high temperatures (> 1200 °C for N-type doping and > 1600 °C for P-type doping) in a high-purity inert gas environment, such as argon.

The distribution of ions injected into the 4H–SiC material follows a Gaussian distribution:

$$N(x) = \frac{Q}{\sqrt{2\pi}\sigma}\exp\left[-\left(\frac{x - R_p}{\sqrt{2}\sigma}\right)^2\right] \tag{4.18}$$

The variable Q represents the amount of injection dose, σ represents the magnitude of injection deviation, and Rp represents the extent of injection depth. For practical applications, it is typically necessary to have a doping distribution that is shaped like a box and achieved through numerous ion injections. To attain a doping distribution resembling a box shape, one can utilize Monte Carlo simulation software like SRIM to calculate crucial factors such the number of injections, energy, and dose. Figure 4.40a displays the simulated Al ionization resulting from the application of specific injection parameters on the 4H–SiC material. The ion distribution is confined within a box-shaped region with a depth of approximately 0.5 μm. Figure 4.40b presents a comparison between the results of the secondary ion mass spectrometry (SIMS) test and the SRIM simulation results following the ion injection experiments. The comparison reveals that the actual injection concentration is in substantial accordance with the simulation results. Additionally, a box-shaped region measuring 0.5 μm has been formed. The SEM profile in Fig. 4.41 displays the injected junction generated with the specified injection parameters. The overall injection depth, including the trailing of Al ions, measures around 0.9 μm, aligning with the results obtained from the SIMS test.

3. Etching

Etching is the technique of selectively removing material from a surface using chemical or physical processes. It can be categorized into wet etching and dry etching. Wet etching involves the utilization of chemical solutions and target materials to effectively remove material. This process has a relatively high corrosion rate, but lacks precise control over the depth of corrosion and is prone to significant lateral corrosion. Dry etching refers to the process of using plasma and the properties of the target material to achieve etching. Depending on the specific mechanism involved,

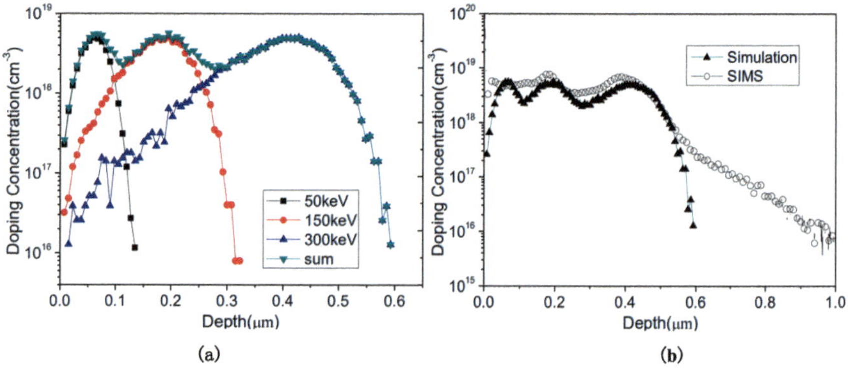


**Fig. 4.40** **a** Simulated Al ion injection doping distribution; **b** Comparison of SIMS test and simulation results [55]

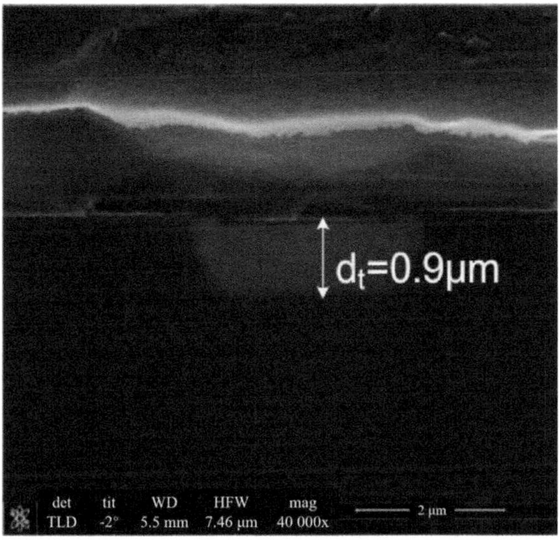


**Fig. 4.41** SEM profile of the actual injection junction [55]

dry etching can be categorized as physical etching, chemical etching, and physical–chemical etching. Physical etching involves the use of plasma and an electric field to blast the surface of a material with sufficient energy, without any reaction between the plasma and the target substance. Chemical etching refers to the process in which plasma interacts with the target material, causing a chemical reaction that leads to the removal of the target material. Physico-chemical etching combines both physical and chemical processes. Dry etching is distinguished by its precise control over etching depth and minimal lateral etching. However, its drawback lies in its slow etching rate and the requirement for costly equipment.

SiC has exceptional chemical inertness towards solvents, making it very resistant to their effects. Consequently, the process of wet etching SiC is notably challenging. SiC single crystals are insoluble in both acids and bases at room temperature. Reactive Ion Etching (RIE) is a dry etching method that is both extremely anisotropic and selective. The process involves utilizing a vacuum system and molecular gas plasma to perform etching. Anisotropic etching is achieved through ion-induced chemical reactions, wherein ion energy is employed to create a surface layer that is susceptible to etching and to facilitate chemical reactions. Additionally, ions are utilized to remove materials from the surface, thereby exposing a clean surface for etching. RIE is commonly employed for the synthesis of SiC in the construction of the countertop framework and indentations. During RIE, the plasma generates reactive radicals that move towards the SiC surface and induce chemical corrosion. Ions with a positive charge are propelled at high speeds within the plasma sheath and collide with the surface of SiC, causing physical deterioration.

The etching gas system can be divided into:

(1) Fluorine-based gases: $SF_6$, $CF_4$, $NF_3$, $BF_3$, $CHF_3$
(2) chlorine-based gases: $Cl_2$, $SiCl_4$, $BCl_3$

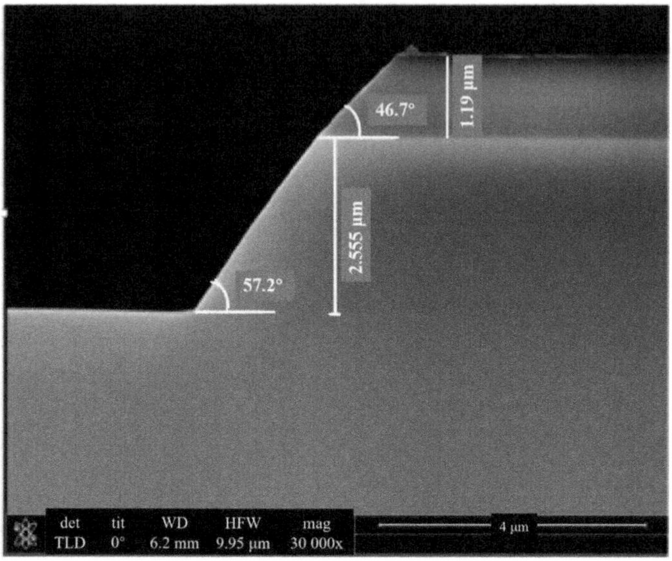


**Fig. 4.42** SEM profile of the Mesa region of the PiN device [3]

(3) Bromine-based gases: $Br_2$, IBr

The addition of either $O_2$ or Ar is typically employed to augment the elimination of carbon atoms or to amplify the active substance, particularly in the realm of fluorinated chemistry. $SiO_2$ was used as the etching mask, and $SF_6/O_2/CHF_3$ was selected as the etching gas. The outcomes of the etching process using this gas combination are illustrated in Fig. 4.42.

4. Oxidation

Silicon carbide (SiC) is the sole compound semiconductor capable of undergoing thermal oxidation to yield $SiO_2$ of superior quality. Thus, SiC may undergo thermal oxidation to effectively address the issues of masking and surface passivation caused by ion implantation and graphic etching. SiC oxidation can be classified into two categories: dry oxygen oxidation and moist oxygen oxidation. The features of the two are as follows: dry oxygen results in the formation of a dense oxide layer with a high dielectric constant, whereas wet oxygen leads to a faster oxidation rate. Water vapor, specifically the hydrogen element, plays a role in passivating the interface state.

5. Passivation

SiC surfaces typically exhibit high concentrations of inherent interfacial states, such as Si\C dangling bonds, C clusters, O vacancies, and other process-induced defects. These defects contribute to elevated device leakage and can even cause premature breakdown, significantly impacting the performance of diode devices. Through the process of oxidizing/passivating the SiC surface under optimal conditions, the number of interfacial states is decreased, resulting in improved isolation and pressure resistance of the passivation system. As a result, the reverse characteristics of SiC diode devices are significantly enhanced. The passivation protection layer system for SiC nuclear radiation detectors typically comprises two components. The first component involves the formation of a first-stage dense passivation layer directly on the SiC surface using the direct thermal oxidation method mentioned earlier. The second component involves the formation of a thicker second-stage composite passivation layer through the CVD deposition and imidization method.

6. Preparation process of 4H–SiC nuclear radiation detector

Taking the PiN diode structure shown in Fig. 4.37 as an example, the specific process preparation steps are as follows:

(1) The process involved preparing thick accidentally doped epitaxial layers (i-layers) and P+ epitaxial layers on N-type 4H–SiC substrates to create PiN-type 4H–SiC epitaxial wafers.
(2) After preparing the substrate, Mesa table etching was carried out.
(3) Termination structure preparation: By utilizing an extended termination structure with an expandable depletion zone, the issue of excessive electric field concentration at the periphery of the main junction can be alleviated, resulting in a significant reduction in leakage current. The terminal P-type area can be created by an ion implantation method, utilizing $SiO_2$ as a mask.
(4) Preparation of the first and second passivation layers: The device surface is subjected to surface thermal oxidation using Plasma Enhanced Chemical Vapor Deposition (PECVD) to create a high-quality $SiO_2$ passivation layer. Selective etching and windowing processes are then performed to prepare the surface for electrode preparation in the next step.
(5) Preparation of backside N-type ohmic contacts and frontside P-type ohmic contacts.
(6) The pad is prepared by sputtering a thicker metal layer on the front side of the device. This is done to reduce oxidative failure of the ohmic contacts, enhance current flow between the electrodes and the external circuit, and improve wire bonding during device packaging.
(7) The third and fourth passivation layers are prepared by applying a silicon nitride and polyimide (Polyimide, PI) passivation layer on the surface of the device. Silicon nitride is a durable material that can withstand high temperatures and has excellent strength and hardness. It is effective in safeguarding devices by preventing contamination, hence enhancing the stability and reliability of the device. Polyimide exhibits properties such as radiation resistance, high thermal stability, chemical resistance, excellent electrical insulation, and the potential for cross-linking upon exposure to ultraviolet (UV) radiation. During the development process, the undeveloped region is dissolved by the developer solution. By subjecting the undeveloped region to heat, the cross-linking reaction is not

**Fig. 4.43** SiC nucleation device on wafer

fully completed, resulting in the formation of a dense and dependable insulating layer as a protective medium. The gadget preparation has been finalized.

### *4.4.2 Detector Electricity Testing*

Figure 4.43 depicts the SiC wafer acquired following the entire preparation procedure, with each silver square on the wafer serving as a SiC nuclear radiation detector. During the scribing, packaging, and nuclear radiation detection test, it is important to first evaluate the pros and cons of the electrical characteristics of the detector. This preliminary testing helps ensure the validity of the process program and allows for the identification of qualified devices that can be used for the final nuclear radiation detection.

1. Schottky contact parameter test—Schottky forward I-V test and parameter extraction

The Schottky barrier height and the ideal factor are the key factors that indicate the performance of a Schottky contact. The determination of the barrier height and ideal factor is carried out for a Schottky contact system created by a certain metal. As an illustration, the optimal barrier height for a Ti/SiC Schottky contact is 1.2 eV, while for a Ni/SiC Schottky contact, it is 1.6 eV. The factor n represents the extent of transport disturbance at the contact interface. The value of n = 1 represents the optimal scenario, which is considered the ideal case. If the contact interface becomes rough, the value of n increases. The barrier height and ideal factor can be determined by analyzing the forward test I-V curves of Schottky diode devices. The equation representing the current–voltage relationship for a Schottky contact is:

$$I = I_S\left(e^{\frac{\beta V}{n}} - 1\right) \tag{4.19}$$

where $I_S$ is the saturation current, $V$ is the applied bias voltage, $n$ is the device ideality factor, $\beta = q/k_B T$, $q$ is the electron charge ($1.6 \times 10^{-19}$C), kB is Boltzmann's constant ($8.62 \times 10^{-5}$eV/K) and $T$ is the absolute temperature (K). The saturation current can be expressed as:

$$I_S = A^* A T^2 \left(e^{-\beta\varphi_B}\right) \tag{4.20}$$

where $h$ is Planck's constant and $m^*$ is the electron effective mass. Taking the logarithm of Eq. (4.19) gives:

$$\log(I) = \frac{\beta V}{n} + \log(I_S) \tag{4.21}$$

Equation (4.21) allows for the determination of the barrier height and ideal factor by analyzing the I-V curve of the Schottky contact. Where $\frac{\beta}{\mathrm{n}}$ represents the gradient of the curve and $\log(I_S)$ represents the point at which the curve intersects the $y$-axis. Figure 4.44 displays an illustrative forward I-V test plot, with the linear section represented by a red dashed line that corresponds to the linear fit curve.

2. Ohmic contact test—TLM method [53]

The metal–semiconductor contact resistance $R_c$ (in Ω) is a crucial physical metric for assessing the quality of an ohmic contact. Contact resistance is the aggregate of resistances associated with the metal–semiconductor contact and is contingent upon the contact's area and geometry. However, in reality, there are highly intricate characteristics at the interface of the MS contact. The resistance value of an ohmic contact is significantly influenced by various parameters, such as the height of the metal barrier, the method used for surface preparation, and the roughness of the interface.

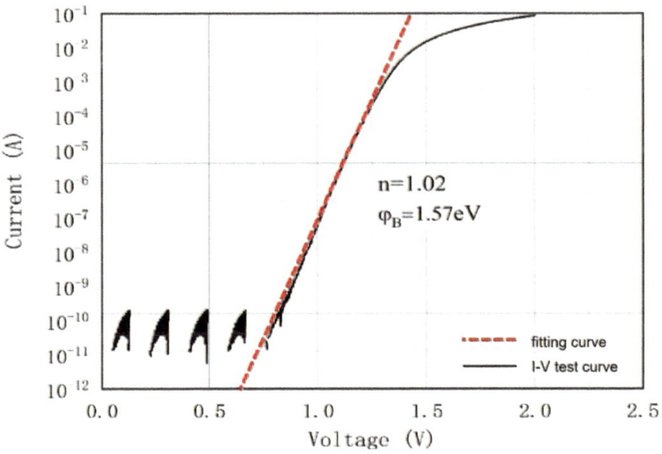


**Fig. 4.44** Schottky contact forward I-V test curve

Given that these parameters are unrelated to the contact's size and shape, the contact resistance does not serve as a reliable indicator of the contact's quality. Hence, it is imperative to incorporate a parameter that accurately quantifies the excellence of the contact, namely the specific contact resistance $\rho_c$ (measured in $\Omega cm^2$). The value of ρc serves as a quantitative indicator to assess the quality of ohmic contact, regardless of the contact area and shape. A lower value of $\rho_c$ indicates better performance of the ohmic contact. $\rho_c$ is a term that is formally defined as:

$$\rho_c = \left(\frac{\partial J}{\partial V}\right)^{-1}_{V=0} \text{ or } \rho_c = \lim_{\Delta S \to 0} R_c \cdot \Delta S \tag{4.22}$$

The following are the main test methods commonly used for specific contact resistance:

(1) Cox and Strack method: The testing procedure involves creating four prototype electrodes with varying radii on the epitaxial side of the semiconductor. On the substrate side, a heavily doped area with a large surface is prepared. The test structure is designed longitudinally. However, a limitation of this method is that it cannot accurately measure the ohmic contact resistance for small resistance values.
(2) The Kuphal method involves using four circular electrodes of equal radius organized in a linear fashion, commonly known as the four-probe method. The current measurement is performed in a lateral manner, which results in limited accuracy. However, it offers the advantage of being user-friendly.
(3) Kelvin method: A four-terminal structure is created by diffusing or implanting ions into a cross-bridge. This approach is very precise but requires significant effort to prepare.
(4) The circular electrode TLM method is a test structure where the electrodes are arranged in concentric rings and the test is conducted in a transverse manner. This method is convenient to produce as it does not require any isolation. Nevertheless, it is necessary for the electrode to have a significantly low square resistance, and the equations used to calculate the parameters are more intricate.
(5) The rectangular electrode transmission line method (TLM) is used. The Rectangular TLM (Transmission Line Methods) structure is a reliable and uncomplicated technique for determining the ohmic contact parameters and assessing the quality of ohmic contact preparation. It has emerged as the primary method for testing ohmic contact and is extensively employed to measure contact resistance. The preparation and testing of the system are less complex and yield more precise results, but, it is necessary to ensure isolation of the test structure. The primary focus of this book is on this particular test methodology.

Figure 4.45 illustrates the fundamental configuration of the TLM test and the model that represents the electrical contact resistance. The TLM structure has a sequence of metal contact patterns that have equal area, with a width of W and a length of L. The graphs exhibit different spacing, denoted as d, and are typically organized in a uniformly spaced sequence. According to Fig. 4.45b, the total resistance ($R_T$)

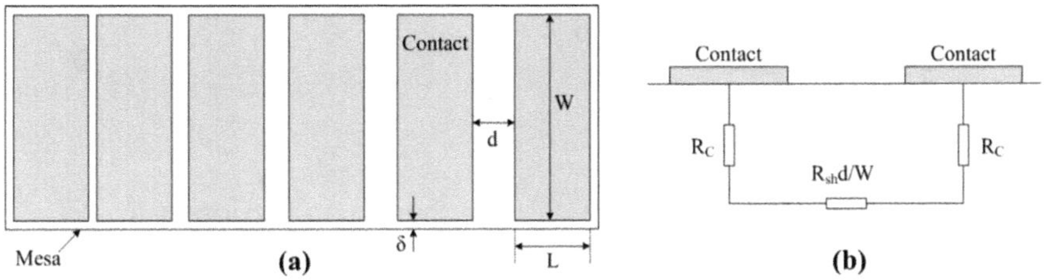


**Fig. 4.45** **a** TLM test structure; **b** ohmic contact equivalent model

between two adjacent contact figures of the TLM can be represented as:

$$R_{\mathrm{T}} = 2R_{\mathrm{c}} + \frac{R_{\mathrm{sh}}d}{W} \tag{4.23}$$

where the contact resistance $R_c$ can be expressed as:

$$R_{\mathrm{c}} = \left(\rho_{\mathrm{c}}\frac{L_{\mathrm{T}}}{W}\right)\coth(L/L_{\mathrm{T}}) \tag{4.24}$$

$R_{sh}$ represents the resistance of the semiconductor material in a thin layer, whereas $L_T$ refers to the transmission length of the current. The transmission length signifies the length of the contact where the majority of the current flows, starting at the contact edge. The abbreviation $L_T$ can be estimated as $\sqrt{\rho_c/R_{sh}}$. $\coth(\mathrm{L}/L_T)$ tends to 1 when $L \geq 1.5L_T$. The LT expression is produced by combining Eqs. (4.23) and (4.24).

$$R_{\mathrm{T}} = 2R_{\mathrm{c}} + \frac{R_{\mathrm{sh}}d}{W} \approx \frac{R_{\mathrm{sh}}}{W}(d + 2L_{\mathrm{T}}) \tag{4.25}$$

The equation above demonstrates the linear relationship between the total resistance $R_T$ and the spacing $d$ of the contact electrodes. Figure 4.46 displays a standard $R_T \sim d$ curve, where the value of $R_{sh}$ can be determined by analyzing the slope of the line, and the value of $L_T$ can be determined by analyzing the intercept of the line with the $x$-axis. The specific contact resistance value can be determined based on this information.

$$\rho_{\mathrm{c}} = R_{\mathrm{sh}}\mathrm{L}_{\mathrm{T}}^2 = R_{\mathrm{c}}L_{\mathrm{T}}W \tag{4.26}$$

3. Interface state/interface charge density test [56]

The SiC N-type epitaxial layers experienced thermal oxidation, and electrodes were applied to the top and bottom sides to create SiC MOS capacitors. The C-V test was utilized to determine various interfacial coefficients, allowing for the assessment of the advantages of the passivated interfaces.

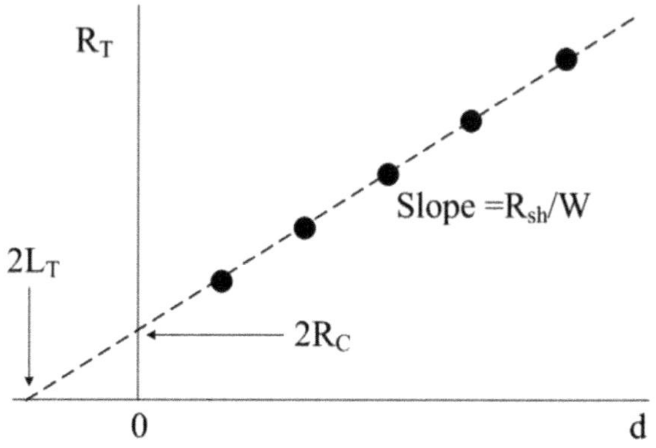


**Fig. 4.46** Schematic of specific contact resistance extraction $R_T \sim d$

(1) The conductance method is used to analyze the interface traps (states) by measuring the conductance ($G_m$) and capacitance ($C_m$) of MOS capacitors. These measurements are taken at different gate voltages and test frequencies. By solving the small-signal equivalent circuit, the parallel conductance ($G_p$) can be calculated. This conductance reflects the energy loss that occurs when carriers are captured and released from the interface traps. The conductance approach aims to obtain the interfacial density of states by carefully measuring the energy loss.

Figure 4.47 is a simple circuit diagram used for evaluating MOS capacitance. The oxide layer capacitance is denoted as $C_{OX}$, the semiconductor capacitance as $C_S$, the capacitance of the interface trap as $C_{it}$, and the resistance of the interface trap that leads to energy loss during the trapping and releasing of carriers as $R_{it}$. By utilizing the circuit representation shown in Figs. 4.47a, b, we can determine the values of parallel capacitance ($C_p$) and parallel conductance ($G_p$).

$$C_P = C_S + \frac{C_{it}}{1 + (\omega\tau_{it})^2} \tag{4.27}$$

$$\frac{G_P}{\omega} = \frac{q\omega\tau_{it}D_{it}}{1 + (\omega\tau_{it})^2} \tag{4.28}$$

where $C_{it}=q \cdot D_{it}$, $\tau_{it}=C_{it} \cdot R_{it}$ and $\omega = 2\pi f$ ($f$) is the test frequency, $\omega$ is the angular frequency), and $\tau_{it}$ is the interface trap time constant. The aforementioned equations describe the energy level traps present in the forbidden bands. However, it is important to note that the energy level distribution of the interface traps at the SiC/$SiO_2$ interface is continuous. When subjected to AC small signals, the capture and release of carriers primarily take place within a few $kT/q$ above or below the Fermi energy level. Thus, the equations for normalized conductance with discontinuous time constants are presented below:

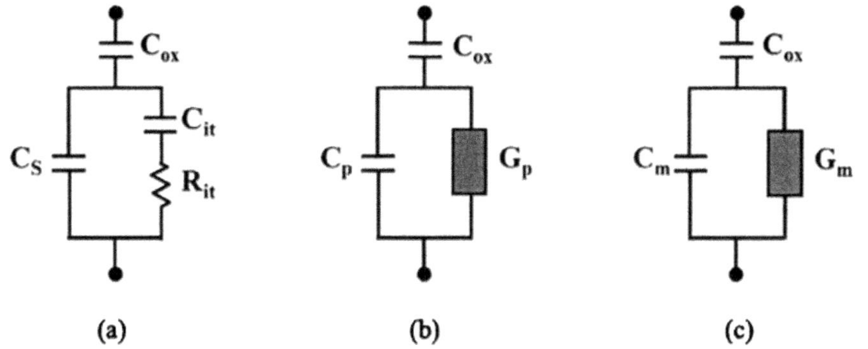


**Fig. 4.47** **a** Simplified circuit diagram of MOS capacitance including interface traps; **b** simplified circuit diagram of parallel conductance; **c** equivalent circuit diagram of measurement

$$\frac{G_P}{\omega} = \frac{qD_{it}}{2\omega\tau_{it}}\ln\left[1 + (\omega\tau_{it})^2\right] \tag{4.29}$$

For practical measurements, the device is regarded as comprising a parallel combination of $C_m$ and $G_m$, as illustrated in the equivalent circuit seen in Fig. 4.47c. The $C_m$ represents the capacitance that has been measured, while $G_m$ represents the conductance that has been measured. By transforming the circuit depicted in Fig. 4.47c into the circuit shown in Fig. 4.47b, the expression for $G_p/\omega$ can be obtained in relation to $C_m$, $G_m$, and $C_{ox}$.

$$\frac{G_P}{\omega} = \frac{\omega G_m C_{OX}^2}{G_m^2 + \omega^2(C_{OX} - C_m)^2} \tag{4.30}$$

To determine the interfacial density of states $D_{it}$, it is necessary to measure and graph the changes in conductance and capacitance as a function of frequency. This can be done by plotting the curve of $G_p/\omega$ vs $\omega$. The highest value of $G_p/\omega$ occurs at $\omega = 2/\tau_{it}$, as indicated by Eq. (4.28).

$$D_{it} = \frac{2.5}{q}\left(\frac{G_P}{\omega}\right)_{max} \tag{4.31}$$

Once the interfacial state density has been acquired, the subsequent task is to determine the distribution of the interfacial traps in relation to their energy levels. At sufficiently high frequencies, the charging and discharging processes of the interface traps cannot keep up with the changes in the AC small signals. Therefore, the effect of the interface traps is disregarded in the equivalent circuit, as depicted in Fig. 4.48. Consequently, the high-frequency C-V curves obtained from the tests can be utilized. These curves represent the relationship between the surface potentials ($\varphi_s$) and the semiconductor capacitance ($C_s$).

**Fig. 4.48** High frequency equivalent circuit diagram

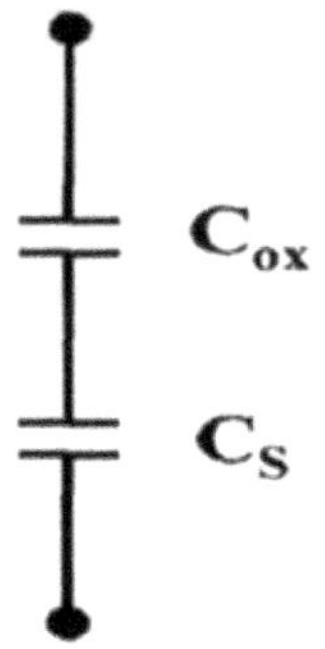


$$C_S = \frac{SqN_d\left|\exp\left(\frac{q\varphi_s}{kT} - 1\right)\right|}{\sqrt{\frac{2kTN_d}{\varepsilon_0\varepsilon_{SiC}}\left\{\exp\left(\frac{q\varphi_s}{kT} - 1\right)\right\}}} \tag{4.32}$$

$$C = \frac{1}{\frac{1}{C_{OX}} + 1/C_s} \tag{4.33}$$

The relationship between the surface potential ($\varphi_s$) and the applied gate pressure is established using the high-frequency C-V curve, where $S$ represents the gate electrode area, $\varepsilon_0$ represents the vacuum permittivity, $\varepsilon_{SiC}$ represents the relative permittivity of 4H–SiC, and $C$ represents the capacitance value in the high-frequency C-V curve. The distribution of the interface state density with energy level position can be computed using the following equation:

$$E_c - E_t = \frac{E_g}{2} - q\left(\frac{kT}{q}\ln\left(\frac{N_d}{n_i}\right) - \varphi_s\right) \tag{4.34}$$

where $N_d$ is the epitaxial doping concentration, $n_i$ is the intrinsic carrier concentration, $k$ is the Boltzmann constant, $T$ is the thermodynamic temperature, and $q$ is the electron charge.

(2) Analysis of charge traps in the oxide layer: The oxide layer traps are mostly found within the oxide layer itself. These traps often catch charge through carrier tunneling, hot carrier injection, or particle radiation. The density of these traps is determined by estimating the quantity of filled charge. Once the charge is captured, the oxide trap typically exhibits a challenging release process and behaves as a stationary charge, resulting in a shift in the flat-band voltage ($V_{fb}$).The procedure for determining the flat-band voltage of the MOS capacitor from the measured C-V curve is as follows. Begin by computing the Debye length ($L_D$):

$$L_D = \sqrt{\frac{\varepsilon_0\varepsilon_{SiC}kT}{q^2N_d}} \tag{4.35}$$

The semiconductor capacitance ($C_{\mathrm{sfb}}$) under flat-band conditions is:

$$C_{\mathrm{sfb}} = \frac{S\varepsilon_0\varepsilon_{\mathrm{SiC}}}{L_{\mathrm{D}}} \tag{4.36}$$

The flatband capacitance ($C_{\mathrm{fb}}$) is a series combination of oxide layer capacitance ($C_{\mathrm{OX}}$) and semiconductor capacitance ($C_{\mathrm{sfb}}$):

$$C_{\mathrm{fb}} = \frac{C_{\mathrm{OX}}C_{\mathrm{sfb}}}{C_{\mathrm{OX}} + C_{\mathrm{sfb}}} \tag{4.37}$$

The maximum capacitance of the accumulation zone in the C-V curve, denoted as $C_{\mathrm{OX}}$, allows for the calculation and extraction of the flat-band capacitance and voltage directly from the measured C-V curve. The quantity of oxide traps ($N_{\mathrm{ot}}$) can be determined by comparing the drift $\mathrm{V}_{\mathrm{fb}}$ of the flat bands of the C-V curves measured prior to and following the capture of the oxide traps:

$$N_{ot} = \frac{C_{OX} \cdot \Delta V_{\mathrm{fb}}}{qS} \tag{4.38}$$

4. Device reverse current test—reverse I-V test

The reverse high voltage of the device during the I-V test is a crucial parameter for assessing the leakage current and voltage characteristics of the detector. By using a high-voltage probe table, the detector is subjected to a voltage scan ranging from 0 to – 1000 V. The resulting I-V curve for each detector on the wafer is typically depicted in Fig. 4.49. Since the 4H–SiC nuclear radiation detector operates under high reverse bias conditions, it requires a low leakage current. Specifically, if the device needs to be in a 1000V reverse bias state, the leakage current should not exceed 100 nA, as indicated by the red line position in Fig. 4.49. Consequently, the wafer device qualification rate is approximately 15%.

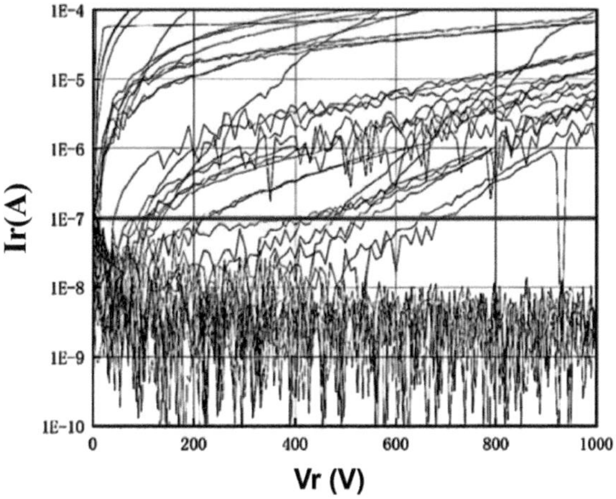


**Fig. 4.49** Reverse I-V test curve

# References

1. Hu Q (2012) Research on SiC neutron detectors. National University of Defense Technology
2. Ruddy FH, Dulloo AR, Seidel JG et al (2004) The fast neutron response of silicon carbide semiconductor radiation detectors. Nuclear science symposium conference record. IEEE, pp 4575–4579
3. Wang S (2018) Research on 4H-SiC PiN diode neutron detectors. Xidian University
4. Giudice AL, Fasolo F, Durisi E et al (2007) Performances of 4H-SiC Schottky diodes as neutron detectors. Nucl Instrum Methods Phys Res 583(1):177–180
5. Kim HS, Ha JH, Park SH et al (2011) Characteristics of fabricated neutron detectors based on a SiC semiconductor. J Nucl Sci Technol 48(10):1343–1347
6. Tairov YM, Tsvetkov VF (1978) Investigation of growth processes of ingots of silicon carbide single crystals. J Cryst Growth 43(2):209–212
7. Junmin G, Jianmin H (2014) Research on the preparation of 4-inch SiC single crystal by PVT method. Modern Instrum Med 1:33–35
8. Yang X, Xu X, Chen X et al (2017) Research progress of wide bandgap SiC single crystal substrate. Power Electron Technol 2017(8):12–16
9. Takahashi J, Ohtani N, Kanaya M (1996) Structural defects in α-SiC single crystals grown by the modified-Lely method. J Cryst Growth 167(3–4):596–606
10. Ha SY, Rohrer GS, Skowronski M et al (2000) Plastic deformation and residual stresses in SiC Boules grown by PVT. Mater Sci Forum 338–342:67–70
11. Yang F, Niu Y, Qian W et al (2016) Overview of homoepitaxial growth of 4H-SiC. Smart Grid
12. Neudeck PG (2000) Electrical impact of SiC structural crystal defects of high electric field devices. Mater Sci Forum 338–342:1161–1166
13. Powell AR, Sumakeris JJ, Khlebnikov Y et al (2016) Bulk growth of large area SiC crystals. Mater Sci Forum 858:5–10
14. Renxu J, Yimen Z, Yuming Z et al (2008) Homoepitaxial growth of N-type 4H-SiC. Acta Physica Sinica 10:6649–6653
15. Hu J (2017) Low-pressure homoepitaxial growth of 4H-SiC and device verification. Xidian University
16. Kordina O, Hallin C, Ellison A et al (1996) High temperature chemical vapor deposition of SiC. Appl Phys Lett 69(10):1456–1458
17. Jia R, Liu S, Xu H et al (2014) Study on the grove model of homoepitaxial growth of 4H-SiC. Acta Physica Sinica 63(3):365–369

18. Lofgren PM, Ji W, Hallin C et al (2000) Modeling of silicon carbide epitaxial growth in hot-wall chemical vapor deposition processes. J Electrochem Soc 147(1):164–175
19. Veneroni A, Omarini F, Moscatelli D et al (2005) Modeling of epitaxial silicon carbide deposition. J Cryst Growth 275(1–2):e295–e300
20. Myers-Ward RL, Shishkin Y, Kordina O et al (2006) High epitaxial growth rate of 4H-SiC using horizontal hot-wall CVD. Mater Sci Forum 527–529:187–190
21. Kimoto T, Cooper JA (2014) Fundamentals of silicon carbide technology: growth, characterization, devices and applications. Wiley-IEEE Press
22. Larkin DJ, Neudeck PG (1994) Site-competition epitaxy for superior silicon carbide electronics. Appl Phys Lett 65(13):1659–1661
23. Larkin DJ (1997) SiC dopant incorporation control using site-competition CVD. Phys Status Solidi 202(1):305–320
24. Yamamoto T, Kimoto T, Matsunami H (1998) Impurity incorporation mechanism in step-controlled epitaxy growth temperature and substrate off-angle dependence. Mater Sci Forum 264–268:111–114
25. Kojima K, Suzuki T, Kuroda S et al (2003) Epitaxial growth of high-quality 4H-SiC carbon-face by low-pressure hot-wall chemical vapor deposition. Jpn J Appl Phys 42(Part 2, No. 6B):L637–L639
26. Yoshida S, Sakuma E, Misawa S et al (1998) A new doping method using metalorganics in chemical vapor deposition of 6H–SiC. J Appl Phys 55(1):169–171
27. Nordell N, Sch?Ner A, Linnarsson MK (1997) Control of Al and B doping transients in 6H and 4H SiC grown by vapor phase epitaxy. J Electron Mater 26(3):187–192
28. Kotamraju S, Krishnan B, Koshka Y (2010) Use of chlorinated carbon and silicon precursors for epitaxial growth of 4H-SiC at very high growth rates. Physica Status Solidi (RRL) Rapid Res Lett 3(5):157–159
29. Zhou Z (2013) Research on the thick-film epitaxial process of 4H-SiC. Xidian University
30. Chen BY, Liu X, Chen ZZ et al (2009) Misoriented domain formation in 6H-SiC single crystal. J Cryst Growth 311(14):3573–3576
31. Lin S, Chen Z, Liu B et al (2010) Identification and control of SiC polytypes in PVT method. J Mater Sci Mater Electron 21(4):326–330
32. Nakashima S, Nakatake Y, Harima H et al (2000) Detection of stacking faults in 6H-SiC by Raman scattering. Appl Phys Lett 77(22):3612–3614
33. Lu YM, Leu IC (2000) Qualitative study of beta silicon carbide residual stress by Raman spectroscopy. Thin Solid Films 377(none):389–393
34. Zhang Z (2014) Research on the growth of 4H-SiC thick epilayers. Xidian University
35. Li Z (2012) Homoepitaxial growth of 4H-SiC and device research. Nanjing University
36. Kimoto T, Miyamoto N (1999) Performance limiting surface defects in SiC epitaxial p-n junction diodes. IEEE Trans Electron Dev Ed 46(3):471–477
37. Chen L, Guy OJ, Jennings MR et al (2007) Study of a novel Si/SiC hetero-junction MOSFET. Solid-State Electron 51(5):662–666
38. Skowronski M, Ha S (2006) Degradation of hexagonal silicon-carbide-based bipolar devices. J Appl Phys 99(1):1–1
39. Kimoto T, Hiyoshi T, Hayashi T et al (2010) Impacts of recombination at the surface and in the substrate on carrier lifetimes of n-type 4H–SiC epilayers. J Appl Phys 108(8):083721
40. Galeckas A, Linnros J, Frischholz M et al (2001) Optical characterization of excess carrier lifetime and surface recombination in 4H/6H–SiC. Appl Phys Lett 79(3):365–367
41. Mannan MA (2015) Defect characterization of 4H-SiC by deep level transient spectroscopy (DLTS) and influence of defects on device performance. Dissertations & theses - Gradworks
42. Hallén A, Fenyö D, Sundqvist BUR et al (1991) The influence of ion flux on defect production in MeV proton-irradiated silicon. J Appl Phys 70(6):3025–3030
43. Auret FD, Deenapanray PNK (2004) Deep level transient spectroscopy of defects in high-energy light-particle irradiated Si. Crit Rev Solid State Mater Sci 29(1):1–44
44. Han C (2016) Development of 4H-SiC PiN power diodes and research on key technologies. Xidian University

45. Hatakeyama T, Suzuki T, Ichinoseki K et al (2010) Impact of oxidation conditions and surface defects on the reliability of large-area gate oxide on the C-Face of 4H-SiC. Mater Sci Forum 645–648:799–804
46. Kimoto T, Miyamoto N, Matsunami H (1999) Effects of surface defects on the performance of 4H– and 6H–SiC pn junction diodes—ScienceDirect. Mater Sci Eng B s 61–62(98):349–352
47. Dudley M, Wang S, Huang W et al (1995) White-beam synchrotron topographic studies of defects in 6H-SiC single crystals. J Phys D Appl Phys 28(4A):A63–A68
48. Neudeck PG, Wei et al (1999) Study of bulk and elementary screw dislocation assisted reverse breakdown in low-voltage (+-n junction diodes. I. DC properties. IEEE transactions on electron devices, vol 46(3), pp 478–484
49. Wahab Q, Ellison A, Henry A et al (2000) Influence of epitaxial growth and substrate-induced defects on the breakdown of 4H–SiC Schottky diodes. Appl Phys Lett 76(19):2725–2727
50. Fan JY, Chu PK (2014) Silicon carbide nanostructures. Springer International Publishing, Cham
51. Kimoto T (2015) Material science and device physics in SiC technology for high-voltage power devices. Jpn J Appl Phys 54(4):040103
52. Sui J (2020) Design and research of P-type 4H-SiC ohmic contact. Hangzhou Dianzi University
53. Guo H (2007) Theoretical and experimental research on ohmic contact of SiC devices. Xidian University
54. Hara S, Teraji T, Okushi H et al (1997) Control of Schottky and ohmic interfaces by unPiNning fermi level. Appl Surf Sci 117–118(none):394–399
55. Yuan H (2017) Design and experimental research on high-performance 4H-SiC Schottky Barrier Diode (SBD)/Junction Barrier Schottky (JBS) devices. Xidian University
56. Jia Y (2018) Research on the trap characteristics of 4H-SiC MOS structures and the reliability of gate oxide layers. Xidian University

# Chapter 5
# Radiation Response Study of SiC Nuclear Radiation Detectors

## 5.1 SiC Nuclear Radiation Detector Energy Spectrum Simulation and Device Optimization Design [1]

To minimize expenses and time spent on modifying the hardware environment, simulation is commonly employed to acquire the performance parameters of the detector prior to determining its geometrical parameters. This allows for guidance in the construction of the detector. Energy spectrum measurement is a fundamental aspect of detector operation. By measuring the energies of incident particles, we can gain insight into the sort of particles involved and the specific nuclear processes that occur, as the energy released during nuclear reactions and decays are distinct and characteristic. Additionally, energy spectroscopy is used to acquire important information regarding essential material and device properties, such as CCE, FWHM, and $\mu$-$\tau$ (carrier mobility lifetime product). The pulse amplitude spectrum in practical measurements is determined by quantifying the amplitude of the output pulse voltage from the shaping circuit, as explained in Sect. 2.5.2. The amplification mechanism employed for the energy spectrum measurements produces a voltage pulse with an amplitude V that is directly proportional to the charge $Q$ detected by the detector. Additionally, the total charge $Q_0$ generated in the detector is directly proportional to the energy E of the charged particles. Under optimal circumstances, the output charge $Q$ of the detector is equivalent to the total charge $Q_0$ created within the detector. The aggregate quantity of charge, denoted as $Q_0$, namely:

$$E \propto Q_0 = Q \propto V \tag{5.1}$$

Assuming perfect conditions (with a charge collection efficiency of 100%), the amplitude V of the output pulse voltage from the shaping circuit should be directly proportional to the energy E of the particles deposited in the detector. Additionally, the energy deposition spectrum, which measures the energy lost by the charged particles in the material, should have the same shape as the experimentally observed



Y. Zhang et al., *Wide Band-Gap Semiconductor Nuclear Radiation Detector*,
Wide Bandgap Semiconductors, https://doi.org/10.1007/978-981-96-0340-4_5

pulse amplitude spectrum. Thus, during the research phase, it is customary to initially compute the energy deposition spectrum of the detector as a theoretical benchmark for its pulse amplitude spectrum. This calculation is then used to evaluate and describe the performance of the detector. Nevertheless, the aforementioned approximation fails to include the operational condition and performance fluctuations of the detecting equipment. To illustrate the distinction between the energy deposition spectrum and the pulse height spectrum, the detector described in reference [2] is employed as an example. Figure 5.1 presents the pulse amplitude spectrum and energy deposition spectrum of a zero-bias PiN diode. Figure 5.1b presents the energy deposition spectra estimated under various settings. Line1 represents the spectrum of energy deposition caused by charged particles throughout the entire SiC material. Line2 represents the spectrum of energy deposition caused by charged particles specifically in the N-epitaxial layer (I-layer) of the SiC. Line3 represents the spectrum of energy deposition caused by charged particles in the space-charged region, which is approximately 4.5 μm in size. It is evident that the energy deposition spectra vary greatly under different situations and are distinct from the recorded pulse amplitude spectrum. Another instance is depicted in Fig. 5.2, which illustrates the fluctuation of the observed pulse amplitude spectrum in relation to the device's operating voltage, as reported in reference [3] in the literature. Merely computing the energy deposition spectrum does not adequately capture these fluctuations. Therefore, to accurately represent the device's performance and obtain a pulse amplitude spectrum that closely matches real-world conditions, we can consider the actual process of generating the pulse amplitude spectrum as described in Sect. 2.5.2. This involves inputting the energy deposition spectrum of the particles that are exposed to the detector device into the device simulation. The output characteristics obtained from the device simulation are then used in the circuit simulation. By combining Monte Carlo simulation, device simulation, and circuit simulation, we can generate a voltage value that corresponds to the incident particle energy, detector structure, and operating status, instead of the original energy deposition value. By employing Monte Carlo simulation to derive the energy deposition spectrum, along with device simulation and circuit simulation, a voltage value is produced that corresponds to the energy of the incoming particles, the structure of the detector, and its operational condition, instead of the original energy deposition value. This chapter will present the radiation response of SiC nuclear radiation detectors, focusing on the study conducted by Dr. Hai-Li Huang in Yuming Zhang's group at Xidian University.

### *5.1.1 Introduction to the Software*

To determine the pulse amplitude spectrum of the particle irradiation detector, one must comprehend many characteristics and their influence on the detection outcomes: (i) The energy and paths of the charged particles that excite the electron–hole pairs in the detector. (ii) The material properties of the detector, including permittivity, carrier mobility, carrier lifetime, and the energy needed to excite electron–hole pairs. (iii)

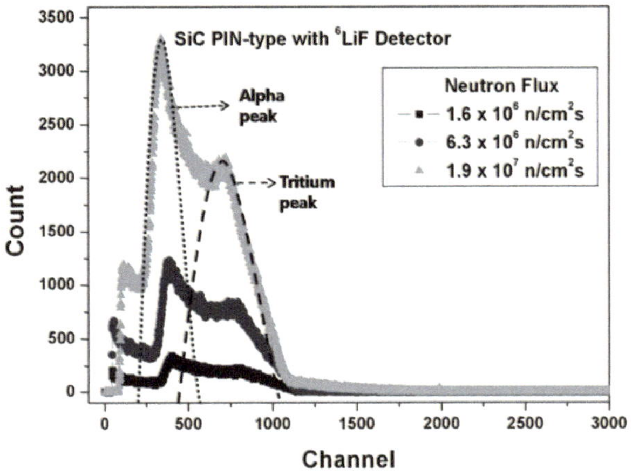


**(a) Measured energy spectrum given in the reference [2]**

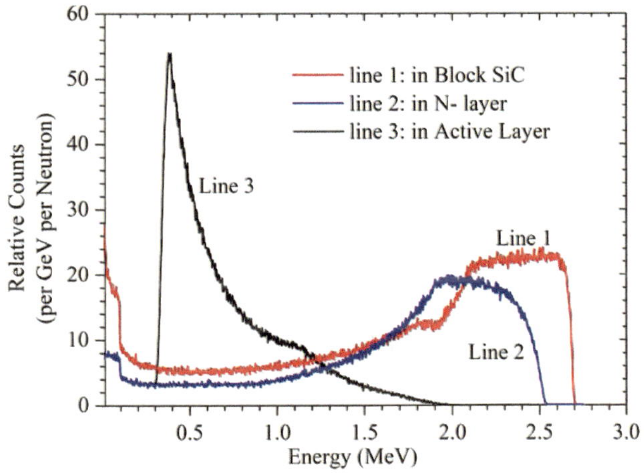


**(b) Energy precipitation spectrum obtained from Monte Carlo simulation**

**Fig. 5.1** Comparison of the measured energy spectrum of the device in the reference [2] with the simulated energy deposition spectrum

The fundamental characteristics of the detector, such as the thickness of the doping concentration in the epitaxial layer and the operating bias. (iv) The electrical structure of the spectral detection system, including important parameters like pulse shaping time and amplification. To acquire and evaluate the aforementioned parameters, the following software options are accessible.

1. Monte Carlo particle transport software

(a) MCNP software [4]

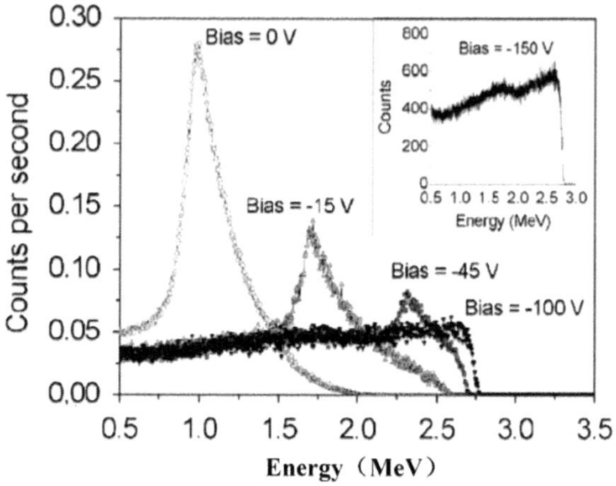


**Fig. 5.2** Variation of the output energy spectrum with bias voltage in the reference [3]

The MCNP software, developed by the Monte Carlo group in the Applied Theoretical Physics Division of the Los Alamos National Laboratory, is a sophisticated Monte Carlo program used for calculating particle transport in intricate three-dimensional geometries. The software has the ability to solve problems involving the transit of neutrons, photons, and electrons, either individually or in combination. It can also calculate the eigenvalue of nuclear critical systems, including both subcritical and supercritical systems. The MCNP program possesses exceptional geometric processing capabilities, enabling it to conduct calculations on materials occupying three-dimensional spaces. These spaces can have geometries comprising one-dimensional, two-dimensional, or even special four-dimensional (e.g., ellipsoidal) surfaces. The program can compute the properties of various nuclides, including isotopes, within the cells of the geometric space. The research encompasses a wide range of fields, including critical device experiments, medicine, and detector design and analysis. It has received unequivocal recognition.

(b) FLUKA software

FLUKA, short for FLUktuierende KAskade, is a Monte Carlo particle transport software developed by CERN, the European Center for Nuclear Research. Its primary use is the computation of particle transportation and reactions within substances, encompassing tasks such as shielding gas pedals and designing targets, analyzing material activation and radiation dosage, designing detectors, and conducting cosmic ray and neutrino analysis, among others. The FLUKA software includes a graphical module that allows users to create material shapes for simulations. Additionally, it may be externally imported to evaluate intricate three-dimensional geometric models. The software additionally enables the importation of diverse visual modules from other origins, hence facilitating the analysis of intricate three-dimensional models. The software has the capability to precisely replicate the behavior of 60 distinct particles, including polarized photons generated by synchrotron radiation and visible light. This is achieved by closely following to the conservation laws governing physical processes, such as particle transportation and interactions between particles. The FLUKA program provides the results of the calculation shown in Fig. 5.1b. FLUKA is utilized in this context to compute the energy and trajectories of charged particles within the detector. Specifically, FLUKA calculates the energy carried by the charged particles upon their incidence on the detector, as well as their angle of incidence.

(c) SRIM software [5]

The SRIM software is a computational tool utilized for determining the stopping mechanism of particles during injection and the penetration depth of injection. The primary premise of this approach is to employ a comprehensive quantum mechanical analysis of ion-atom collisions. Typically, SRIM considers the atoms in motion as "ions" and regards all the atoms in the target as "atoms". SRIM comprises two primary programs and incorporates several other components. SRIM comprises two primary programs, along with several specialized case programs. One of the software tools used is SR, primarily employed for basic cases to calculate the stopping process and injection depth. The other tool is TRIM, mainly utilized to simulate the transfer process of ions in the material and calculate the energy transfer of atoms during transfer and collision using the Monte Carlo simulation method. These calculations are based on the following principles: The Monte Carlo method involves tracking the trajectory of the incident ions using computer simulation. Throughout the simulation, the positions of the ions, the amount of energy lost, and the characteristics of the secondary ions are recorded. Ultimately, the anticipated values of the necessary physical properties and their related statistical errors are examined. The SRIM software lacks the capability to model electrons, neutrons, and photons. However, it is capable of simulating the movement of heavily charged particles commonly found in nuclear radiation detection, such as protons, deuterons, tritium nuclei, and alpha particles, within a medium. The interface is user-friendly and the initial setup process is straightforward.

In the study of SiC nuclear radiation detectors, SRIM software is commonly employed to determine the depth at which charged particles penetrate and the distribution of energy deposition. On the other hand, MCNP and FLUKA software are frequently utilized to calculate the paths of electrons and neutrons in semiconductor materials, as well as the energy deposition. In specific applications that involve a front conversion layer, such as thermal and fast neutron detectors (as explained in Sect. 4.1.4), these software tools are often used to compute the energy carried by incident charged particles and their angle of incidence.

2. Sentaurus Tcad semiconductor device simulation software [6]

Sentaurus TCAD is a powerful software developed by Synopsys, Inc. that is used for optimizing and simulating the development of semiconductor processes and devices. It has a diverse range of applications in the field of semiconductor process and device development. The software utilizes mathematical techniques to solve fundamental physical partial differential equations, enabling the analysis and computation of the structure and performance of semiconductor devices. Additionally, the integration of simulation in simple circuit systems allows for the analysis of the interaction between several devices. Furthermore, the software offers interconnect modeling and extraction techniques that furnish data on crucial parasitic factors required for optimizing device performance. The SDE module is utilized for constructing intricate 2D and 3D device structures and analyzing their fundamental electrical characteristics, optoelectronic properties, high-voltage behavior, irradiation effects, and other relevant properties. The SDEVICE module is employed for simulating the electrical characteristics of these devices. Both modules offer different computational accuracies and allow users to utilize pre-existing physical models or create customized ones. The simulation research of SiC nuclear radiation detectors involves utilizing Sentaurus to compute the pulse current produced by charged particles when they interact with the detector. Additionally, Sentaurus will be employed for the structural optimization of detector diode devices.

3. Energy spectrum detection system circuit simulation software

Multisim is a simulation software developed by National Instruments (NI) Ltd. The program is mostly utilized for the creation of analog and digital circuit boards. The software offers robust modeling and analysis capabilities, allowing users to conveniently edit circuit schematics using a graphical interface. Multisim offers an extensive collection of components and a range of virtual instruments for circuit design and verification. Users can obtain circuit nodes and electrical signals over time without requiring extensive knowledge of general analog circuit simulation technology, such as Simulation Program with Integrated Circuit Emphasis (SPICE).

### *5.1.2 Device Simulation Model and Material Parameters [6]*

The semiconductor diode device serves as the central component in semiconductor nuclear radiation detectors. The diode device's structural parameters have a direct impact on the selection of the detector's operating voltage, depletion region width, depletion region electric field strength, and peak electric field distribution. These parameters, in turn, affect the detector's detection efficiency, charge collection efficiency (CCE), energy resolution, leakage current, and response time, which are key performance indicators. Hence, Sentaurus Tcad device simulation plays a crucial role in precisely transforming the energy deposition spectrum into a pulse amplitude spectrum. Additionally, it aids in guiding the ideal design of the detection device and determining the selection of structural parameters for the actual device flow sheet.

Numerical calculations are essential for computer modeling of semiconductor devices. The numerical computations for simulation with Sentaurus Tcad software rely on a sequence of physical models and equations derived from known theories or empirical formulas in the fields of semiconductor and solid state physics. Device simulation is a method of virtually testing the electrical and physical properties of semiconductor devices. The software used for this requires a significant quantity of information, which needs to be discretized, meaning it is calculated based on mesh division. The software incorporates a program (SDE module) that encodes the semiconductor structure with a well-defined grid division. Subsequently, the device characteristics simulation (SDEVICE) module employs numerical calculations on a customized grid to simulate the behavior of the semiconductor device. Hence, the partitioning of the mesh is of utmost significance, as it directly impacts the velocity, precision, dependability, and convergence of the outcomes obtained from the simulation of semiconductor devices. It is essential to understand the precision of grid subdivision based on varying criteria. The accuracy of device simulation relies heavily on the selection of physical models and parameters, as well as the utilization of different equations and physical models to achieve accurate simulation values, given the specific simulation objectives.

1. **Drift–diffusion model**

The fundamental objective in simulating semiconductor devices is to determine the motion status of carriers within the semiconductor space region. In the most elementary model, specifically the fundamental drift–diffusion transport model, the movement of the carriers can be determined by simultaneously solving three fundamental equations:

(1) Poisson's equation:

$$\nabla^2\varphi = -\frac{q}{\varepsilon}\left(p - n + N_{\mathrm{D}}^{+} - N_{\mathrm{A}}^{-}\right) \tag{5.2}$$

The Poisson equation establishes the relationship between the total charge distribution throughout the semiconductor and the electric field. In this equation, $\varphi$ represents the potential, $\varepsilon$ is the capacitance coefficient of the material, $q$ denotes the charge of the electrons, $n$ and $p$ represent the densities of electron and hole carriers respectively, $N_{\mathrm{D}}^{+}$ represents the concentration of ionizing donors, and $N_{\mathrm{A}}^{-}$ represents the concentration of ionizing acceptors.

(2) Current continuity equation:

$$\nabla\vec{\mathbf{J}}_{\mathbf{n}} = qR + q\frac{\partial n}{\partial t} \tag{5.3}$$

$$-\nabla\vec{\boldsymbol{J}}_{\mathbf{p}} = qR + q\frac{\partial p}{\partial t} \tag{5.4}$$

The electron–hole continuity equation represents the preservation of the overall flow of carriers within the semiconductor, taking into account the generation of complexes and the change in concentration over time. In this equation, $R$ denotes the net rate of complexes, $\vec{\boldsymbol{J}}_{\boldsymbol{n}}$ represents the vector for electron current density, and $\vec{\boldsymbol{J}}_{\boldsymbol{p}}$ represents the vector for hole current density.

(3) Expression of current density:

$$\vec{J}_{\mathrm{n}} = \mathrm{nq}\mu_{\mathrm{n}}\vec{E} + qD_{\mathrm{n}}\nabla n \tag{5.5}$$

$$\vec{J}_{\mathrm{p}} = \mathrm{nq}\mu_{\mathrm{p}}\vec{E} + qD_{\mathrm{p}}\nabla p \tag{5.6}$$

where $D_{\mathrm{n}}$ and $D_{\mathrm{p}}$ are the electron and hole diffusion coefficients, respectively.

2. **Incomplete dissociation model**

Given the significant ionization energy of the 4H–SiC material and the inability of impurities to fully ionize at elevated temperatures, it is imperative to incorporate an incomplete ionization model for impurities in the simulation. The model equation for this purpose is as follows:

$$N_{\mathrm{A}}^{-} = \frac{N_{\mathrm{A}}}{1 + g_{\mathrm{A}}\frac{p}{p_1}}, N_{\mathrm{D}}^{+} = \frac{N_{\mathrm{D}}}{1 + g_{\mathrm{D}}\frac{n}{n_1}} \tag{5.7}$$

$$p_1 = N_{\mathrm{v}}\exp\left(-\frac{E_{\mathrm{A}} - E_{\mathrm{v}}}{KT}\right), \quad n_1 = N_{\mathrm{c}}\exp\left(-\frac{E_{\mathrm{c}} - E_{\mathrm{D}}}{KT}\right) \tag{5.8}$$

$$E_{\mathrm{A}} - E_{\mathrm{v}} = E_{\mathrm{A},0} - E_{\mathrm{v},0} - a_{\mathrm{A}}(N_{\mathrm{A}} + N_{\mathrm{D}})^{1/3} \tag{5.9}$$

$$E_{\mathrm{c}} - E_{\mathrm{D}} = E_{\mathrm{c},0} - E_{\mathrm{D},0} - a_{\mathrm{D}}(N_{\mathrm{A}} + N_{\mathrm{D}})^{1/3} \tag{5.10}$$

$$p = \exp\left(\frac{E_{\mathrm{A}} - E_{\mathrm{Fp}}}{KT}\right) \tag{5.11}$$

$$n = \exp\left(\frac{E_{\mathrm{Fn}} - E_{\mathrm{D}}}{KT}\right) \tag{5.12}$$

$g_{\mathrm{A}}$ and $g_{\mathrm{D}}$ are coefficients that describe the degradation of impurity energy levels. $N_{\mathrm{D}}$ and $N_{\mathrm{A}}$ represent the concentrations of effective donor and acceptor impurities. $E_{\mathrm{Fn}}$ and $E_{\mathrm{Fp}}$ represent the quasi-Fermi energy levels of electrons and holes. $E_{\mathrm{D}}$ and $E_{\mathrm{A}}$ represent the energy levels of donor and acceptor impurities. $E_{\mathrm{c}}$ and $E_{\mathrm{v}}$ represent the energy levels of the conduction and valence bands. $N_{\mathrm{c}}$ and $N_{\mathrm{v}}$ represent the effective densities of states in the conduction and valence bands, respectively. The Boltzmann constant, denoted as $K$, and the absolute temperature, represented by $T$, are related to the effective reduction of the activation energies of the acceptor and donor in the semiconductor. These reductions are described by Eqs. (5.9) and (5.10) respectively. The total doping concentration in the semiconductor influences this reduction. The specific values of each parameter can be found in the 4H-SiC parameter file integrated into the software.

3. **Composite model**

The compounding mechanisms of semiconductors encompass direct compounding, indirect compounding (known as Shockley–Read–Hall or SRH compounding), Auger compounding, and surface compounding. Due to its large forbidden bandwidth, 4H–SiC, an indirect band gap semiconductor material, does not easily facilitate the direct combination of electrons and holes. Instead, the compounding process primarily relies on impurity or defect centers located near the center of the band gap. As a result, the simulation employs an indirect compounding mechanism. Furthermore, given the presence of highly doped regions of semiconductor diodes, such as the substrate region and the P-type region in PiN diodes, the Auger composite mechanism is reintroduced. To summarize, the simulation primarily utilizes the SRH composite and Auger composite models, which are characterized by the following model equations:

$$R_{SRH} = \frac{np - n_i^2}{\tau_p\left[n + n_i \exp\left(\frac{E_t - E_i}{KT}\right)\right] + \tau_n\left[p + n_i \exp\left(-\frac{E_t - E_i}{KT}\right)\right]} \tag{5.13}$$

$$R_{auger} = \left(C_n n + C_p p\right)\left(np - n_i^2\right) \tag{5.14}$$

The variables in question are $ni$ (the intrinsic carrier concentration), $E_t$ (the composite center energy level), $E_i$ (the intrinsic energy level), $\tau$ (the carrier lifetime), and $C_n$ and $C_p$ (quantities associated with the carrier concentration).

4. **Band Gap Narrowing Model**

The width of the band gap is contingent upon temperature and the concentration of doping. The semiconductor material has a negative temperature coefficient for its band gap width, meaning that as the temperature increases, the band gap width decreases. The equation provided describes the relationship between the band gap width ($E_g$) of 4H–SiC and the temperature ($T$).

$$E_g(T) = E_g(0) - \frac{\alpha T^2}{T + \beta} \tag{5.15}$$

where α and β are the relevant material parameters. In the case of heavy doping, the larger the impurity concentration is, the smaller the band gap width of the semiconductor material is instead, i.e., the band gap narrowing effect:

$$\Delta E_g = E_{ref}\left[\ln\left(\frac{N_A + N_D}{N_{ref}}\right) + \sqrt{\left(\ln\left(\frac{N_A + N_D}{N_{ref}}\right)\right)^2 + 0.5}\right] \tag{5.16}$$

where $E_{ref}$ and $N_{ref}$ are the relevant material parameters and $E_g$ is the reduction of $E_g$.

5. **Mobility model**

Mobility is a measure of the conductivity of carriers and is influenced by factors such as effective mass, scattering probability, doping concentration, temperature, and applied bias. It is inversely related to effective mass and scattering probability. Under low electric fields, the primary factors causing scattering in 4H–SiC materials are ionized impurity scattering and lattice vibrational scattering. In this case, the mobility may be accurately characterized using the Masetti model. At high electric field, the movement speed of the carrier particles reaches a maximum and stops increasing further. This phenomenon is known as saturation, and the mobility of the particles in this state is characterized using the Canali model.

(1) Constant mobility model:

The default activation of this model occurs when utilizing the mobility model, in which channel free electrons solely interact with phonons by scattering. Consequently, the mobility is solely determined by the lattice temperature, and its value may be calculated using the following equation:

$$\mu_{\mathrm{const}} = \mu_{\mathrm{L}}\left(\frac{T}{300K}\right)^{-\delta} \tag{5.17}$$

where $\mu_{\mathrm{L}}$ is the default value for body mobility at 300 K.

(2) Masetti model:

$$\mu_{dop} = \mu_{min1}\exp\left(\frac{P_c}{N_A + N_D}\right) + \frac{\mu_{const} - \mu_{min2}}{1 + ((N_A + N_D)/C_r)^{\alpha}} - \frac{\mu_1}{1 + (C_s/(N_A + N_{\mathrm{D}}))^{\beta}} \tag{5.18}$$

For the 4H–SiC material, Eq. (5.18) can be expressed under the default material parameters of the software:

$$\mu_{\mathrm{n}} = 40 + \frac{950 \times \left(\frac{T}{300K}\right)^{-2.4} - 40}{1 + \left((N_{\mathrm{A}} + N_{\mathrm{D}})/1.94 \times 10^{17}\right)^{0.61}} \tag{5.19}$$

$$\mu_{\mathrm{p}} = 15.9 + \frac{125 \times \left(\frac{T}{300K}\right)^{-2.15} - 15.9}{1 + \left((N_{\mathrm{A}} + N_{\mathrm{D}})/1.76 \times 10^{19}\right)^{0.34}} \tag{5.20}$$

(3) Canali model:

$$\mu_{\mathrm{highfield}} = \frac{\mu_{\mathrm{low}}}{\left(1 + \left(\frac{\mu_{\mathrm{low}}E}{v_{\mathrm{sat}}}\right)^{\beta}\right)^{\frac{1}{\beta}}} \tag{5.21}$$

where $\mu_{low}$ is the mobility at low electric field, which depends on the combined effect of the other mobility models that have been activated. $E$ denotes the electric field strength, and $v_{sat}$ denotes the saturation drift velocity of the carriers.

6. **Correlation model for non-equilibrium carrier generation under irradiation**

Importing the corresponding nonequilibrium carrier production distribution under a specific energy deposition spectrum into the Sentaurus Tcad device simulation is a crucial factor that affects the accuracy of the energy spectrum simulation of SiC nuclear radiation detectors. This process involves transforming the energy deposition spectrum and pulse height spectrum. This text provides a concise introduction to the many models of nonequilibrium carrier creation available in the Sentaurus Tcad device simulation software. It aims to assist users in selecting the most appropriate model for their individual simulation investigations.

(1) Constant carrier generation model:

This model is the most basic method for calculating the continuous generation of carriers, and it can be applied universally to semiconductor devices or tailored to specific regions or materials.

(2) γ-irradiation generation model:

The user can specify the total dose (in rad) or dose rate (in rad/s) of the irradiation, the duration of the irradiation, and the standard deviation of the Gaussian rising and falling edges.

(3) α-particle model:

Carrier generation induced by α particle irradiation is applicable solely for transient simulations. The software incorporates the count of electron–hole pairs generated prior to the starting transient period into the initial carrier density at the start of the simulation. The model may be configured with the energy, position, direction, and angle of the incoming particles, and the energy deposition along the trajectory aligns with the Bragg curve.

(4) Heavy ion model:

Upon entering a device, a heavy ion undergoes energy loss and produces electron–hole pairs. The key variables associated with the heavy ion model include the particle's energy and type, the angle at which the particle enters the device (with respect to the plane of the particle at the moment of incidence on the SiC), and the correlation between the particle's energy loss within the device (linear energy transfer) and the quantity of electron–hole pairs produced. In contrast to the α-particle model, the heavy ion model allows users to customize the energy deposition distribution along the trajectory. This makes it more flexible than the α-particle model. Consequently, when studying the incidence of charged particles, the statistical distribution of single-particle energy deposition obtained from MC simulation can be imported into the Sentaurus Tcad software with a higher degree of compatibility.

The subsequent section will provide a more comprehensive explanation of the heavy ion model, specifically focusing on the computation of the pulse amplitude spectrum of the particle irradiation detector. Due to the complexity of using models in Tcad software simulation, it is necessary for the user to consult the software's user manual for detailed instructions on settings. This book provides a brief overview of commonly used device simulation models and references additional models that may be relevant. In addition, the energy deposition in semiconductors can be more complicated due to particle irradiation, such as the energy deposition spectrum of D-T neutrons (14 MeV) directly incident on a 4H–SiC material as shown in Fig. 1. 4. Currently, it is highly challenging to incorporate the energy deposition spectra into the Sentaurus Tcad software for the purpose of simulating the pulse amplitude spectra. Simultaneously, it is evident that the pulse amplitude spectra of the two nuclear reactions, $^{12}C(n, \alpha)^{9}Be$ and $^{28}Si(n, p)^{25}$ Mg, closely correspond to the peak locations and counts in the energy deposition spectra obtained from the simulation. Consequently, the simulation of the energy deposition spectra of the incident D-T neutrons can partially capture the characteristics of the pulse amplitude spectra of the real detectors. To determine the energy of the incident neutrons and analyze the neutron energy spectrum, SiC nuclear radiation detectors are commonly used. These detectors are particularly effective in detecting fast neutrons involved in two series of nuclear reactions: $^{28}Si(n, p)^{25}$ Mg and $^{12}C(n, \alpha)^{9}Be$ [7]. In summary, the energy deposition spectrum simulation and the subsequent simulation of the pulse amplitude spectrum can be utilized to investigate the characteristics of the detector output. Additionally, the pulse amplitude spectrum simulation using Sentaurus Tcad software considers the detector device's operational state and any performance variations, making it more intricate. Furthermore, the Sentaurus Tcad software simulation is essential for optimizing the design of the detector diode device structure, in addition to simulating the pulse amplitude spectrum.

### *5.1.3 Principle of Calculating the Pulse Amplitude Spectrum of the Particle Irradiation Detector*

For the case of charged particles incident from the outside of the detector as shown in Fig. 5.4, the flow of the particle irradiation detector energy spectrum simulation is shown in Fig. 5.5.

Let us assume that each output pulse of the detector corresponds to a single incident particle. The pulse output is primarily determined by three factors:

(1) The angle $\theta$ and energy $E$ of the particle at the time of incidence;
(2) The operational state of the detector, particularly the amount of the bias voltage;
(3) The structural parameters and material qualities of the detector.

The aforementioned three points have been established.

The pulse amplitude spectrum is derived by quantifying the voltage amplitude $V$ of the pulse generated by the particle upon reaching the detector. To accomplish this, it is imperative to initially determine the energy $E$ and the angle of incidence $\theta$ at which the particle approaches the interface between the front contact metal and the SiC material (as indicated by the red dots in Fig. 5.4). When a charged particle approaches, in the case where there is only one type of charged particle with a specific incident energy, we can generate a random number $\theta rand$ between 0 and $2\pi$ to represent the angle of incidence $\theta$. The incident energy $E$ of the particle is directly related to $\theta$ based on the thickness of the metal layer on the device's surface. In situations involving multiple particles and a complex energy composition, it is essential to employ Monte Carlo (MC) software to generate the desired particles randomly. Additionally, a relevant module should be utilized to calculate the distribution of energy ($E$) and angle ($\theta$) of the incident particles upon reaching the interface between the metal on the SiC surface and the SiC body material, as depicted in Fig. 5.6. Subsequently, the Monte Carlo method is employed to generate a pair of random numbers ($Erand$, $\theta rand$) that represent the energy and angle of the charged particles upon their interaction with the SiC material, based on the distribution of energy $E$ and angle $\theta$.

Performing one device and circuit simulation for each ($Erand$, $\theta rand$) pair is not feasible due to the large number of pairs generated. Thus, it is postulated in this context that the ultimate magnitude of the output voltage V of the detector undergoes continuous variation with respect to $E$ and $\theta$. Subsequently, once a specific number of ($E$, $\theta$) pairs and their accompanying $V$ values are determined, the output voltage amplitude Vrand for any given ($Erand$, $\theta rand$) can be calculated using interpolation. Each $Vrand$ count corresponds to one valid particle incidence in the experiment. Repeating the above process in large numbers yields the desired pulse amplitude spectrum, as shown in Fig. 5.7.

The values of ($E$, $\theta$) can be adjusted artificially to achieve the desired level of computation precision. To calculate $V$ for a given ($E$, $\theta$) pair, as per the operating principle of the energy spectrum measurement system outlined in Sect. 2.5.2, it is essential to have knowledge of the pulse current $I(t)$ produced by the charged particle incident detector at that moment. Additionally, the voltage pulse $V(t)$ generated by the input of pulse current $I(t)$ into the subsequent signal processing system, along with the magnitude of $V(t)$, which is the required $V$ in this context, must also be known.

To calculate $I(t)$, transient current calculations must be conducted using the SDEVICE module in Sentaurus. In order to determine the impulse current produced by a charged particle striking the detector, it is essential to incorporate a Heavy Particle (Heavy Ions) model with the fundamental model that will be discussed in Chap. 4.

Figure 5.8 in Sentaurus displays the occurrence of a particle with significant mass. This particle's behavior is determined by several model parameters, resulting in a straight trajectory through the material in a consistent direction. Within the range of $L \sim L + \Delta L$, the movement of the particle results in the production of a total charge of $Q$. The charges are dispersed in the plane perpendicular to the trajectory following either Gaussian or exponential distributions, with a characteristic radius $R$. When the particle reaches the detector at time $t$, Sentaurus uses the given parameters to calculate the distribution of extra excited carriers in the device. This resulting excess carrier distribution is then considered as part of the device's initial state and is directly included in the simulation calculation. Computations for simulating a process or system.

The SRIM software can be used to analyze the two-dimensional energy deposition and acquire the necessary parameters for describing heavy particles. It is important to mention that in the actual scenario, the characteristic radius *R* is a parameter in a three-dimensional space. However, in order to simplify the simulation calculations, a two-dimensional simulation model will be employed in the Sentaurus simulation. The two-dimensional energy deposition analysis provided by the SRIM software accurately represents the projection of the energy deposition in a three-dimensional space onto a two-dimensional plane. Furthermore, previous investigations [8, 9] have focused on the characteristic radius *R* using Si materials, yielding characteristic radii at the nanoscale level. The small size of the nanoscale features in the Sentaurus simulation results in a high concentration of excited carriers within a small area and a short period of time. This leads to a significant increase in the number of simulation meshes. Additionally, the mesh division is not uniform and is highly concentrated in certain areas, which negatively impacts the simulation's efficiency and the convergence of the iterative process. Hence, we choose to derive the distribution of *R*-values along the particle's track by statistically averaging the energy deposits produced around the trajectory when a substantial number of particles are directed into SRIM.

Using the SRIM software, we simulated the energy deposition of a 2.73 MeV $^3H$ particle as it vertically hits the SiC target. The computed result is displayed in Fig. 5.9a. The left side of the figure provides a direct representation of the relationship between the trail position (*L*) and the charge (*Q*). On the other hand, the shaded area on the bottom side represents the two-dimensional projection of the average energy deposition resulting from the incident of a large number of particles. During the SRIM simulation, the software divides the target into multiple segments and computes the energy deposition within each segment when a particle passes through it. This allows us to determine the energy deposition $E \perp(L)$ of the particle in each segment that is perpendicular to the direction *L*, near the position of the particle's trajectory. The energy deposition is determined by employing either an exponential or Gaussian fit. In this case, we note that the trend of $E \perp(L)$ aligns more closely with a Gaussian distribution, and hence a Gaussian fit is employed. Once the desired distribution of excess carriers is chosen, a Gaussian or exponential fit is employed to determine the characteristic radius *R* that corresponds to the point *L* of the observed trend $E \perp(L)$. In this case, a Gaussian fit is used as the trend closely resembles a Gaussian distribution. Figure 5.9b displays the ultimate retrieval of *Q* and the change in *R* as a function of *L*.

Furthermore, it is worth mentioning that by examining the clearly deviated track of particle energy deposition in Fig. 5.9a, it becomes evident that, in the SRIM simulation, the energy of each individual incoming particle is predominantly concentrated in the block it traverses. Although the SRIM software does not provide a specific definition for the characteristic radius *R*, the simulation results indicate that when the material scale is reduced, the energy from a single incident particle is still only deposited in the block that the particle passes through. It is important to note that the size of the block in SRIM is fixed in proportion to the material size. Consequently, the SRIM software does not consider the characteristic radius $R$, and it is not possible to determine the feature radius $R$ based on the energy deposition of a single incident particle in SRIM.

Once the aforementioned parameters are acquired, Sentaurus can be utilized to compute the pulse current $I(t)$ produced by a charged particle when it strikes the detector at various energies $E$ and angles $\theta$, while operating under a specific bias. This calculation can be performed using Fig. 4.4 as a reference, and the results are displayed in Fig. 5.10. The particle depicted in the picture is a $^{3}H$ particle with an energy of 2.73 MeV, while the device is being operated in a reverse biased state with a voltage of 200 V. The output current is suitably adjusted on the time axis to facilitate visualization. Modifications to the device structure, material properties, and model parameters will result in corresponding changes in the obtained $I(t)$, which will finally manifest as a variation in $V$. Figure 5.10 displays the current output.

Once the pulse current $I(t)$ has been acquired, it is imperative to calculate the associated voltage pulse $V(t)$. For instance, the Multisim software is employed to create a simulation of the circuit [10, 11] depicted in Fig. 5.3. In the diagram, the detector is simplified to a pulsed current source (Current Source) connected in parallel with the detector reverse bias capacitor $C_{\text{detector}}$. The input, $I(t)$, is fed into the amplifier A250, which is a charge sensitive amplifier explained in Sect. 2.5.2. The integrating capacitance, $C_f$ is used. $R1$-$C1$ acts as a high pass filter to speed up the decay of the integrated signal. The amplifier A275 and the LRC loop are then used for signal amplification and low-pass filtering. Finally, the output OUT generates a pulse voltage signal, $V(t)$, as shown in Fig. 5.11a. The required variation of $V$ with $E$ and $\theta$ can be produced by performing the foregoing computations for all selected ($E$, $\theta$) pairs and using the resulting $I(t)$ values to obtain $V(t)$ and extract the peak $V$. This is illustrated in Fig. 5.11b. The particle depicted in the image is a 3H particle with an energy of 2.73 MeV, and the device is subjected to a reverse bias of 0 V.

It is worth mentioning that the conventional forming circuit depicted in Fig. 5.3 includes an isolation capacitor, $C_n$, positioned between the front end of the amplification link and the bias circuit of the detector. This capacitor is connected in parallel with the reverse bias capacitance of the detector, known as $C_{\text{detector}}$, which changes

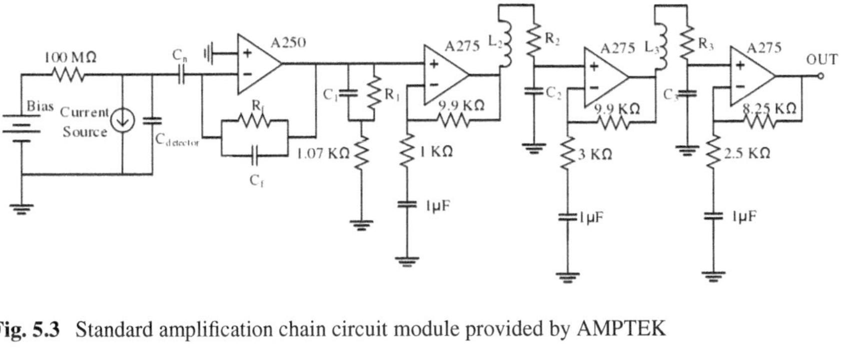


**Fig. 5.3** Standard amplification chain circuit module provided by AMPTEK

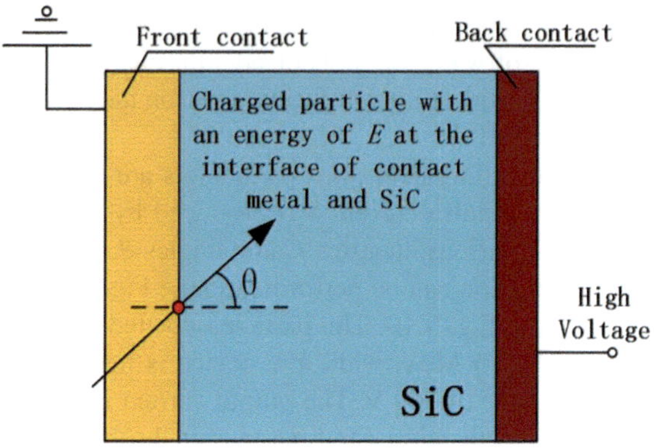


**Fig. 5.4** Schematic diagram of a charged particle carrying energy E and incident on the detector at an angle θ

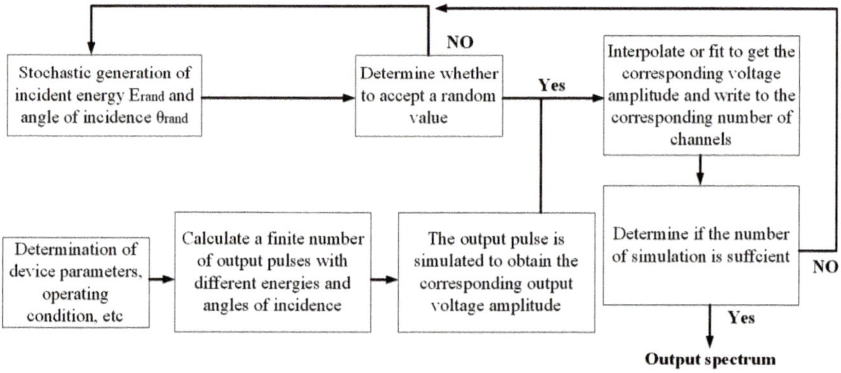


**Fig. 5.5** Simulation flowchart of the detector pulse amplitude spectrum

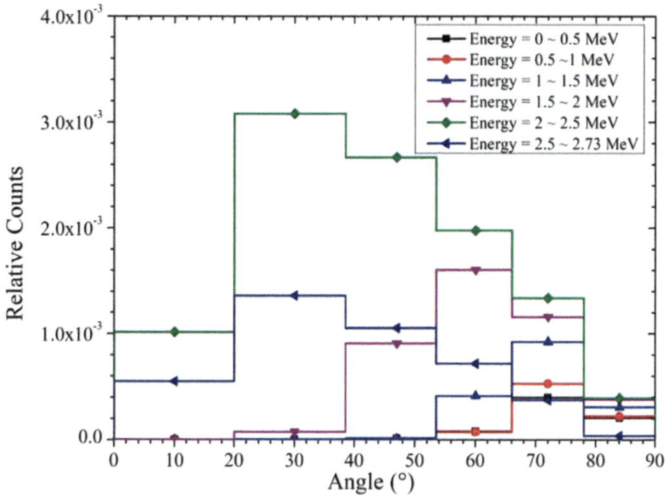


**Fig. 5.6** Plot of the relative number of incident charged particles in different energy-angle intervals

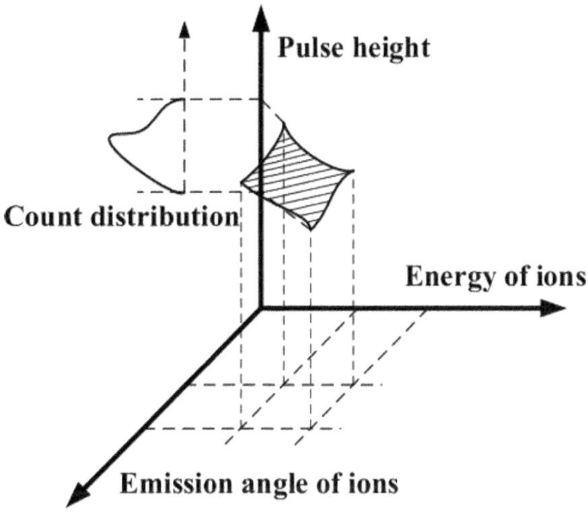


**Fig. 5.7** Schematic diagram of the pulse height spectrum calculation process

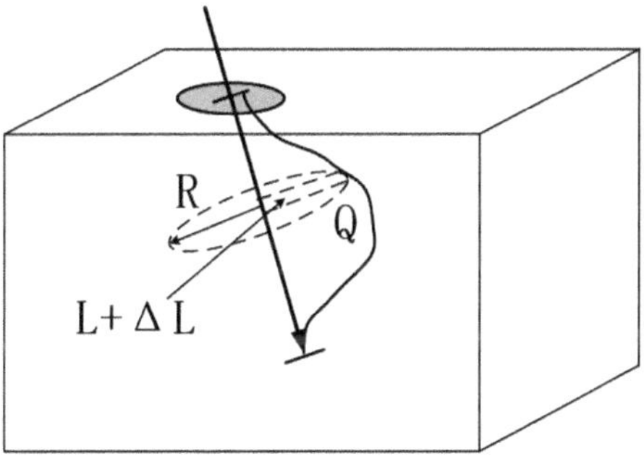


**Fig. 5.8** Descriptive parameters of a heavy particle incident semiconductor device

depending on the reverse bias voltage of the device. Hence, in order to determine the change in the pulse amplitude spectrum in relation to the device's bias, particularly when the size of $C_{\text{detector}}$ is similar to that of $C_{\text{n}}$, it is crucial to describe the dimensions of the aforementioned capacitors. Furthermore, this analysis will not consider the impact of circuit configuration and amplifier parameters, particularly frequency characteristics, on the calculation results. The focus will solely be on adjusting the component sizes in the amplification link based on literature references to determine the output pulse peaking time, $T_{\text{p}}$. The amplification connection is thereafter fine-tuned to match the values of the corresponding components. To achieve the necessary pcaking time, the capacitance and inductance sizes in Fig. 5.3 are determined by the following relationships with $T_{\text{p}}$ measured in μs: $C1 = 360$ pF $\times T_p$; $C2 = 592$ pF $\times T_p$ $L2 = 150$ μH $\times T_p$; $C3 = 564$ pF $\times T_p$ $L3 = 82$ μH $\times T_p$。

The output pulse *Vrand*, which corresponds to the values (*Erand*, *θrand*), is determined and quantified using interpolation on a surface where the amplitude of the voltage pulse $V$ varies with both the energy $E$ and the angle $\theta$ (as shown in Fig. 5.11b). Currently, each *Vrand* count corresponds to a single occurrence of

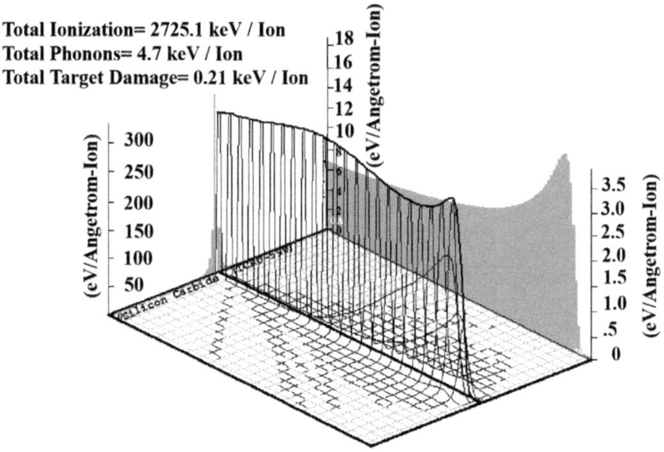


(a) Simulation results given by SRIM for vertical particle incidence

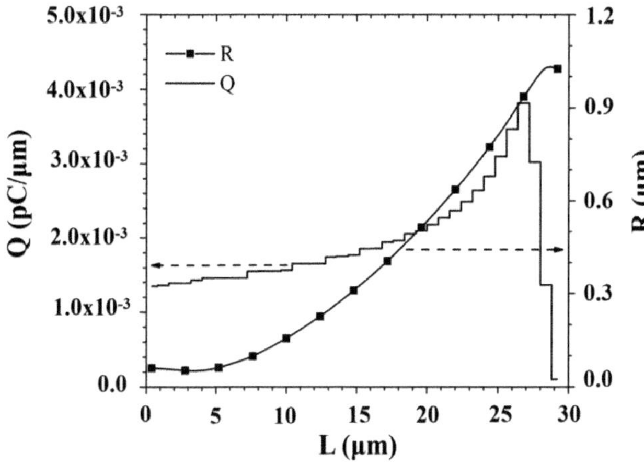


(b) Extraction of parameters to obtain the variation of $Q$ and $R$ with $L$

**Fig. 5.9** Extraction of heavy particle parameters required for the simulation using SRIM, with incident particles of 2.73 MeV$^3$ H particles

a genuine charged particle in the experiment. To get the appropriate pulse amplitude spectrum, the aforementioned procedure is iterated for a significant quantity of randomly produced (*Erand*, *θrand*) pairs.

The computational methods mentioned above can also be used to situations where charged particles are produced and travel within the detector, as depicted in Fig. 1. 4. However, additional parameters are necessary to describe the formation of a pulse in such cases. Due to the involvement of nuclear reactions, the location and motion direction of charged particles generated inside the detector are often random and variable. Additionally, the generation of multiple charged particles must be considered,

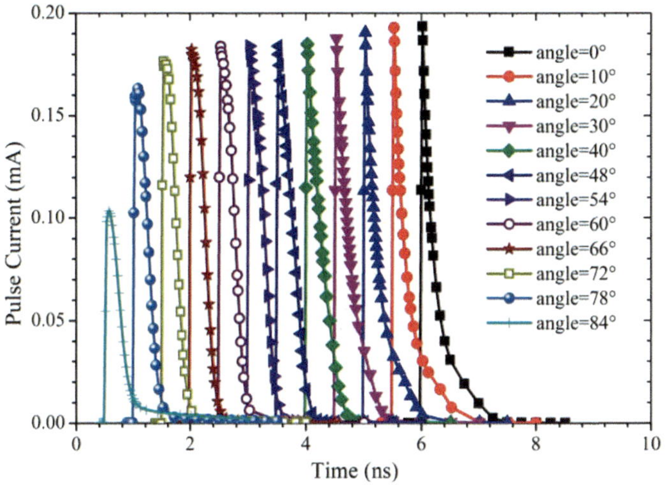


**Fig. 5.10** Calculation of the pulse output current $I(t)$ under different conditions using Sentaurus

which complicates the required pulse amplitude space for interpolation. Furthermore, if the motion direction of different particles is not in the same plane, a three-dimensional model is necessary for calculating $I(t)$ in the Sentaurus simulation. This significantly increases the simulation time and cost. Due to the aforementioned factors, there is a significant and rapid increase in the time required for simulation. Therefore, only the straightforward scenario of a single particle entering the detector from the outside is taken into account. For more complex scenarios, it is advised to use the energy deposition spectrum as an approximation for the actual pulse amplitude spectrum.

### 5.1.4 *Optimized Design at the Device Level [12–14]*

1. The selection of epitaxial layer thickness and working voltage

Using the α-particle SiC nuclear radiation detector as a case study, we can observe the software-generated cross-section in Fig. 5.12. This diagram displays the range distribution of charged α-particles in SiC material with an energy of 5.05 MeV, as simulated by the SRIM software. The simulation reveals that the statistical range of α-particles with this energy is 16.6 μm. Figure 5.13 illustrates the primary energies of α-particles, measuring 5.156 and 5.486 MeV, as they enter the 4H–SiC material in a vertical direction. The amount of energy lost by the α-particles depends on the path they take upon entering. The highest distances traveled by α-particles when they deposit all of their energy are 17.25 μm and 19.30 μm, respectively. The curves showing the energy deposition align with the Bragg curves of the α-particles depicted in Fig. 2.1.

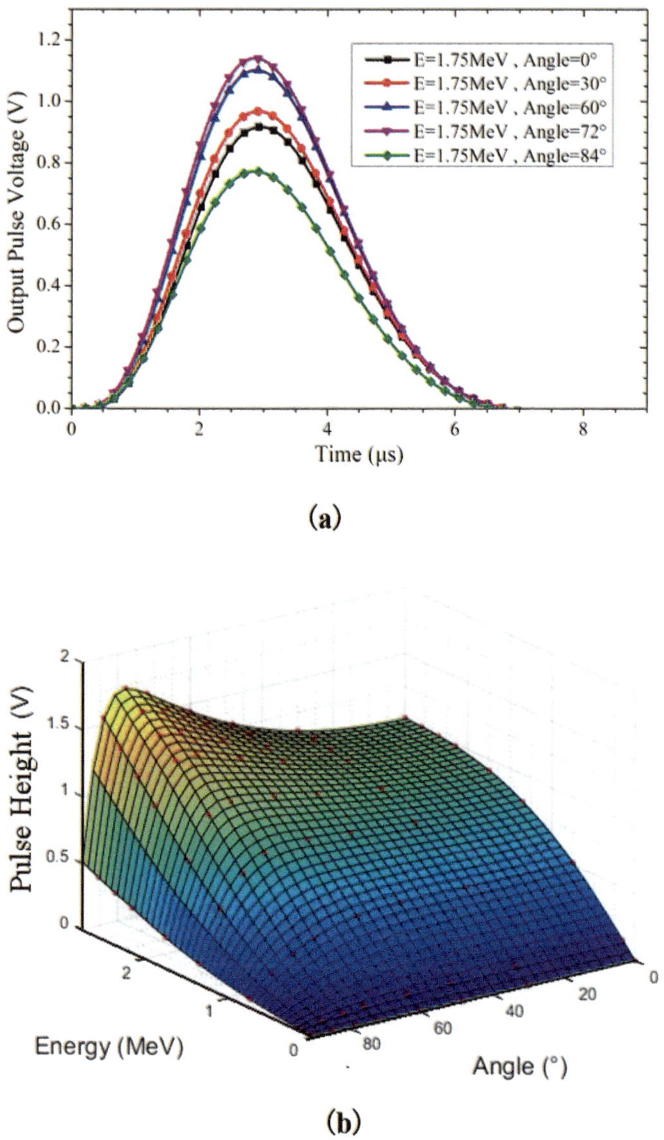


**Fig. 5.11** Output pulse voltage and its amplitude extraction. **a** Output pulse voltage at the OUT end in Fig. 5.3 for different incidence conditions, Tp = 3 μs; **b** variation of pulse voltage amplitude $V$ with incident particle energy and angle

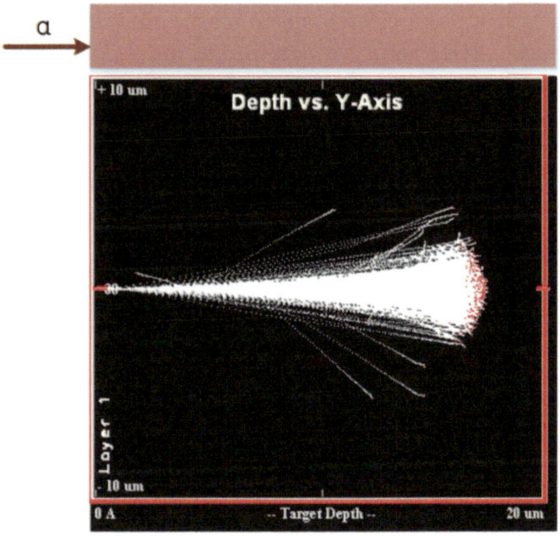


**Fig. 5.12** Range of α-particles of 5.05 MeV in SiC material

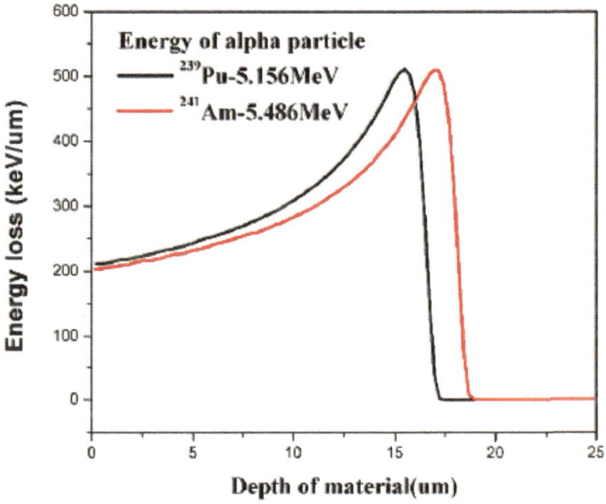


**Fig. 5.13** 5.156 and 5.486 MeV α-particle Energy deposition in 4H–SiC materials

To enhance the detector's effectiveness, it is crucial to guarantee that the entire energy of the α-particle is deposited within the sensitive zone. In order to meet the requirements for detecting nuclear radiation particles, it is necessary to ensure that the thickness of the high-quality SiC epitaxial layer, which acts as the sensitive region of the detector, is greater than or as close as possible to the range of the α-particles. This should be done within the limits allowed by the process conditions. Additionally, to enhance the efficiency of charge collection in the detector, it is necessary to design the thickness of the depletion region in the epitaxial layer in a way that facilitates the quick separation of electron–hole pairs generated by ionization of incident particles. The PN junction diode, namely the epitaxial layer in the N-type area, does not experience any significant effects until it reaches full depletion. According to Eq. (5.22), the extent of depletion is determined by the $N_A X_A$, with a smaller value of $N_D$ resulting in a greater value of $X_D$. To minimize reverse leakage, the device should achieve the desired thickness of the depletion region at the lowest reverse bias voltage possible, or fully deplete the epitaxial layer. This is important because the thickness of the epitaxial layer is typically designed to match the maximum range of incident particles. Hence, it is necessary to minimize the doping concentration of the epitaxial layer. Based on the current process level, the doping concentration of the epitaxial layer can be reduced to the order of magnitude of $10^{13}$ cm$^{-3}$.

$$N_A X_A = N_D X_D \tag{5.22}$$

In the given equation, $N_A$ represents the doping concentration and $X_A$ represents the depletion thickness of the P-type area. Similarly, $N_D$ represents the doping concentration and $X_D$ represents the depletion thickness of the epitaxial layer. The value of the reverse bias voltage that causes full depletion of the epitaxial layer at a specific epitaxial layer doping concentration can be calculated by combining Eq. (5.23) for the built-in potential of a non-simplex semiconductor and Eq. (5.24) for the depletion layer width of a unilateral mutant junction under an applied reverse bias voltage. Figure 5.14 illustrates the correlation between the thickness of the depletion layer and the reverse voltage for Schottky junctions with varying doping concentrations, as determined from calculations in the literature using TCAD software. Additionally, Eq. (5.24) is also valid for Schottky junctions.

$$\varphi_{bi} = \frac{kT}{q}\ln\frac{N_A N_D}{n_i^2} \tag{5.23}$$

$$W_D = \sqrt{\frac{2\varepsilon_s\left(\varphi_{bi} - V - \frac{2kT}{q}\right)}{qN}} \tag{5.24}$$

The symbol φbi represents the inherent potential of the PN junction, $n_i$ denotes the intrinsic concentration of carriers in the 4H–SiC device at room temperature, and $\varepsilon_s$ represents the dielectric constant of the 4H–SiC material. Furthermore, once the epitaxial layer is fully depleted, increasing the reverse bias voltage allows for a greater influx of carriers, resulting in an enhanced drift rate. Consequently, this leads to an improved reaction time of the detector. Figure 5.15 illustrates the drift velocity of carriers at various reverse bias voltages, as obtained through sentaurus simulation. It is evident that the drift velocity of carriers at lower voltages experiences a rapid increase as the reverse bias voltage increases. Additionally, the drift velocity of holes is significantly lower than that of electrons at low voltages, accurately indicating that the low-field mobility of holes is considerably inferior to that of electrons. As the reverse bias voltage increases, the high-field mobility decreases, resulting in a stronger effect on the electron. This causes the hole drift velocity in the high field to gradually approach that of the electron. For example, in Fig. 5.15, when the reverse

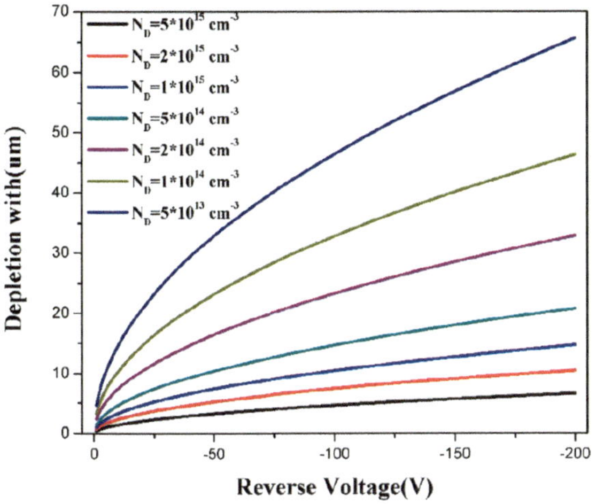


**Fig. 5.14** Variation of depletion layer thickness with reverse voltage for different doping concentration conditions [14]

bias voltage is 800 V, the electron and hole drift velocities are approximately 1.6 $\times 10^7$ cm/s, which is about 80% of the saturation drift velocity. If the reverse bias voltage continues to increase until it reaches the avalanche breakdown point (2600 V), the carrier drift velocity will increase slowly. The velocity of carrier drift exhibits a gradual increase. By examining the lower right graph in Fig. 5.15, it is evident that when a reverse bias voltage of 800 V is applied, the epitaxial layer beneath the main junction of the device becomes fully depleted. This condition satisfies the operational criteria of the detector and promotes efficient separation and collection of the electron–hole pairs generated by the incident radiation. Additionally, the electron and hole drift velocities in this state are both high and comparable.

2. Junction capacitance and intrinsic reverse leakage current control [15]

The reference presents a schematic picture, displayed in Fig. 5.16, illustrating the device structure used for simulating and optimizing the electrical performance of a 4H–SiC SBD device for radiation detection applications [15]. The detector structure consists of a substrate made of 4H–SiC, which has a thickness of approximately 370 μm and a doping concentration of $5 \times 10^{18}$ cm$^{-3}$. It also includes a 0.5 μm thick 4H–SiC buffer layer with a doping concentration of $1 \times 10^{18}$ cm$^{-3}$. The epitaxial layer, with a thickness of 21 μm, completes the structure. The device structure of the 4H–SiC Schottky diode α-particle detector consists of Schottky contact electrodes on the 4H–SiC epitaxial layer and ohmic contact electrodes on the strongly doped side of the substrate.

The electrical characteristics of the device are influenced by the doping concentration of the epitaxial layer of the detector. In the literature, the doping concentration of the epitaxial layer of a 4H–SiC Schottky diode nuclear radiation detector has been simulated using five sequential values: $1 \times 10^{13}$ cm$^{-3}$, $5 \times 10^{13}$ cm$^{-3}$, $1 \times 10^{14}$ cm$^{-3}$, $5 \times 10^{14}$ cm$^{-3}$, $1 \times 10^{15}$ cm$^{-3}$. A reverse bias voltage of – 200 V was applied in the simulation. The I-V data obtained from the experiment is displayed in Fig. 5.17. It

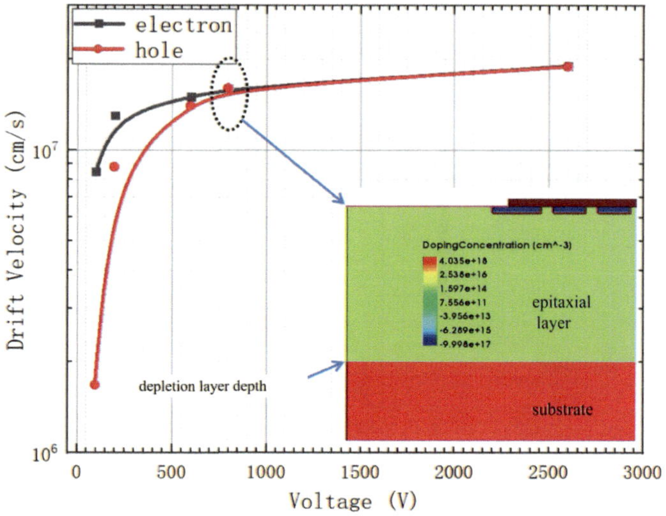


**Fig. 5.15** Carrier drift rate in depletion region

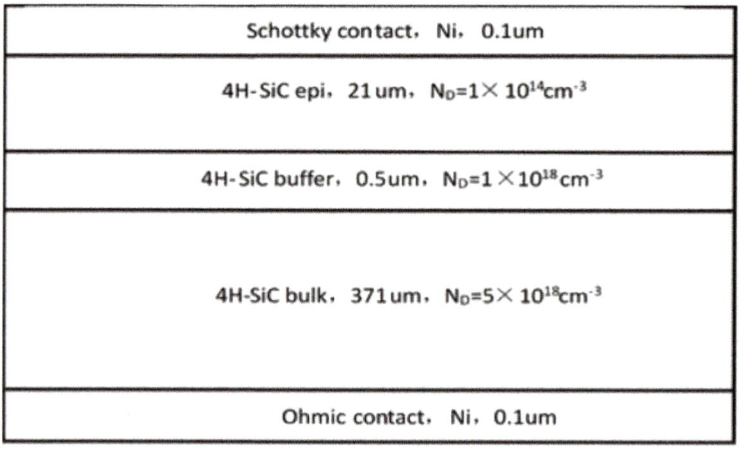


**Fig. 5.16** Schematic diagram of 4H–SiC Schottky diode detector structure

is evident that when the doping concentration increases, the reverse leakage current also increases at the same voltage. An elevation in reverse leakage current will result in a wider energy resolution of the detector, hence impacting the device's performance. Consequently, reducing the doping concentration of the epitaxial layer leads to a decrease in reverse leakage current and an improvement in energy resolution of the device at high voltage.

Now, we are taking into account the impact of junction capacitance on the design of the detector. Figure 5.18 displays a schematic illustration illustrating the functioning

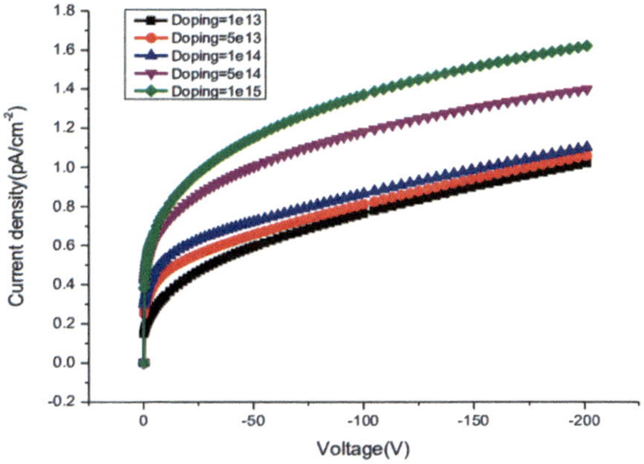


**Fig. 5.17** Comparison of inverse I-V with different epitaxial layer doping concentrations [15]

of a semiconductor detector. The amplitude u of the output pulse signal produced by the ray is directly proportional to the amount of energy that the incoming ray loses in the sensitive area of the detector.

$$u = \frac{Eq}{\omega C} = \frac{Q}{C} \tag{5.25}$$

The variables in the equation are as follows: $E$ represents the energy of the incident ray, $q$ represents the electron charge, $\omega$ represents the average energy needed to generate one electron–hole pair, $Q$ represents the total ionization charge produced by the incident ray in the depletion layer, and $C$ represents the sum of the detector's junction capacitance (or barrier capacitance), $C_d$, and the stray capacitance, $C_s$. It is worth noting that the value of the stray capacitance, $C_s$, is considered insignificant,

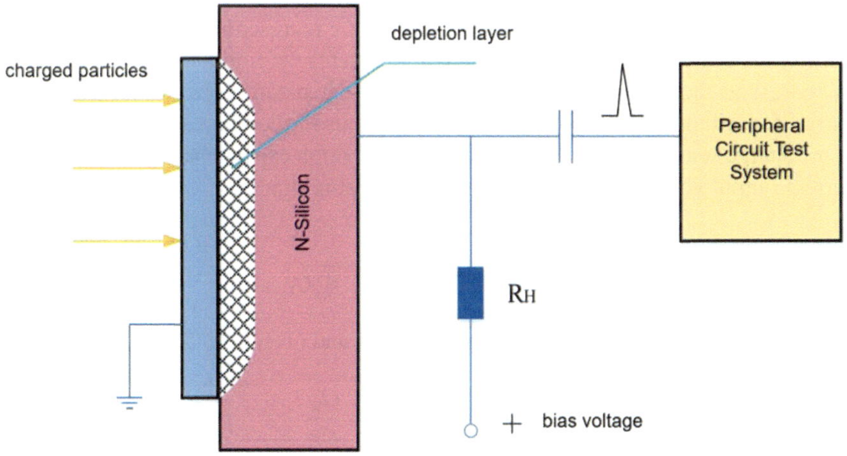


**Fig. 5.18** Operating principle of the semiconductor detector [16]

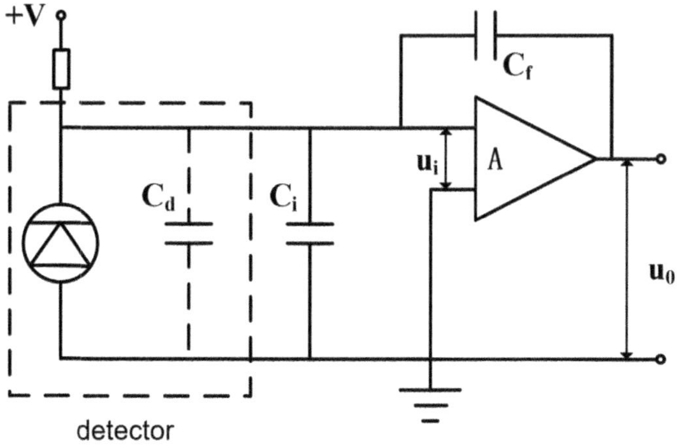


**Fig. 5.19** Equivalent circuit diagram of the detector and charge-sensitive preamplifier coupling

and the detector capacitance, $C$, is approximately equal to the barrier capacitance, $C_\mathrm{d}$.

The amplified output pulse from the semiconductor detector will be processed using a charge-sensitive preamplifier. The circuit diagram of this preamplifier is depicted in Fig. 5.19. The incident ray in the sensitive zone of the detector generates a total ionization charge, denoted as $Q$. A fraction of the charge $Q_0$ is captured by the charge-sensitive preamplifier, while the remaining charge $Q_\mathrm{d}$ stays in the detector, with a collection efficiency denoted as $\eta$:

$$\eta = \frac{Q_0}{Q} = \frac{Q_0}{Q_0 + Q_\mathrm{d}} = \frac{(1 + A)C_\mathrm{f}}{C_\mathrm{d} + C_\mathrm{i} + (1 + A)C_\mathrm{f}} \tag{5.26}$$

The variables in the equation are as follows: $C_\mathrm{f}$ represents the feedback capacitance of the charge sensitive preamplifier, $A$ represents the amplifier's open loop gain, $Q_0$ represents the input charge of the charge sensitive preamplifier, $C_\mathrm{d}$ represents the barrier capacitance of the detector, and $C_\mathrm{i}$ represents the combined input capacitance of the charge sensitive preamplifier and the distribution capacitance of the input. The current output voltage $u_0$ of the charge sensitive amplifier is:

$$u_0 = \frac{AQ_0}{C_\mathrm{d} + C_\mathrm{i} + (1 + A)C_\mathrm{f}} \tag{5.27}$$

When the amplifier has open-loop gain $A \gg 1$ and $(1 + A)C_\mathrm{f} \gg (C_\mathrm{d} + C_\mathrm{i})$, then:

$$u_0 = \frac{AQ_0}{(1 + A)C_\mathrm{f}} \approx \frac{Q_0}{C_\mathrm{f}} \tag{5.28}$$

Equation (5.27) states that when capacitance increases at a constant voltage, the collection efficiency of the detector decreases. This causes an increase in the amount of charge $Q_d$ retained inside the detector and a decrease in the input charge $Q_0$ in the charge-sensitive preamplifier. As a result, the output pulse amplitude of the preamplifier decreases, causing a shift in the peak of the output pulse amplitude spectrum towards the low-channel address. Furthermore, the augmentation of $C_d$ will impair the temporal responsiveness of the detection system.

The C-V characteristics of the device epitaxial layer for various doping concentrations have been simulated in literature [15]. The corresponding data is presented in Fig. 5.20. Figure 5.20 illustrates that when a reverse bias voltage of − 60V is applied, the capacitance of the 4H–SiC material decreases as the doping concentration increases. Specifically, a lower doping concentration results in a smaller capacitance under the same bias voltage. Additionally, for a certain concentration, the capacitance decreases as the reverse bias voltage increases. However, the capacitance remains constant once the reverse bias voltage reaches a certain level. Hence, the findings depicted in Figs. 5.17 and 5.20 indicate that reducing the doping concentration of the epitaxial layer in the detector leads to lower reverse leakage current and capacitance values. Consequently, this results in a higher ideal energy resolution for the detector.

Similarly, the TCAD software can be employed to examine the electrical properties of PiN-structured detectors. This analysis involves investigating various structural parameters, including the thickness and concentration of the P-region and the thickness and concentration of the I-region. The purpose is to determine the optimal structural parameters for the devices.

3. Control of non-ideal reverse leakage [12]

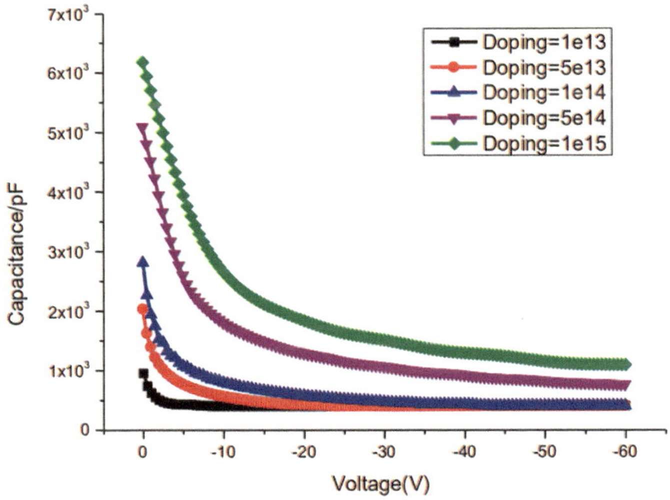


**Fig. 5.20** Comparison of C-V at different epitaxial layer doping concentration reverse bias

(a) Leakage mechanism of 4H–SiC-based diode devices

Addressing device reverse leakage is a crucial challenge in the development of SiC radiation detectors that operate at high reverse bias voltages. Device leakage typically encompasses two primary components: body leakage and surface leakage. Regarding diode-type devices, the body leakage adheres to the fundamental principles of PN junction and Schottky junction. Specifically, for a PN junction diode, the leakage current is the combination of reverse diffusion current and depletion region current. On the other hand, for a Schottky diode, the reverse leakage is influenced by the collective impact of thermionic emission, carrier diffusion, mirror force, and tunneling effect. The observed device leakage is significantly greater than the calculated theoretical value due to two main factors. Firstly, under reverse high-voltage conditions, the electric field distribution across the detector depletion layer is not uniform horizontally. This non-uniformity leads to a concentration of electric field lines at the junction region, known as the electric field concentration effect. Consequently, this effect increases the local reverse leakage current and decreases the breakdown voltage. Secondly, the presence of various types of defects (such as point defects, dislocations, and layer faults) in the actual device material further contributes to the increased leakage. However, the actual device material has many sorts of flaws such as point defects, dislocations, and layer defects. These faults not only cause extra body leakage by creating a composite center, but also introduce additional surface effects.

For SiC material, the ideal reverse current of a PN-type device is often very low, at $10^{-20}$ A/cm$^2$. However, in actual SiC devices, the reverse leakage current is often far larger than this theoretical value. This is mostly due to the unresolved crystal material imperfections in SiC material, which result in a pronounced leaking body effect and surface effect.

The body leakage in SiC devices is primarily influenced by the presence of material body flaws. The dislocation body effect is characterized by the distortion or rupture of lattice chemical bonds near the dislocation nucleus. This results in alterations in the electronic energy state or bandgap width in that region. When a reverse bias is applied, the high electric field triggers the activation of these electronic energy states, which are situated in deep energy level regions, leading to the formation of leakage points. Consequently, the reverse leakage is increased. Hence, effectively addressing the device body effect mostly relies on the methodology of managing the material imperfections. The body leakage effect can be deliberately mitigated using techniques such as substrate preparation and improved epitaxial growth.

Surface leakage is a significant factor in the overall leakage of actual SiC devices. It is caused by defects on the material surface, defects induced during the manufacturing process, and high-density SiC/$SiO_2$ interfacial states. Surface leakage has a greater impact on the reverse leakage of the device compared to defects within the device itself, making it more critical and detrimental. Currently, the main factor contributing to reverse leakage in SiC-based devices is surface leakage, which is particularly challenging to optimize due to the increased complexity of materials and processes. Many literature findings indicate that experimental curve fitting can be used to extract body leakage and surface leakage. The analysis often finds a significant proportion of surface leakage. The high surface leakage of the devices can be attributed to two primary factors: firstly, the presence of elevated surface states and surface damage resulting from the fabrication process, which is primarily influenced by process optimization; and secondly, the inherently high surface electric field of the device, which is closely linked to the structural design of the device. Hence, in order to minimize leakage in diode devices, device designers must optimize the device structure to decrease the elevated surface electric field.

In the case where surface effects dominate, the concentration of high electric field on the device surface becomes a significant cause of surface leakage. This leads to a substantial increase in reverse leakage current, premature breakdown, and potential damage to the passivation dielectric layer. Consequently, the device becomes inoperative, rendering neutron detection impossible. The detector primarily receives a signal from the particles that are exposed to radiation in the depletion region, causing carrier excitation. This leads to the creation of a finite inductive charge, which is then collected and amplified by the circuit system. However, the reverse leakage of the device is also collected, which affects the detector's response signal. This leakage causes an increase in the deviation between the detected signal amplitude and the actual signal amplitude. Detector applications necessitate higher amounts of surface leakage compared to power devices.

To prepare 4H–SiC-based junction detectors, the first step is to address the issue of electric field concentration in the diode device structure during the design phase. This is important in order to minimize surface leakage and body leakage resulting from high field intensity at specific points. By doing so, the intrinsic noise of the detector device can be reduced, and a reliable and counter-biased working electrical environment can be ensured. Nevertheless, the presence of surface defects and material flaws in the device is limited by the device preparation process and material quality. Consequently, it is challenging to establish an accurate defect model that accurately represents the leakage current of the actual device in device simulation. Therefore, in simulation studies, the common approach is to optimize the electric field distribution at the boundary using the terminal structure in order to reduce the leakage current.

(b) Terminal structure design of low leakage 4H–SiC-based diode devices

This book presents the utilization of TCAD software in designing terminal structures for SiC nuclear radiation devices, using PiN-type detectors based on 4H–SiC as a specific case study. In actual devices, their boundaries are always there, and there is consistently a varying degree of electric field concentration at the device boundaries. Figure 5.21 is a schematic figure depicting the electric field concentration effect. In this effect, the field strength within the detector body measures $9.58 \times 10^4$ V cm$^{-1}$, whereas at the corner of the etching stage, it reaches $3.0 \times 10^6$ V cm$^{-1}$, which is equivalent to the critical breakdown field strength of SiC. Currently, the device's bias voltage is at 483 V. To enhance the device's ability to withstand power and ensure its reliability, it is imperative to incorporate a terminal protection structure into the PiN. The device terminals consist of planar and mesa structures, which are the two

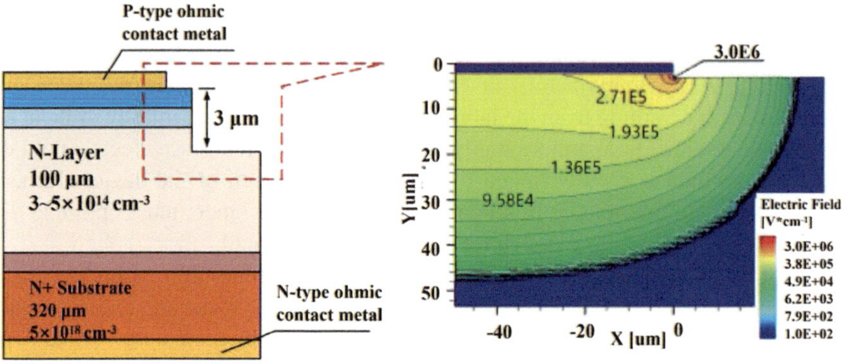


**Fig. 5.21** Schematic diagram of the electric field concentration phenomenon at the edge step of the device

primary structures of the device. The 4H–SiC material possesses high hardness, a small thermal diffusion coefficient, and stable chemical properties. Consequently, impurities cannot be selectively doped through diffusion effectively. Therefore, in the preparation of the planar structure of the 4H–SiC PiN device, the main junction P+ region can only be formed using high-energy ion implantation. However, this method of forming the P+ region involves high-energy ion implantation. Nevertheless, the primary intersection created during the process of high-energy ion implantation exhibits significant lattice damage and a low rate of ion annealing activation. Consequently, this results in a diminished conductivity modulation parameter for the device. Hence, PiN detectors based on 4H–SiC frequently employ Mesa-based PiN main structures, which efficiently regulate the maximum body electric field and surface electric field.

By optimizing the process of a single Mesa termination structure, the concentration of electric field at the surface corners is reduced. However, it is still not possible to completely eliminate the occurrence of a single point with a high electric field at the PN junction edge. By utilizing extended termination structures capable of enlarging the depletion zone, the single-point high-field issue occurring at the periphery of the primary junction can be alleviated, resulting in a significant reduction in leakage. Field Plate (FP) [18], Guard Ring (GR) [19], and Junction Termination Extension (JTE) [20, 21] are commonly used approaches for terminal protection. Each technique is depicted in Fig. 5.22. In order to enhance the sensitivity range and optimize the energy deposition and electrical signal output, SiC detectors require operation at a high reverse bias. The use of a thick passivation layer is necessary to accommodate the high surface electric field in the case of high reverse bias. Consequently, the thin oxygen-based field plate approach is unsuitable for the fabrication of high voltage detectors. The optimization of the protective ring approach involves various design factors, including ring spacing, ring width, and the number of rings. These parameters are intricate and require careful consideration. The distribution of the electric field and the ability to tolerate voltage are highly sensitive to the spacing of multiple rings around the main junction. Hence, because to the intricate nature of the optimization design, the tolerance of the process, and the enhancement of voltage withstand efficiency, this book selects JTE technology as a case study to illustrate terminal optimization design. Figure 5.23 displays the fundamental configuration and characteristics of the primary junction area of the device.

(c) Simulation of JTE terminal length

The size of the junction-terminated extended terminal structure directly impacts the maximum electric field peak of the device. To minimize the inherent noise of the detector, it is crucial to minimize the highest electric field peak of the device. The

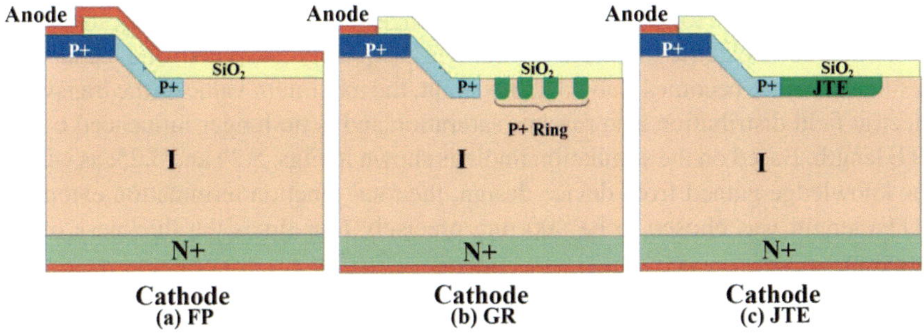


**Fig. 5.22** Schematic diagram of each terminal structure profile of PiN-based devices, where Anode denotes the anode and Cathode denotes the cathode [17]

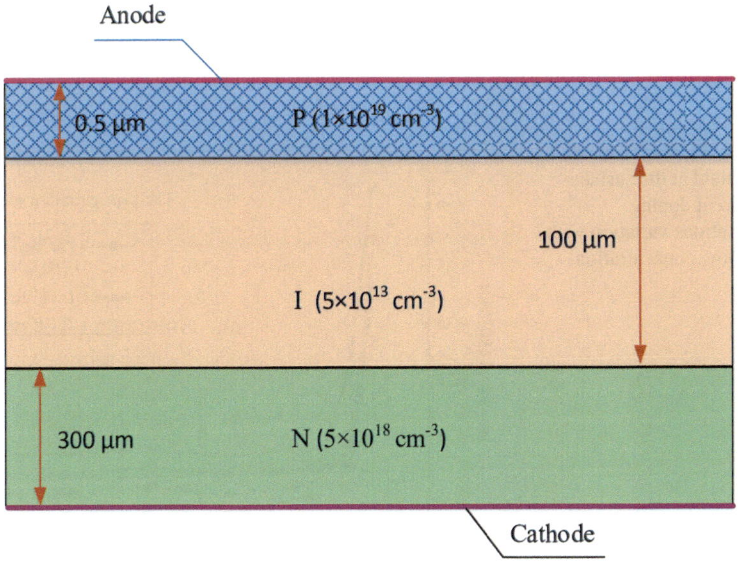


**Fig. 5.23** Basic structure and parameters of PiN devices

simulation investigates the relationship between the length of the junction-terminal extended terminal structure and the peak value of the transverse maximum electric field in single-region JTE (Single-JTE)-based PiNs (as shown in Fig. 5.22c) under various doping concentrations. Based on the simulation results presented in Figs. 5.24 and 5.25, it is evident that the trends of the device's surface maximum electric field peak (the highest value of the transverse electric field intercepted at the surface of the main junction) and the in vivo electric field peak (the highest value of the transverse electric field intercepted 0.5 μm below the surface of the main junction) remain consistent under different doping concentrations. Specifically, the peak electric field decreases as the length of the JTE increases, eventually reaching a saturation point. This is primarily due to the fact that the length of the Junction Termination Extension (JTE) can effectively mitigate the concentration of electric field at the edge of the main junction. However, once the JTE length exceeds a certain threshold, the phenomenon of electric field concentration is adequately alleviated, and the electric field distribution becomes stable. At this point, the maximum value of the transverse electric field distribution also reaches saturation and is no longer influenced by the JTE length. Based on the simulation findings shown in Figs. 5.24 and 5.25, as well as the knowledge gained from device design, the total junction termination extension (JTE) length was chosen to be 400 μm, precisely four times the thickness of the I-region.

(d) Optimization of JTE terminal doping concentration

The maximal electric field peaks on the surface and in the body of a Single-JTE-based PiN with JTE doping concentration under a reverse voltage of 500 V are presented, assuming a total length of JTE of 400 μm. Based on the simulation presented in Fig. 5.26, the trends of the peak surface maximum electric field and the peak maximum electric field in the body of PiN with Single-JTE with JTE doping concentration are similar. Both of them initially decrease and then increase as the

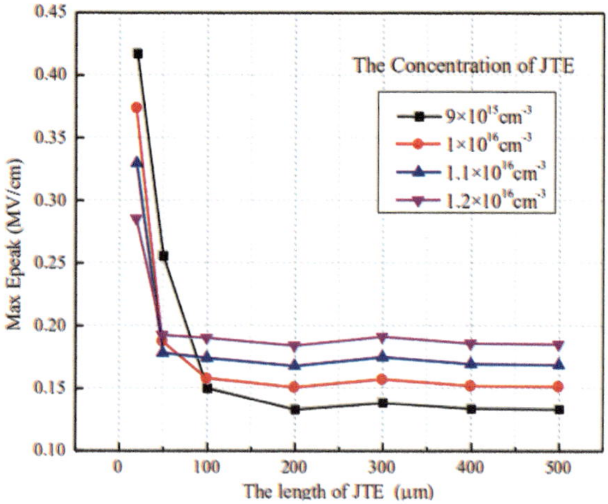


**Fig. 5.24** Peak maximum electric field at the surface for different doping concentrations variation with JTE doping concentration

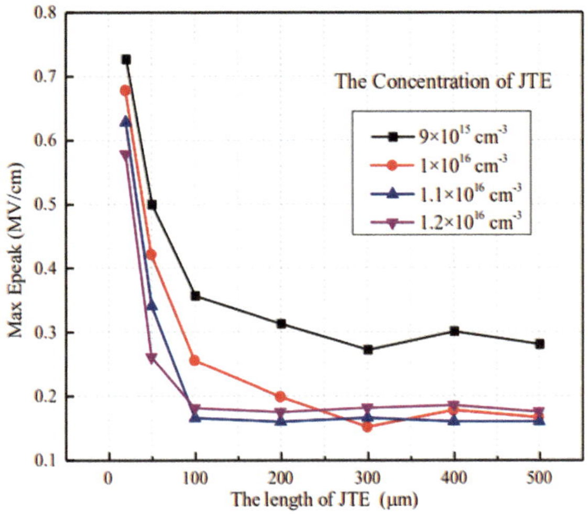


**Fig. 5.25** Maximum peak electric field in vivo for different doping concentrations variation with JTE doping concentration

JTE doping concentration increases. When the doping concentration is below the optimal level, the maximum value of the peak surface electric field is noticeably smaller than that in the body of the peak electric field. Similarly, when the doping concentration exceeds the optimal level, the peak surface electric field is smaller than that in the body of the peak electric field. When the doping concentration exceeds the ideal level, the peak surface electric field reaches its highest value at the same point as the peak electric field in the body. The ideal doping concentration of JTE is $1 \times 10^{16}$ $cm^{-3}$.

Additionally, the multi-zone Junction Termination Extension (JTE) is more efficient in enhancing device performance. The JTE has a total length of 400 μm, with the Double-JTE also having a total length of 400 μm. The JTE1, located close to the

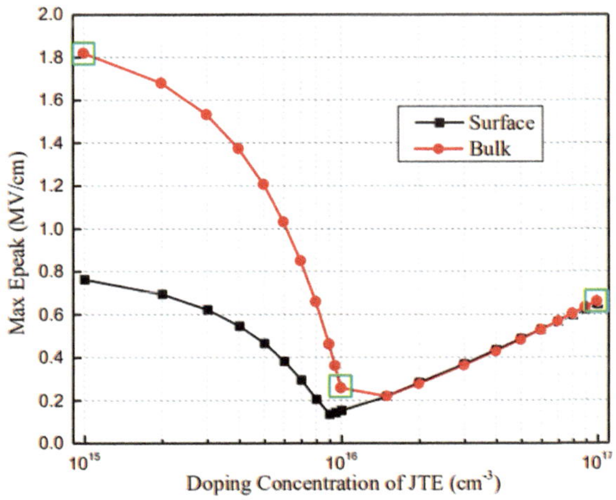


**Fig. 5.26** Variation of maximum electric field peak with JTE doping concentration in PiN with Single-JTE

main junction, and JTE2, located further from the main junction, both have lengths of 200 μm. This means that the JTE region, as depicted in Fig. 5.22c, is divided into two parts (left and right) with different doping concentrations but equal lengths. The simulation also considered the variations in the surface maximum electric field peak and the in vivo maximum electric field peak based on the JTE doping concentration, while maintaining a reverse voltage of 500 V on the base PiN. The JTE doping concentration mentioned refers to the doping concentration in the JTE1 region. In both circumstances, the doping concentration in the JTE2 region is 0.5 times that of the JTE1 region. Figures 5.27 and 5.28 provide a comparison of the performance benefits and drawbacks of PiN with Single-JTE and Double-JTE. It is evident that when the JTE doping concentration is below the optimal level, the maximum electric field peaks on the surface and in vivo of PiN with Single-JTE and Double-JTE coincide. Conversely, when the doping concentration exceeds the optimal level, the PiN with When the doping concentration exceeds the optimal level, the peak electric field strength on the surface and in the body of the PiN with Double-JTE are significantly lower than those of the PiN with Single-JTE. Consequently, the PiN with Double-JTE is more effective in reducing reverse leakage and minimizing detector noise.

The TCAD device simulation software is a robust tool for simulating semiconductor devices. In addition to the simulation study mentioned earlier, the software offers a comprehensive selection of physical models for users to utilize. Users can conduct a thorough simulation of device characteristics to meet the requirements of nuclear radiation detector applications. This will aid in the structural design and parameterization of SiC nuclear radiation detectors.

## 5.2 Response of SiC Nuclear Radiation Detectors to Charged Particles

### *5.2.1 Simulation Results of Charged Particle Irradiation Response Analysis [22]*

The $^{241}$Am source is widely utilized as a radioactive source, with the 5.486 MeV α-particles it emits extensively employed for testing the efficacy of detectors in pulse amplitude spectroscopy. As the particle penetrates deeper into the SiC material, it deposits more energy and generates a greater number of electron–hole pairs. This implies that particles with varying energies produce varying quantities of electron–hole pairs in the SiC, while particles with the same energy generate an equal number of electron–hole pairs when they strike the SiC at the same angle. Moreover, the quantity of electron–hole pairs gathered by the detector fluctuates in accordance with the thickness of the detector's sensitive region. Hence, the output characteristics of the detector are contingent upon various elements, such as the particles' angle of incidence, energy, and bias voltage. The SRIM software is employed to acquire the energy distribution of secondary particles in SiC.

To ensure the comparability of the simulation results with the experimental data, we employ the 4H–SiC Schottky diode detector described in reference [23] for our simulation investigation. Figure 5.27 displays the structure. The substrate used is N+ type 4H–SiC, which is doped with nitrogen at a concentration of $1 \times 10^{18}$ cm$^{-3}$. It has a thickness of 360 μm. On top of the substrate, there is a 13 μm-thick N-type epitaxial layer that is lightly doped with nitrogen at a concentration of $1.5 \times 10^{15}$ cm$^{-3}$. This layer is depleted to function as the main sensitivity region of the detector when a reverse bias voltage is applied. The cathode is a Ni-/4H–SiC Schottky contact with a Ni metal. The cathode has a thickness of 0.1 μm and a barrier height of approximately 1.75 eV. The anode consists of a 4H–SiC/Ni/Au ohmic contact, with a Ni metal thickness of 0.1 μm and an Au metal thickness of 6 μm.

Figure 5.28 displays the energy distribution resulting from a vertically incident α-particle with an energy of 5.486 MeV on SiC. Figure 5.28 illustrates that the ionization energy of α-particles in SiC rises as the incident distance increases. However, once it reaches a maximum value, the ionization energy rapidly declines to zero. The range of 5.486 MeV α-particles in SiC is approximately 18.26 μm, and the maximum distance they can travel before being incident is roughly 19 μm.

Figure 5.29 displays the current pulses produced by the detector when 5.486 MeV alpha particles are encountered at various angles, with a device bias of 20 V. In Fig. 5.29, the amplitude and duration of the output current's pulse both rise as the incident angle increases. The TCAD program simulated the thickness of the sensitive zone of the 4H–SiC SBD detector at varied bias voltages, as shown in Table 5.1. By applying a bias voltage of 20 V, it is determined that the thickness of the sensitive zone is around 3.56 μm. This value is significantly smaller than the range of α-particles in SiC, which is 18.26 μm for particles with an energy of 5.486 MeV. Consequently, as the angle of incidence increases, a greater amount of energy from α-particles is detected. As a result, additional electron–hole pairs are produced, which causes the output current pulse amplitude of the detector to rise. Under a constant bias voltage,

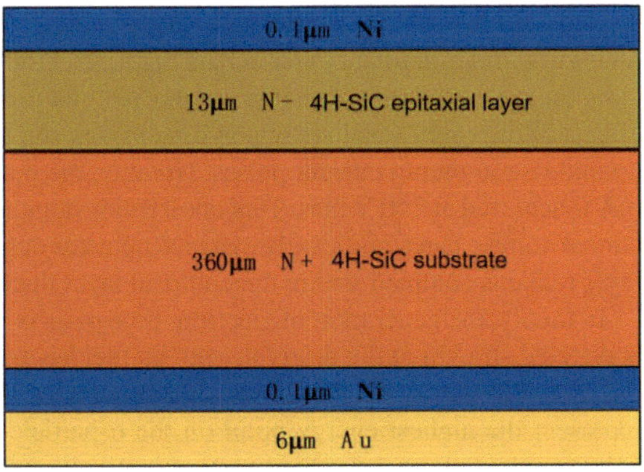


**Fig. 5.27** Schematic diagram of 4H–SiC Schottky diode detector structure

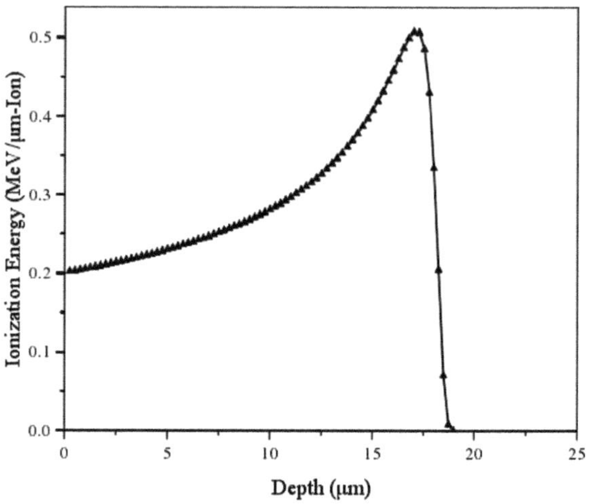


**Fig. 5.28** Energy deposition of 5.486 MeV alpha particles in SiC

the duration for charge collection lengthens as the number of electron–hole pairs increases. Consequently, the time it takes for the pulse to decrease also increases as the angle of incidence rises. At an incidence angle of 0°, the peak current measures approximately 0.38 mA, whereas at an incidence angle of 78°, the peak current is around 0.49 mA. If the thickness of the sensitive zone is significantly smaller than the range of the incident particles, the detection current can be raised by appropriately increasing the incidence angle.

Under different reverse bias voltages, the output current pulses of the detector under 5.486 MeV α-particle vertical incidence are illustrated in Fig. 5.30. As evidenced by the experimental data, the height of the output current pulses increases markedly with the applied reverse bias voltage. Table 5.1 indicates that when the reverse bias reaches 240 V, the depletion layer width approximates 12.75 μm. Considering the epitaxial layer thickness of 13 μm, it can be inferred that the depletion layer ceases to expand beyond 240 V. For biases below 240 V, the Schottky diode's depletion region enlarges with increasing voltage, effectively broadening the detector's sensitive thickness. This expansion facilitates the generation of additional electron-hole pairs while simultaneously enhancing the electric field strength, thereby accelerating the collection rate of charge carriers per unit time. Collectively, the synergistic effects of increased sensitive region dimensions and intensified electric field lead to a rapid rise in output current pulses. Notably, the maximum output current of 1.28 mA is achieved at 240 V bias. This observation demonstrates that elevating the operational voltage can significantly enhance detector output current when the depletion layer remains confined within the epitaxial layer thickness.

In theory, the generation of electron–hole pairs is maximized when an α-particle is directed straight at the detector, with an incidence angle of 0°. Consequently, the output voltage pulse obtained from the output current pulse at 0° should accurately represent the highest energy point on the α-particle energy spectrum. The output voltage pulses of the detector at various bias voltages (20, 80, 140, and 240 V) were

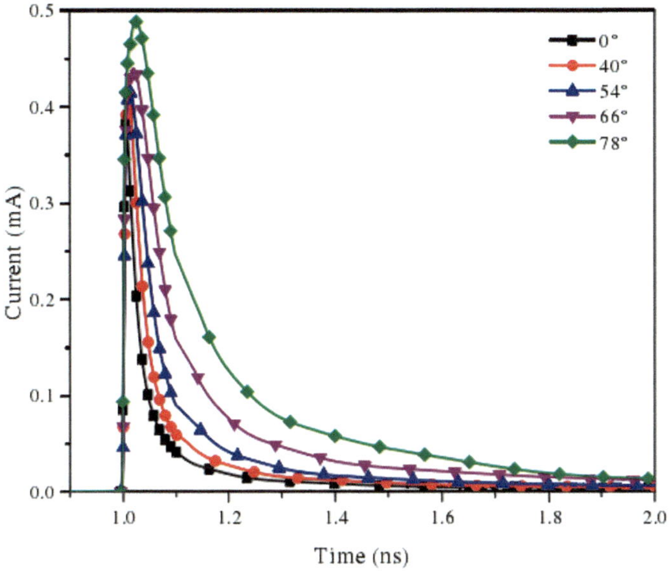


**Fig. 5.29** 5.486 MeV alpha particles at 20 V at different angles of output current pulse at incidence

**Table 5.1** Sensitive zone thickness of 4H–SiC SBD detectors at different bias pressures

| Reverse bias voltage (V) | 20 | 80 | 140 | 240 |
|---|---|---|---|---|
| Sensitive zone thickness (μm) | 3.56 | 7.18 | 9.74 | 12.75 |

calculated based on the current pulses observed under collimated conditions. The peak voltages of these output pulses were then plotted on a graph. Additionally, the number of channels where these peaks were located were extracted from the experimentally measured energy spectrum (e.g., Fig. 5.31) and included in the graph (Fig. 5.32). Figure 5.32 demonstrates that the calculated output voltage pulse peaks at various bias voltages align with the peak positions of the experimental energy spectrum, indicating a strong correlation between their relative positions. By associating the peak of the experimental energy spectrum with an output voltage pulse peak of 240 V, the remaining output voltages are adjusted to align with the number of channels. This method results in the generation of Fig. 5.33. Figure 5.33 shows that the peak voltage of the output voltage pulse deviates little from the peak of the energy spectrum, with a maximum variation of only 1.4%. Accounting for the impact of noise and other variables, the variance is insignificant.

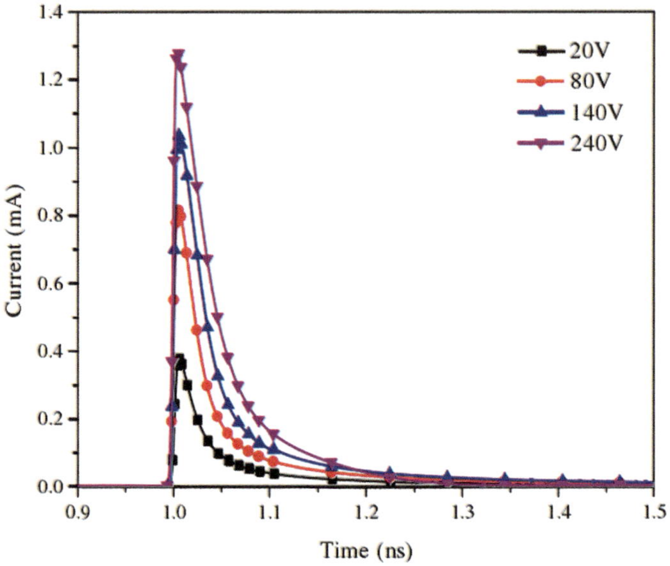


**Fig. 5.30** Vertical α-particles of 5.486 MeV at different bias voltages output current pulse at incidence

**Fig. 5.31** Measured energy spectrum of the detector in reference [23] for α particles at different bias pressures

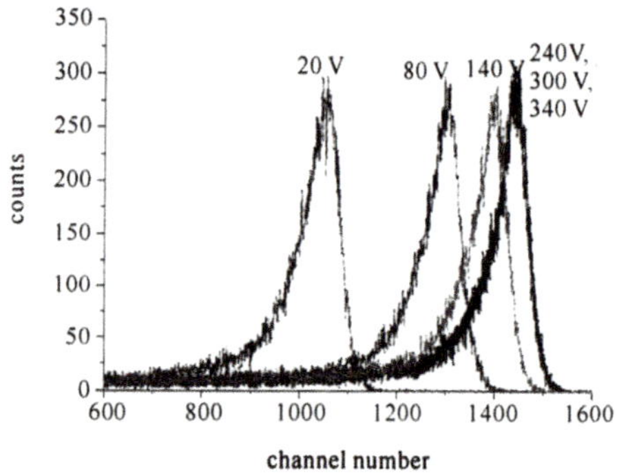


## 5.2.2 *Experimental Study and Analysis of Charged Particle Irradiation Response*

Figure 5.34 displays the measurement platform for the pulse amplitude spectrum. The measurement platform for pulse amplitude spectrum utilizes a comprehensive array of ORTEC electronic modules. The high-voltage power supply supplies a bias voltage of up to 1500 V to the detector through the type 142 front-end amplifier (142 pre-amplifier) [24]. The output current pulse of the detector is initially integrated by the charge sensitive 142 pre-amplifier. The integrated signal is then sent to the type 855 shaping amplifier (855 main amplifier) [25] for shaping and amplification.

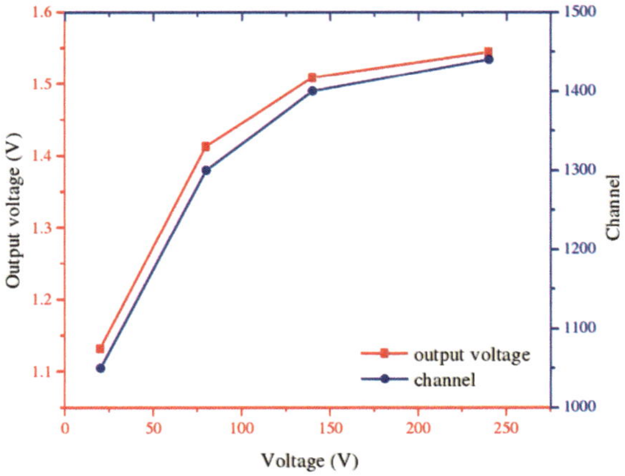


**Fig. 5.32** Comparison of the calculated output voltage pulse peaks with the peaks of the experimental energy spectrum reported in reference [23]

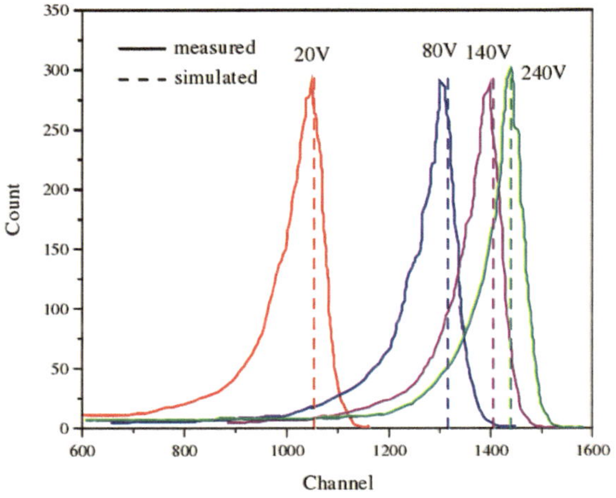


**Fig. 5.33** Calculated peak voltage on the experimental energy spectrum from reference [23]

The resulting voltage signal, which has been shaped into pulses, is then inputted into the 927 multichannel analyzer for analysis. The analyzer outputs the results of the energy spectrum statistics to the computer. All ORTEC modules are powered by a special power supply module, which provides them with a 12/24 V DC operating voltage. In this case, the "E-channel" of the 142 preamplifier is chosen because it has a longer decay time for the current integrated signal. To ensure a high signal-to-noise ratio, it is important to keep the connection line between the detector and the 142 preamplifier as short as possible, around 15 cm in this experiment. When detecting α-particles, the shaping time of the 855 preamplifier is set to 1.5 μs, but the actual peak time should be 3.3 μs. The α-particle detection process involves setting the formation time of the 855 main amplifier to 1.5 μs. However, the actual peak time is measured to be 3.3 μs, and the actual half-width of the forming pulse is found to be 4.95 μs. To enhance the signal-to-noise ratio and prevent the pulse voltage signal amplitude

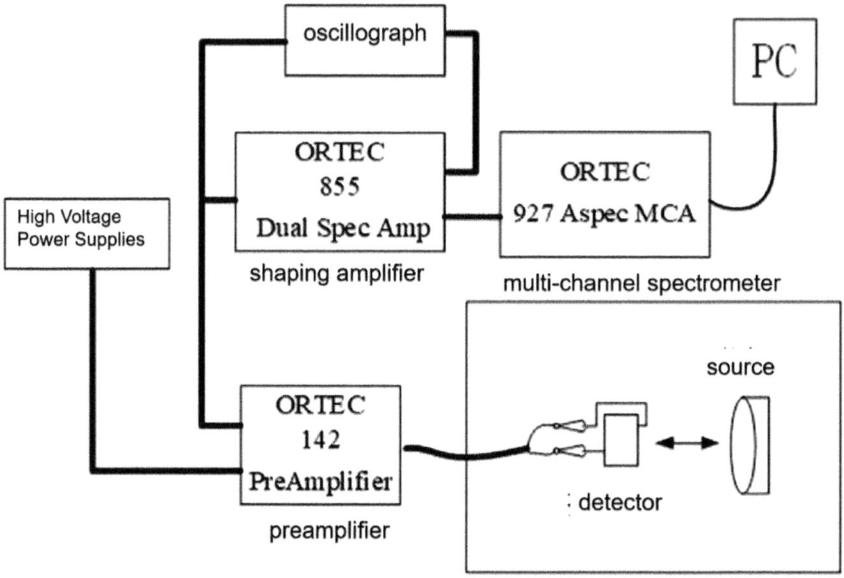


**Fig. 5.34** Schematic diagram of the experimental platform used for energy spectrum measurements

from exceeding the range, the amplification for the 241 Am α-particle is adjusted to 50. The output signal of the 855 main amplifier can be directly transferred to the 927 multi-channel analyzer for pulse amplitude analysis, which subsequently generates the needed pulse amplitude spectrum by statistical methods. To measure the $^{241}$Am α-particle spectrum, the $^{241}$Am source is positioned inside a spacious aluminum alloy shielded container together with the detector.

The detector's longitudinal structure, depicted in Fig. 5.35, is established using the current attainable level of low-doped thick epitaxy. Prior to the experiment, the energy spectrum of α-particle irradiation in the detector is simulated. The technique described in Sect. 5.1.3 is employed to compute the pulse amplitude spectrum of the device. Additionally, the energy deposition spectrum will be computed for the purpose of comparison. To simplify the calculations, certain basic assumptions are made:

(1) The $^{241}$Am source has a negligible thickness, meaning that the energy absorbed by the source itself from alpha particles is not taken into account.
(2) Aside from the 2 μm Al layer, there is also a 500 nm Au layer that serves to establish a connection between the ohmic contact and the copper surface of the housing during the packing of the device.
(3) Following placement, the source and detector are precisely aligned at the center, with the emitting surface of the source and the receiving surface of the detector being parallel. This arrangement is depicted in Fig. 5.36 and mirrors the conditions used during the actual experiment. The source can be modeled as a circular surface with a diameter of 1 cm, while the detector receiving surface is a square

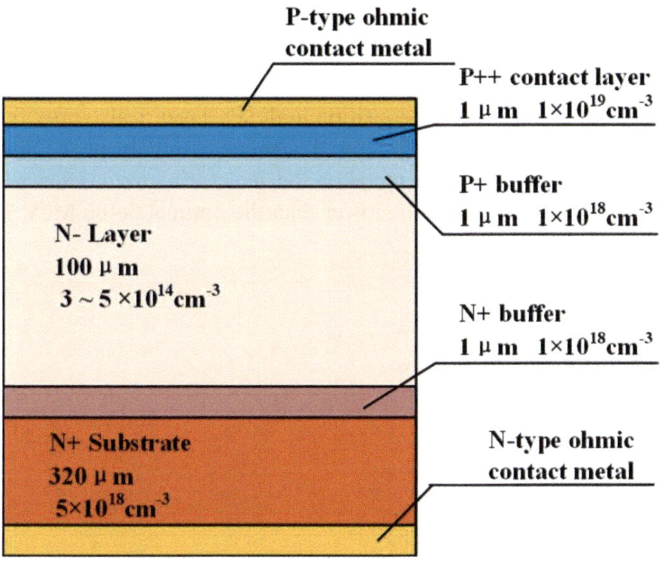


**Fig. 5.35** Schematic diagram of the longitudinal structure of the SiC-PiN type irradiation detector

with a side length of 2 mm. The distance between the source and detector is around 2 mm.

Based on the above parameters, the SRIM program was used to determine the energy deposition of 5.486 MeV α-particles as they hit the detector at various angles. This calculation is illustrated in Fig. 5.37a. The energy deposition spectrum is estimated by combining the data presented in Fig. 5.37a with interpolation and the Monte Carlo approach, as demonstrated in Fig. 5.37b. Due to the high doping concentration of the 2 μm P+ ohmic contact layer on the device surface, which is $1 \times 10^{18}$ cm$^{-3}$, the presence of a space charge area is negligible even when the device is reverse-biased

**Fig. 5.36** Schematic diagram of the relative position of the device and the source

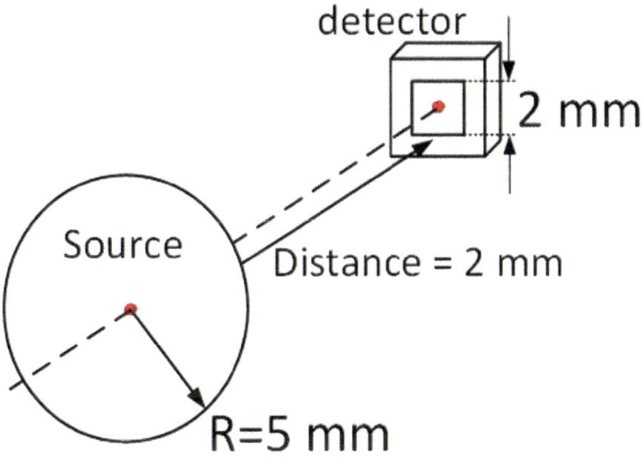

at 3000 V. Furthermore, based on the carrier lifetime against doping concentration depicted in the aforementioned SRH model, it is evident that the carrier lifetime in the P+ layer is exceptionally diminished. Hence, the energy deposition spectrum acquired by disregarding energy deposition in the P+ layer is also provided here.

The energy spectrum depicted in Fig. 5.37b has a solitary peak. However, the presence of a specific thickness of metal on the detector's surface causes the energy of α-particle deposition in SiC to be lower than the optimal 5.486 MeV. Due to the

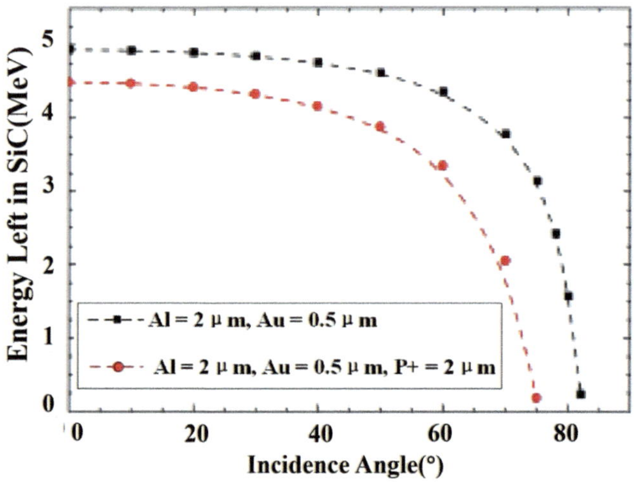


(a)

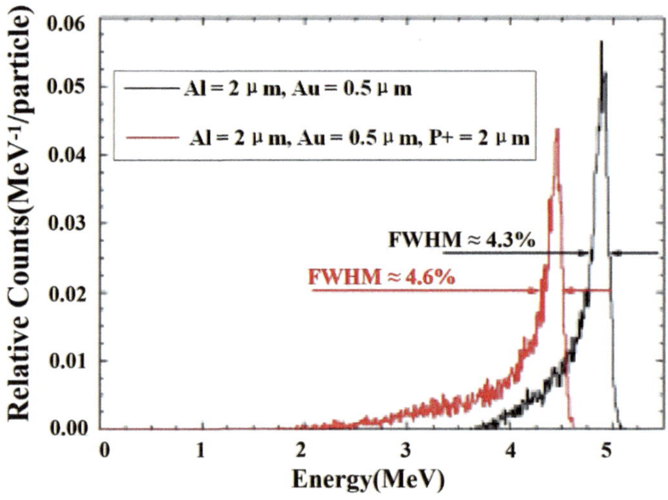


(b)

**Fig. 5.37** Schematic diagram of the relative position of the device and the source uncertainty in the direction of α-particles generated by the radioactive source, the energy attenuation caused by the metal surface varies at different angles of incidence. This leads to an aberration in the energy deposition spectrum, namely towards the lower end of the energy range. The detector spectrum's theoretical energy resolution for the 5.486 MeV α-particle is around 4.6% and 4.3% when considering the presence and absence of the P+ layer, respectively. The detector's pulse amplitude spectrum can be recalculated by utilizing the energy spectrum calculation method described in Sect. 5.1.3. This involves combining the measured epitaxial layer concentration parameter of the device (referred to as E1) with the energy deposition distribution of α-particles obtained from the SRIM software simulation. The simulation and test voltages are configured at 30 and 200 V, respectively, to represent scenarios where the depletion layer thickness is smaller than the range of charged particles and where the depletion layer thickness is larger than the range of charged particles. The simulation results are displayed in Fig. 5.38, which exhibits a distinct shape compared to the energy deposition spectrum depicted in Fig. 5.37b. This discrepancy arises due to variations in the bias voltage and the depletion layer width of the device. Specifically, with a bias voltage of 30 V, the depletion layer width is around 10 μm, whereas at 200 V, it expands to around 26 μm. The peak voltage of the spectral line does not exceed 200 V when the bias voltage is set at 30 V. The expansion of the energy spectrum caused by the device construction and operational bias is restricted in current device designs. Below a pulse voltage of 2.6 V, the relative counts in the energy spectrum remain constant. This suggests that the improved energy resolution at lower voltages is caused by incomplete charge collection. According to theory, particles that hit the SiC material at a small angle should deposit more energy, leading to a larger output pulse. However, as the bias voltage decreases, these particles deposit more energy in the neutral region, resulting in a larger output pulse. The neutral area experiences energy deposition, leading to a reduction in the actual output voltage amplitude, resulting with final counts gathering at around 3 V.

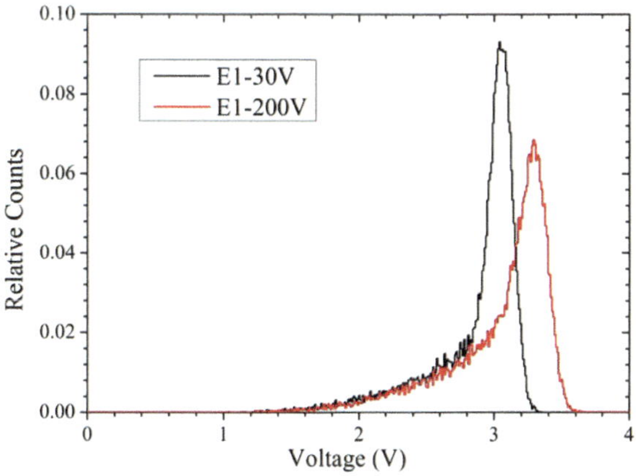


**Fig. 5.38** Simulation results of the pulse amplitude spectrum of device E1 at different voltages

To mitigate electromagnetic interference noise in space, the $^{241}$Am source and detector were positioned in close proximity within an aluminum alloy shielding cavity during the measurement of the alpha irradiation spectrum. Figure 5.39 displays the pulse amplitude spectrum as measured by the detector. The analysis includes comparisons between measurements taken in vacuum and non-vacuum circumstances, as well as measurements taken at various bias voltages in a vacuum environment. Within the non-vacuum spectrum, the device exhibits a solitary peak that possesses specific statistical attributes. This peak is observed above the cut-off threshold and below 100 channels. It is believed that this peak is a result of the crosstalk caused by the external voltage supply line. Given that the bias voltage for the 142 preamplifier and detector must be directly provided by an external voltage supply line, it is quite improbable to prevent the occurrence of this type of noise once it manifests. The pulse amplitude spectra measured in non-vacuum exhibit a notable widening effect and a noticeable shift towards the lower channel locations, when compared to the spectra measured in vacuum. Within the vacuum spectrum, when the device bias voltage reaches 600 V, a notable change occurs. The shape of the spectrum undergoes significant alterations. Specifically, a considerable increase in noise counts is observed in the low-channel-count region (below 100 channels). Additionally, the characteristic single peak vanishes, being replaced by a smaller peak that exhibits fewer counts and is distorted towards the high-energy end. The spectral line shape of the device under normal operation closely resembles the ideal energy deposition spectral line shape, albeit with considerable broadening. At both 30 and 200 V bias levels, the device exhibits a conventional single peak. The energy resolutions at these bias conditions are 5.45% and 10% respectively. The simulation results of the pulse amplitude spectrum are generally in agreement with the experimental findings. However, the major peak of the 200 V bias in the simulation is more distinct than the observed value. This discrepancy is due to the absence of consideration for spectral line broadening caused by noise in the simulation.

The energy resolution of SiC nuclear radiation detectors for α-particles in the energy range of 4.8–7.7 MeV, as studied by Wu Jian and Jiang Yong's research team at the Chinese Academy of Engineering Physics [26, 27], is found to be between 0.61 and 0.90%. This energy resolution is comparable to that of high-resolution SiC α-particle detectors reported globally. The 4H–SiC nuclear radiation detector is a Schottky structure, as depicted in Fig. 5.40. The epitaxial layer is formed on a 4H–SiC single crystal substrate using the CVD homogeneous epitaxial technique. The ohmic contact is applied using electron beam evaporation and then annealed at a temperature of 1000 °C. Simultaneously, the thickness of the $Si_3N_4$ and $SiO_2$ protective layers is carefully regulated to 0.1 μm in order to minimize energy loss and divergence of charged particles within the detector's dead layer, which is formed by the protective layer and electrodes. This precise control results in a calculated energy loss of only 78.3 keV for α-particles with an initial energy of 6.0 MeV within the detector's dead layer. The 4H–SiC device has a sensitive zone with an area of 25mm$^2$. The remaining specific specifications are indicated in Fig. 5.40. According to the data presented in Fig. 5.41, the leakage current measures at 14.92 nA/cm$^2$ under a reverse bias voltage of 200 V.

**Fig. 5.39** Energy spectrum measured by detector E1. **a** Energy spectrum at 200 V, vacuum and atmospheric pressure; **b** variation of the energy spectrum with bias pressure under vacuum conditions

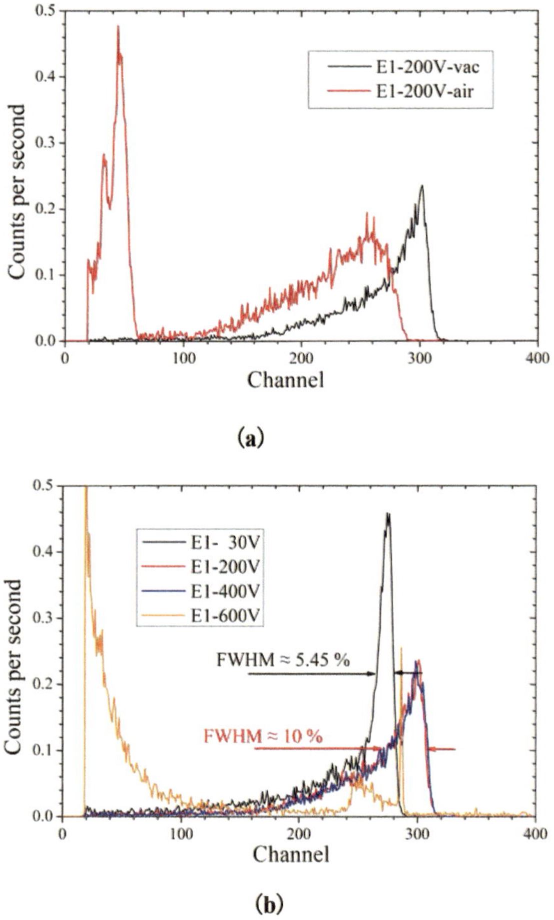


The experimental setup is depicted in Fig. 5.42, which resembles the composition module illustrated in Fig. 5.34. However, in this setup, a collimator is positioned between the detector and the α source to facilitate the straightforward collimation of the α particles. The experiment utilizes a $^{226}$Ra source with five primary α particle energies to investigate the energy resolution of the 4H–SiC radiation detector at various α particle energies. It also examines the energy linearity of the detection system while mitigating potential errors in incident particle energy caused by variations in coating thickness among different radioactive sources. Additionally, the energy spectrum of α particles emitted by the $^{226}$Ra source was measured using a GM20VA AuSi surface barrier detector under identical conditions, in order to make a comparison.

Figure 5.43 displays the energy spectrum of alpha particles emitted by the $^{226}$Ra source in the 4H–SiC nuclear radiation detector. Table 5.2 displays the precision of the 4H–SiC detection system in measuring the energy of alpha particles emitted by

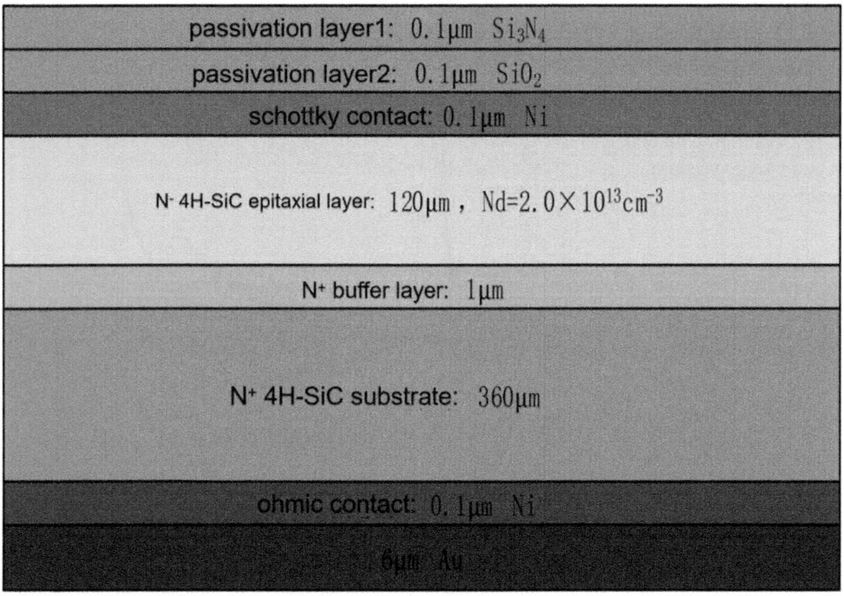


**Fig. 5.40** Schematic cross-section of 4H–SiC Schottky diode

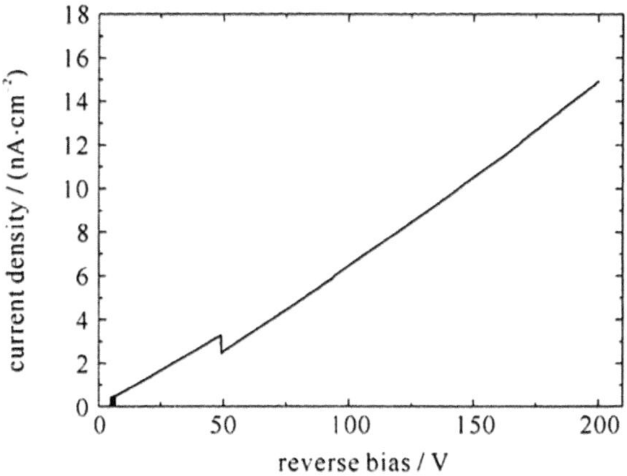


**Fig. 5.41** Variation of 4H–SiC leakage current density with bias voltage

the $^{226}$Ra source. This precision is determined from the energy spectrum depicted in Fig. 5.43. Upon comparing the data of the Si detection system (Table 5.2), it is evident that the energy resolution index of the 4H–SiC detection system is already on par with the resolution index of the well-established Si detection system.

For the α-particle energy spectrum measurement system composed of 4H–SiC nuclear radiation detectors, the α-particle energy spectrum broadening is caused by the system electronic noise, detector leakage current, carrier statistical rise and fall, detector dead layer, radiation source α-particle energy broadening, α-particle

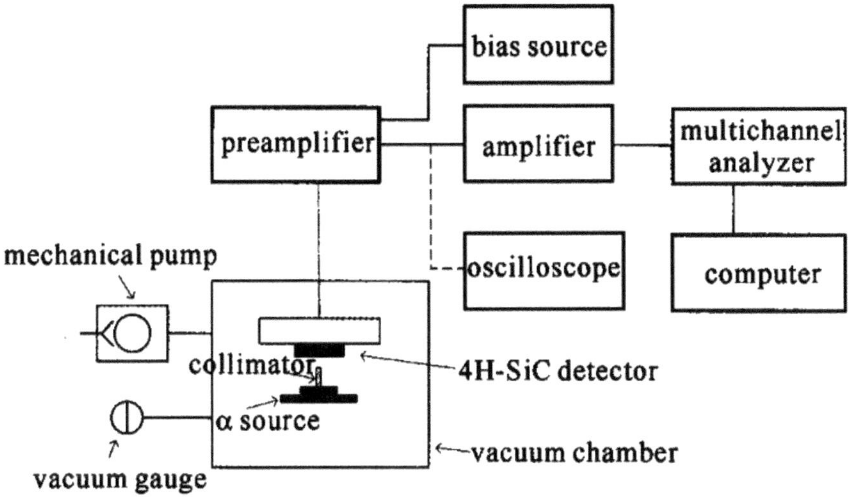


**Fig. 5.42** Experimental layout of 4H–SiC nuclear radiation detector alpha particle energy spectrum test

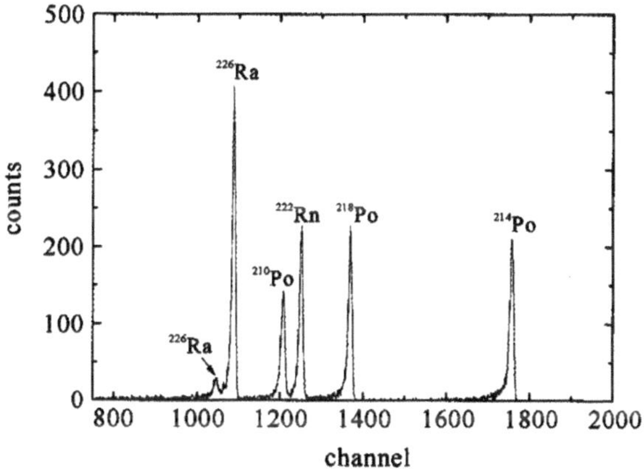


**Fig. 5.43** Alpha particle energy spectrum of the$^{226}$ Ra source obtained by the 4H–SiC nuclear radiation detector

**Table 5.2** Energy resolution of the main 5 alpha particles of the 4H–SiC detection system for the$^{226}$ Ra source and comparison with Si detectors

| Isotopes | Energy resolution in 4H–SiC (%) | Half height width$_{SiC}$ (keV) | Energy resolution in Si (%) | Half height width$_{Si}$ (keV) |
|---|---|---|---|---|
| $^{226}$Ra | 0.90 | 42.41 | 1.55 | 77.97 |
| $^{210}$Po | 0.87 | 45.72 | 0.72 | 40.20 |
| $^{222}$Rn | 0.72 | 39.03 | 0.59 | 34.28 |
| $^{218}$Po | 0.61 | 36.34 | 0.60 | 37.88 |
| $^{214}$Po | 0.68 | 51.93 | 0.50 | 38.82 |

incidence angular distribution, etc., i.e., the half-height width of the α-particle energy spectrum ($\delta_{FWHM}$) has the following expression [28]:

$$(\delta_{\mathrm{FWHM}})^2 = \left(\delta_{\mathrm{FWHM}}^{\mathrm{e}}\right)^2 + \left(\delta_{\mathrm{FWHM}}^{\mathrm{l}}\right)^2 + \left(\delta_{\mathrm{FWHM}}^{\mathrm{s}}\right)^2 + \left(\delta_{\mathrm{FWHM}}^{\mathrm{d}}\right)^2 + \left(\delta_{\mathrm{FWHM}}^{\mathrm{a}}\right)^2 \quad (5.29)$$

$\delta_{\mathrm{FWHM}}^{\mathrm{e}}$ refers to the widening of the energy spectrum caused by the electronic noise of the system. $\delta_{\mathrm{FWHM}}^{\mathrm{l}}$ is the widening of the energy spectrum caused by the leakage current of the detector. $\delta_{\mathrm{FWHM}}^{\mathrm{s}}$ is the widening of the energy spectrum caused by the statistical fluctuations in the number of carriers produced by α-particles in semiconductors. $\delta_{\mathrm{FWHM}}^{\mathrm{d}}$ is the widening of the energy spectrum caused by the presence of a dead layer in the detector, which includes the protective layer of the 4H–SiC and electrodes. $\delta_{\mathrm{FWHM}}^{\mathrm{a}}$ is the widening of the energy spectrum caused by the distribution of energy levels of the α-particles emitted by the radiation source and the variation in the angle at which the radiation source is incident.

Wu et al. examined the impact of different parameters on the widening of the energy spectrum. They used the emission of 6 MeV α-particles from the $^{218}$Po nuclide in Table 5.2 as a specific case study. The ORTEC 419 precision pulse generator was employed to produce a pulse signal that mimicked the 6 MeV α-particle. This signal was then fed into the preamplifier in the system depicted in Fig. 5.42. The purpose of this setup was to experimentally measure the impact of the first two terms of Eqs. (5.29) on the widening of the energy spectrum. The measured root-mean-square (RMS) value for this broadening was 13.02 keV. The equation shown below describes the expansion of the energy spectrum caused by the stochastic fluctuations of the carriers [29]:

$$\delta_{\mathrm{FWHM}}^{\mathrm{s}} = 2.35\sqrt{F\varepsilon E} \quad (5.30)$$

Let $F$ represent the Fano factor, which is determined to be 0.128 [30]. ε represents the average energy needed to generate a pair of electron–hole pairs in 4H–SiC, which is determined to be 7.28 eV [30], the same as the energy of the incident α-particle. The resulting 6 MeV α-particle $\delta_{\mathrm{FWHM}}^{\mathrm{s}}$ in the end of the 4H–SiC nuclear radiation detector is estimated to be 5.56 eV. The quantitative $\delta_{\mathrm{FWHM}}^{\mathrm{a}}$ term in Eq. (5.29) can be disregarded due to the implementation of a collimated system in the measurement setup. By analyzing the experimental data presented in Table 5.2, we can estimate that the broadening of the energy spectrum $\delta_{\mathrm{FWHM}}^{\mathrm{d}}$ is 33.47 keV. This broadening is primarily caused by the detector dead layer, which includes the protective layer, electrodes, and materials such as $Si_3N_4$, $SiO_2$, and Ni. The broadening of the spectrum of α-particles measured by the 4H–SiC nuclear radiation detector is mainly influenced by this dead layer. Furthermore, utilizing Fig. 5.43 as a reference, an examination was conducted to assess the energy linearity of the 4H–SiC detector system for α particles within the energy range of 4.8–7.7 MeV. The findings are presented in Fig. 5.44. Figure 5.44 demonstrates that the measuring system formed by this detector has excellent linearity (with a linear correlation coefficient R = 0.99999) within the specified energy range. This is a notable advantage of semiconductor

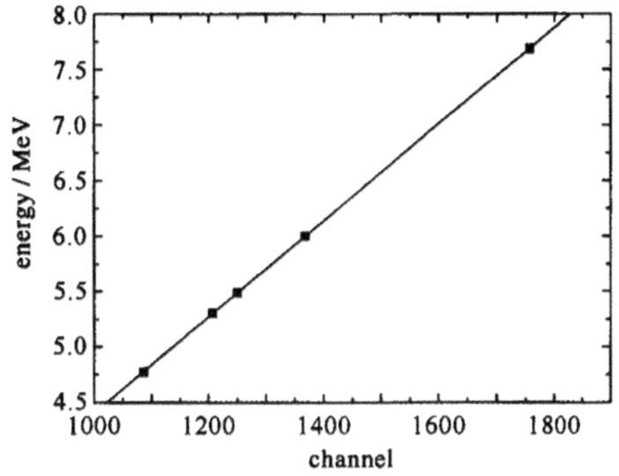


**Fig. 5.44** Energy linearity of the 4H–SiC nuclear radiation detector for alpha particles

detectors. The experimental results and analysis of the high-resolution α detector, as presented by Wu et al., can be attributed to its exceptionally low energy resolution, mostly stemming from the following factors: ① Low-doped thick epitaxial layer: The epitaxial layer becomes fully depleted at a lower bias voltage. The range of α particles is smaller than the thickness of the depletion zone, allowing for complete collection of the energy deposited in the detector. This, in turn, further lowers energy divergence. (2) The implementation of an energy-collimation system; (3) Exceptionally low leakage current; (4) Thin dead layer thickness.

In 2015, the Slovak Academy research team [31] led by Zat'ko B conducted foreign research on 4H–SiC Schottky diodes. These diodes were specifically used as chambers for detecting α-particles. To limit the impact of α-particle energy dispersion, a thin contact layer of Ni/Au Schottky measuring only 15 nm was employed. At ambient temperature and with a bias of 700 V, the detector exhibits a reverse leakage current density of merely 0.3 nA/cm$^2$, as indicated by its current–voltage characteristics. The analysis of the spectral performance reveals that the energy resolution for 5.486 MeVα emitted from $^{241}$Am in a pulse height spectrum with a reverse bias of 200 V is only 0.25%. The energy resolution for 5.486 MeVα released from 241 Am is barely 0.25%. The comparison with the investigation conducted by Wu et al. provides additional evidence that the dead layer of the detector is the primary component responsible for the widening of the energy spectrum observed in the 4H–SiC nuclear radiation detector while measuring α particles.

## 5.3 Response of SiC Nuclear Radiation Detectors to X-Rays

### *5.3.1 Pulsed X-Ray Transient Response Experiments*

As science and technology advance, new equipment like the Fractured Neutron Source (FNS) and pulsed reactors (PRs) have been implemented. A notable characteristic of these devices is their ability to generate a substantial quantity of particles

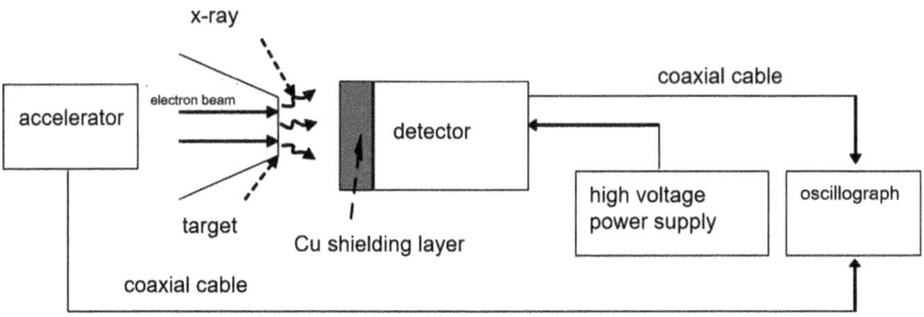


**Fig. 5.45** Schematic diagram of the pulsed X-ray experimental setup

within a brief timeframe [32]. As stated in Sect. 2.5.2, it is necessary for the detector output signal to precisely represent the whole duration of the recorded particle population. Contemporary detection of pulsed irradiation necessitates detectors with time performance in the range of nanoseconds or even sub-nanoseconds [33]. The pulsed X-ray transient response experiment utilizes the detector configuration depicted in Fig. 5.35, while the experimental arrangement is illustrated in Fig. 5.45.

The gas pedal generates a high-energy electron beam that strikes the Be target, resulting in the emission of X-ray radiation in the approximate $2\pi$ direction. The energy of the X-ray rays is around 60–70 keV. The SiC nuclear radiation detector and the electrical components responsible for supplying the detector bias loop are enclosed within a shielded box. This box is open at the front and covered with multiple layers of thin Cu sheets to prevent interference from other energetic charged particles. The gas pedal trigger signal and the detector output pulses are transmitted to the oscilloscope using separate coaxial cables. Figure 5.46 displays the external circuit of the detector. When measuring pulsed radiation, the detector signal strength is very strong at the moment of measurement. To maintain consistent voltage at both ends of the detector, an extra capacitor is typically attached to the high-voltage power supply as a charge buffer. The detector's current pulses pass through the oscilloscope's 50 Ω resistor $R_s$ and are transformed into voltage pulses that are seen on the oscilloscope. Due to the extended length of the cables transmitting the signals during the test, an inductor L is incorporated into the circuit model.

Due to the existence of process fluctuations, the experiments were performed on devices numbered E1, G & F. The average doping concentration $N_D$ and the built-in potential $\varphi_{bi}$ of the devices obtained from the C-V test are shown in Table 5.3.

The pulse response of device E1, biased at 400 V, is illustrated in Fig. 5.47a. The pulse accurately represents the overall traits of this particular set of devices: The detector output pulse exhibits notable variations at its peak. The primary pulse response lasts approximately 15 ns, with a half-height width of around 7.5 ns. The rising edge of the pulse takes about 3 ns, while the falling edge lasts approximately 10 ns. Additionally, there is some overshoot present. Following the cessation of the pulse, despite the signal's ability to promptly revert back to the initial level, there persists a prolonged oscillation of the signal in close proximity to the baseline,

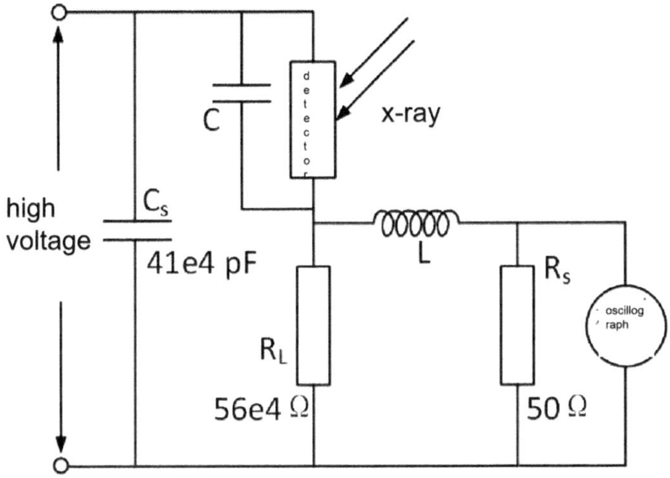


**Fig. 5.46** Schematic of the external circuit of the device when measuring pulsed X-rays

**Table 5.3** Linear fit of V-1/$C^2$ curves extracting $N_D$ and $\varphi_{bi}$

| Device Number | E1 | G | F |
|---|---|---|---|
| $N_D$ $(\times 10^{14}\ \mathrm{cm})^{-3}$ | 3.544 | 3.098 | 2.641 |
| $\varphi_{bi}$ (eV) | 3.734 | 3.145 | 2.893 |

enduring for around 200–300 ns. Figure 5.47b depicts the concurrent representation of the gas pedal's trigger monitoring signal and the corresponding output pulse of device E1 for reference. The trigger signal exhibits a rising edge with a duration of approximately 0.75 ns, a falling edge with a duration of approximately 2 ns, and a pulse half-height breadth of around 1 ns. The transmission of the detector signal is delayed by approximately 25 ns due to the utilization of a lengthy coaxial wire. The aging of the equipment interface causes noticeable signal jitter near the conclusion of the main pulse of the trigger signal, however it does not impact the detection experiment.

Figure 5.48 displays the relative amplitudes of the output pulse signals from different devices at a reverse bias of 400 V. The amplitude of the monitored signal pulses was utilized as a normalization condition for comparison. The relative pulse amplitude of device G is the largest, while the relative pulse amplitude of device F is the lowest. However, these results do not align with the average doping concentration of the three devices mentioned above. Typically, when the effective carrier concentration declines, the space charge area expands, which is advantageous for the separation and collection of electron–hole pairs, leading to a greater output pulse. There is a hypothesis that suggests a possible connection between the variation in the effectiveness of each electron beam pulse and the lifespan of the device. Figure 5.49 displays the output voltage pulses of device F under various biases. As the bias of device F increases, the output pulse amplitude experiences a notable

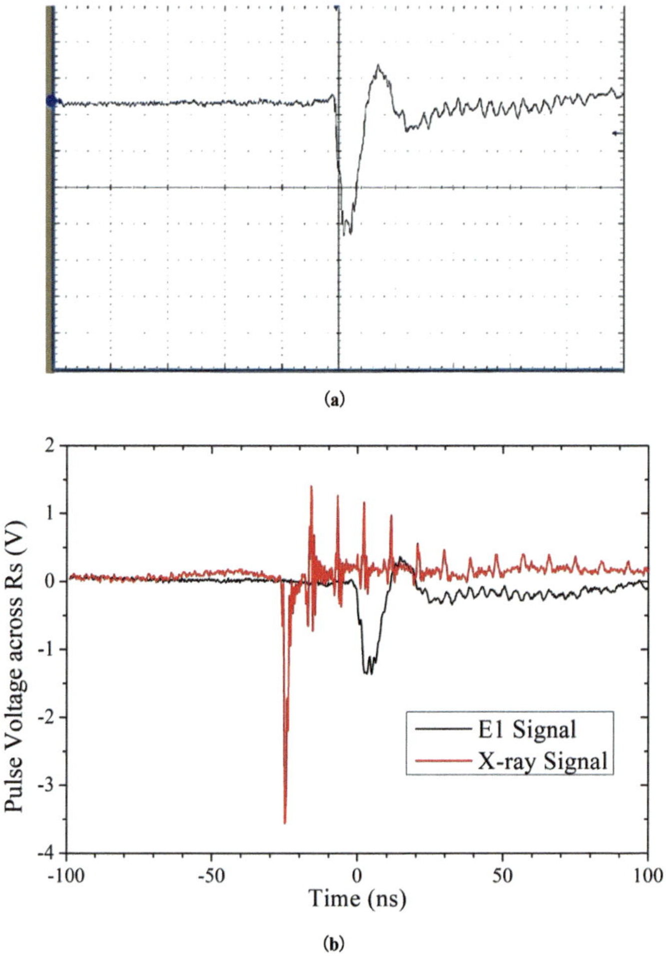


**Fig. 5.47** Pulsed X-ray response measured at 400 V bias of E1 device

increase. However, parameters such as rise time, fall time, and half-height width remain relatively unchanged. Additionally, the amplitude of subsequent signal oscillations, particularly when a reverse bias of 600 V is applied, demonstrates an increase. This suggests that decreasing the capacitance of the device may have resulted in a greater mismatch in impedance. Figure 5.50 shows the output pulse signal of a Si-PiN detector at the product level. It is important to mention that the pulse signal is reduced by a factor of 10. The Si-PiN detector exhibits a substantially higher output pulse intensity compared to the SiC device, mostly due to the following factors: Si atoms have more X-ray absorption capability compared to C atoms. Additionally, Si-PiN features a bigger surface area, a thicker semi-insulating i-layer, and lower doping, resulting in a larger effective volume for X-ray absorption in Si-PiN compared to

SiC radiation detectors; Furthermore, Si, being a narrow-band semiconductor, necessitates a lower average energy (about 3.6 eV) to stimulate a pair of electron–hole pairs compared to SiC, which is a wide-band semiconductor (around 7.8 eV). Consequently, a solitary X-ray photon can stimulate a greater number of electron–hole pairs and generate a more substantial pulse current in Si. The pulse waveform of the Si-PiN detector exhibits a nearly triangular shape, and the primary pulse demonstrates minimal jitter; The rising and falling edges are essentially symmetrical; The rise time is 3 ns, the pulse half-height width is 4 ns, and the main pulse response is about 7 ns; Unlike the SiC radiation detector, which exhibits excessive oscillations and protracted deviation from the baseline following the main pulse, the Si-PiN detector only displays minor oscillations towards the conclusion of the main pulse, followed by a swift recovery to the baseline level. In this case, by ensuring that the external circuitry employed in the SiC device is compatible with that of the Si-PiN detector, additional circuit refinement for the SiC device should effectively reduce the signal oscillations and expedite the restoration to the initial state. Furthermore, to enhance the detection effectiveness of SiC-based detectors and amplify the output pulse signal strength, it is imperative to fabricate SiC-based detectors with bigger surface areas. Simultaneously, the expansion of the device's surface area results in an elevation of its capacitance, potentially impacting the time-related properties of the pulse. This can lead to an elongation of both the rise time and fall time of the device. Consequently, it becomes necessary to decrease the doping concentration of the epitaxial layer even more, enhance the quality of the epitaxial material, and further reduce the density of dislocations.

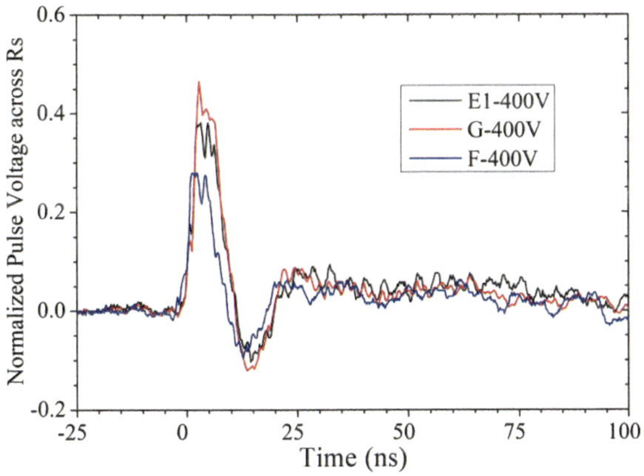


**Fig. 5.48** Taking the peak value of the trigger monitoring signal pulse as the standard, the normalized pulse signal

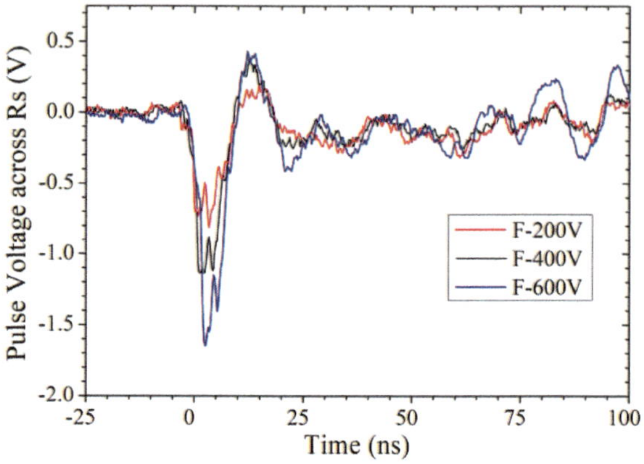


**Fig. 5.49** Device F at different bias voltages pulse signals obtained by probing

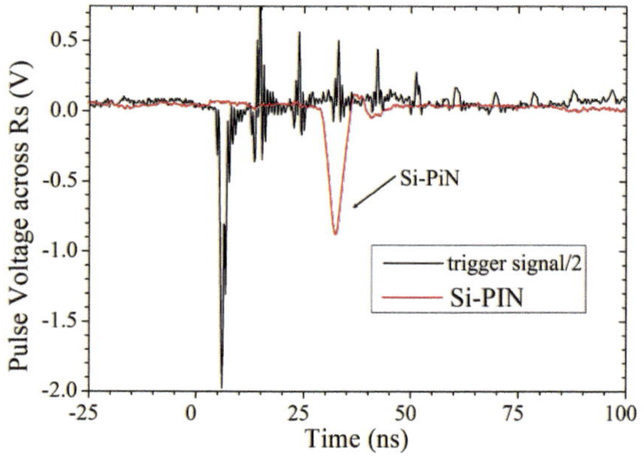


**Fig. 5.50** Product-level Si-PiN detector with tenfold attenuated output pulse signal

## *5.3.2 Pulsed X-Ray Transient Response Simulation Analysis*

Due to the unknown correlation between the trigger monitoring signal and the actual pulse intensity, and the fact that the output amplitude of the pulse voltage is significantly smaller than the bias voltage of the detector, the simulation of Sentaurus does not consider the impact of the bias voltage at both ends of the device on the output signal. After adding the excitation signal, our focus is solely on the waveform change of the output pulse current signal, without considering the absolute magnitude of the signal. Initially, the monitoring signal's frequency spectrum is evaluated using Matlab's Fast Fourier Transform function FFT due to the existence of periodic noise. This analysis is conducted on the triggered monitoring signal Fig. 5.51a in various time intervals, as depicted in Fig. 5.51b.

After filtering out a portion of the noise appropriately:

$$V = \begin{cases} 0 & t < 0 \\ 5.24\,\mathrm{e}^{-0.85\times 10^{9t}} & t \geq 0 \end{cases} \tag{5.31}$$

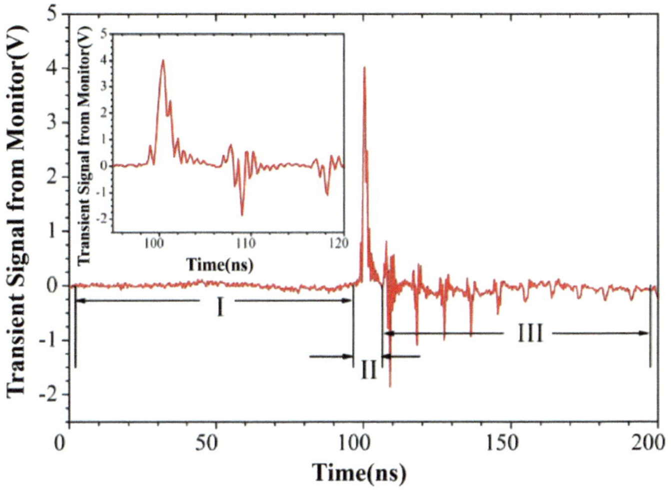


(a)

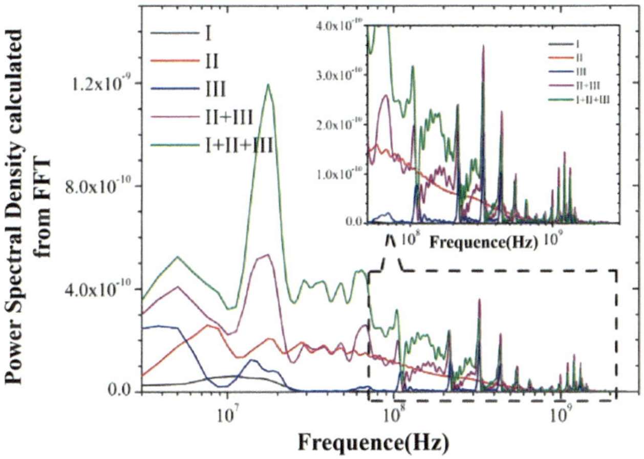


(b)

**Fig. 5.51** Analysis of accelerator trigger pulse signal. **a** Accelerator pulse signal waveform shown by the trigger monitor signal; **b** spectrum analysis of the trigger monitor signal

The equation above represents the waveform of the monitoring signal's pulse signal, as depicted in Fig. 5,52. The X-ray pulse signal waveform of the Sentaurus input is in agreement with this waveform. Given that Sentaurus characterizes X-ray irradiation as a consistent generation of carriers across a specific duration, it is necessary to discretize the pulse waveform in this context. The depicted loop in Fig. 5.46 was created using Multisim software, where the detector is regarded as a current source. The output current pulse, shown by the black line in Fig. 5.53, was computed using Sentaurus software and replaced. The experimental signal transmission employs a coaxial cable with a parasitic inductance ($L$) of around 250 nH. The simulation results indicate that the detector has a capacitance of around 29 pF when subjected to a bias voltage of 400 V. The voltage pulse waveform derived from the simulation is depicted as the red line in Fig. 5.53. The signal's oscillation period is slightly shorter than the measured signal. Once the capacitance in the Multisim simulation hits 40 pF, as indicated by the blue line in Fig. 5.53, the simulated outcomes generally align with the experimental results. The discrepancy could be attributed to the parasitic capacitance of either the package enclosure or the coaxial wire.

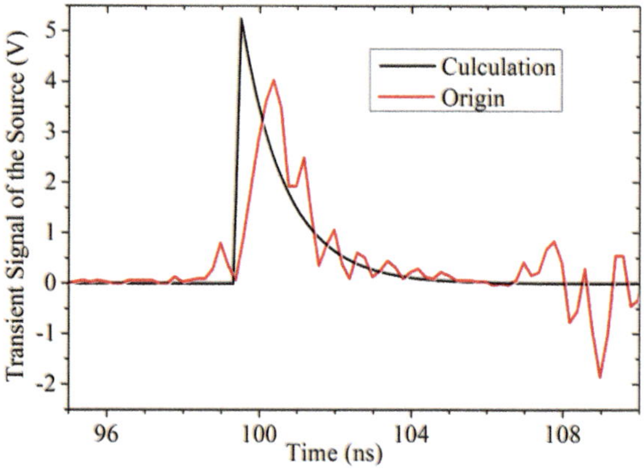


**Fig. 5.52** Monitoring pulse waveform after filter fitting

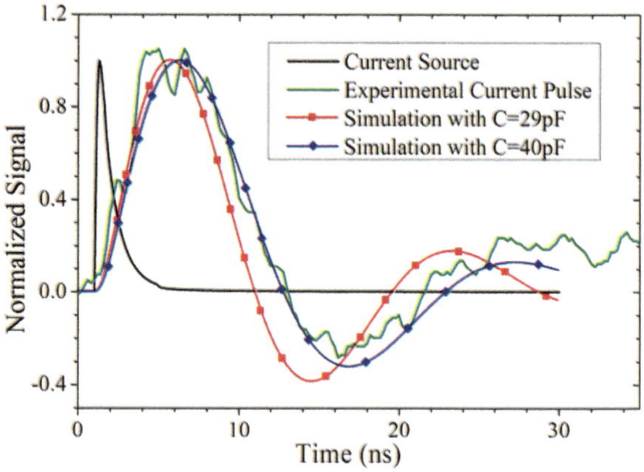


**Fig. 5.53** Comparison of experimental and simulation results of X-ray output pulse waveform

## 5.4 Response of SiC Nuclear Radiation Detectors to Neutrons of Different Energies

There has been a rise in the need for novel neutron detectors in recent years. Due to their lower operating voltage, faster charge collection, and compact size, semiconductor neutron detectors have become highly appealing as an alternative to other types of detectors. They are utilized in various applications, including nuclear reactor power control, nuclear weapon or fissile material detection, and drug screening [34]. SiC has gained recognition as a viable material for neutron detection, particularly in challenging conditions, due to its broad bandgap (4H–SiC, 3.26 eV), high thermal conductivity (1.3 $cm^{-1}$ K), and high shift threshold energy ($Ed_{si}$ = 35 eV, $Ed_c$ = 25 eV). Additionally, SiC has remarkable irradiation tolerance [35]. SiC nuclear radiation detectors have been extensively described for measuring thermal neutron, superthermal neutron, fast neutron flux rate, and spectral analysis. Moreover, SiC nuclear radiation detectors have utilized thin neutron converter layers, such as $^{6}$LiF or $^{10}$B, to enhance their detection performance [2, 36–39].

### 5.4.1 *Thermal Neutron Detection Performance Study*

Previous research on semiconductor neutron detectors has primarily concentrated on enhancing detection efficiency through structural design [40]. Only a limited number of studies have examined the pulse output of detectors that detect single charged particle incidents [35]. As mentioned earlier, although it is feasible to compute the energy deposition spectrum for an estimation of the pulse output, the influence of the device's operating parameters has not been considered, and there has been minimal effort to simulate the pulse amplitude spectrum. The forthcoming section will provide a comprehensive explanation of the process involved in determining the pulse amplitude spectrum of a SiC thermal neutron detector. This approach will be based on the methodology outlined in Sect. 5.1.3, which relies on analyzing the pulse output of individual particles produced by the neutron capture reaction within the detector.

In order to detect neutrons, it is essential to initiate nuclear reactions to generate secondary charged particles. These charged particles then undergo ionization, resulting in the production of numerous electron–hole pairs. Finally, the collection of these electron–hole pairs enables the detection of neutrons, as the neutrons themselves are uncharged and incapable of directly exciting carriers. Due to the low reaction cross section of Si and C atoms with thermal neutrons, it is imperative to employ atoms such as $^{6}$Li and $^{10}$B, which possess a significant thermal neutron capture cross section, to initially capture neutrons. This capture leads to a subsequent nuclear reaction that releases charged secondary particles, thereby enabling the detection of thermal neutrons. By doing this, the detector's ability to detect thermal neutrons can be significantly enhanced. The data used in this study are sourced from the literature [2]. The device structure is illustrated in Fig. 5.54. It consists of a P+ layer with

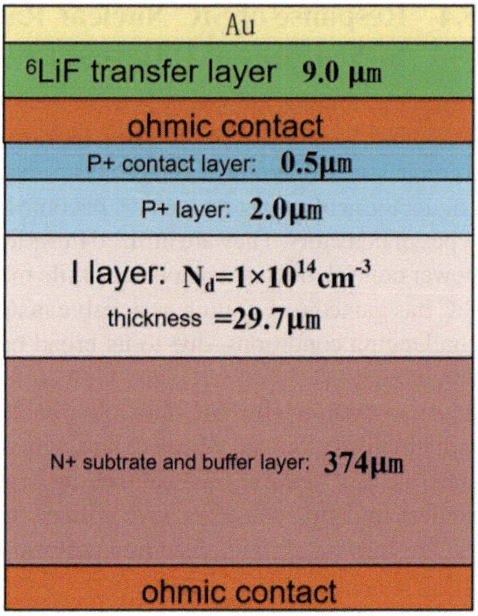


**Fig. 5.54** Schematic of PiN junction type SiC thermal neutron detector structure [2]

a thickness of 2.5 μm, which corresponds to a doping concentration exceeding $1 \times 10^{18}$ cm$^{-3}$. Additionally, an I layer with a thickness of approximately 30 μm is present, with a doping concentration of $1 \times 10^{14}$ cm$^{-3}$. The P-type ohmic contact metals consist of a 30 nm layer of nickel (Ni) and a 200 nm layer of gold (Au). A 9 μm conversion layer made of $^6$LiF was applied externally. The detector was used with no bias and its energy spectrum is shown in Fig. 5.1a.

A nuclear reaction between a thermal neutron and a $^6$Li atom produces two charged particles with the following reaction equation:

$$^6\mathrm{Li} + \mathrm{n} \rightarrow \alpha(2.05\,\mathrm{MeV}) +^3 \mathrm{H}(2.73\,\mathrm{MeV})(\mathrm{Q} = 4.78\,\mathrm{MeV}) \tag{5.32}$$

Due to the significant difference in energy levels, the thermal neutron (0.0253 eV) can be treated as a nuclear reaction caused by the emission of α-particles and $^3$H-particles in random directions, with two particles being emitted in the exact opposite direction. This is illustrated in Fig. 5.55. Thus, the requirement for the presence of only one charged particle entering the detector from the external environment is fulfilled. The desired pulse amplitude spectra can be generated by individually computing the pulse amplitude spectra of α-particles and $^3$H-particles, and then combining them together. To obtain the energy and angle distributions of α-particles and $^3$H-particles for the generation of (*Erand*, *θrand*), it is necessary to draw the target material required for the simulation in FLUKA. This is because the thickness of the conversion layer is significant compared to the range of these charged particles. The size of the device and the thickness of the conversion layer, which are provided in the literature [2], should be taken into account when determining the target material.

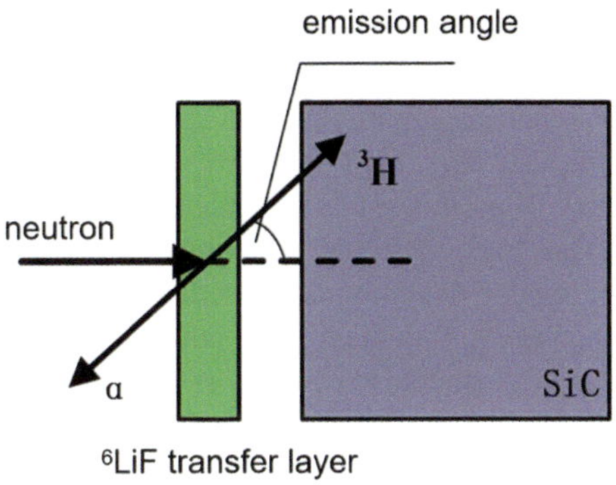


**Fig. 5.55** Schematic of thermal neutron detection using $^6$LiF conversion layer

Afterwards, the Materials and Compounds module is employed to specify the necessary $^6$LiF, contact metal, and SiC materials, while the Low-Mat module is called upon to adjust the diameters of neutron capture cross-sections for different target materials. Afterwards, a statistical region is established at the boundary between SiC and the contact metal. The URSBDX module is then employed to measure the energy and emission directions of the two charged particles as they travel through the predetermined statistical region. Figure 5.1b displays the energy deposition spectra acquired from charged particles hitting SiC material at different energies and angles (*Erand*, *θrand*) using the FLUKA code. In contrast to Fig. 5.1a, which exhibits a pulse height spectrum with two peaks, the calculated energy deposition spectra do not show this characteristic. The findings indicate that the bias condition significantly influences the configuration of the output energy spectrum.

1. Counting

The analysis focused on the relative neutron sensitivity, which is determined by dividing the total number of reaction products calculated in SiC by the total number of incident neutrons. This analysis examined how the thickness of the $^6$LiF neutron converter layer affects the relative neutron sensitivity. The number of $^3$H and α particles emitted from the $^6$LiF material and entering the SiC material were measured and adjusted to account for each neutron. The total of these two counts represents the relative neutron sensitivity. The results are depicted in Fig. 5.56a and closely resemble the findings reported in reference [41]. The sensitivity reaches its maximum between 25 and 30 μm, which closely corresponds to the $^3$H counts. Subsequently, the sensitivity diminishes as the thickness of the conversion layer increases. The data shows that 2.73 MeV $^3$H has a range of approximately 30 μm in $^6$LiF. Figure 5.56b illustrates the flux of 2.73 MeV $^3$H in $^6$LiF when it is incident vertically. The $^6$LiF layer was adjusted to a thickness of 25 μm in order to guarantee a suitably elevated sensitivity. Subsequently, maps of the charged particles were generated, taking into account the emission angle. Fig. 5.57 displays the energy spectrum of the charged particles at the metal electrode-SiC contact, as obtained by FLUKA simulation. The α particles were measured within the energy range of 0.15–2.05 MeV. Particles with energies below 0.15 MeV were disregarded since their distances traveled were similar to the thickness of the P+ layer, resulting in insignificant pulse heights. The counting of $^{3}$H particles is limited to the energy range of 0.15–2.75 MeV. Similarly to α particles, particles with energies below 0.15 MeV are disregarded for the same purpose. The trend of counts within a specific angular range is identical for both α and 3H particles. It is possible to hypothesize that the counts exhibit an upward trend as the angle increases, reaching a maximum between 20° and 40°, and then declining. A notable deviation is noticed for 3H particles between the energy interval of 0.15–0.35 MeV, with the highest point appearing between 40° and 54°. This phenomenon can be attributed to two factors: firstly, when charged particles deposit a given amount of energy, an increase in the emission angle θ results in a decrease in the distance $L$ between the reaction point and the $^{6}$LiF/SiC border, thus leading to a drop in the number of counts. Conversely, the steric angle in the vicinity of $\theta$ can be expressed as $\sin(\theta)d\theta$, therefore the steric angle near $\theta$ rises as $\theta$ increases, leading to an increase in counts. Consequently, there is an observed pattern of counts inside a specific range of angles. The counts exhibit a varying trend depending on the energy deposition, which is believed to be associated with the alteration in energy deposition per unit length over the particle range (Bragg curve). The simulated thickness of $^{6}$LiF is smaller than the range of $^{3}$H in $^{6}$LiF, resulting in lower counts of low energy particles. Conversely, the range of α particles in $^{6}$LiF is only 7 μm, leading to larger counts of low energy particles.

2. Pulse height calculation

According to the predictions made by FLUKA and SRIM, it has been determined that 2.73 MeV $^{3}$H has a range of 30 μm in SiC. This indicates that in order to ensure full charge collection, the active layer thickness should be more than 30 μm. To create depletion layers with a Ni doping concentration of $2 \times 10^{14}$ cm$^{-3}$, a reverse bias of 200 V is necessary. A 4H–SiC PiN diode, as seen in Fig. 5.58a, underwent simulation utilizing a TCAD platform that utilized a 2D structure. The p-layer was doped with aluminum, whereas both the i-layer and N-layer were doped with nitrogen. The doping concentration of each layer is: P+ = $1 \times 10^{19}$ cm$^{-3}$; i = $2 \times 10^{14}$ cm$^{-3}$; N+ = $5 \times 10^{18}$ cm$^{-3}$. The static and transient analyses include many fundamental models, including (1) the Shockley–Read–Hall (SRH) and Auger composite models, and (2) models for high-field saturation, doping-dependent behavior, and carrier scattering mobility. Subsequently, the Poisson equation as well as the continuity equations for electrons and holes are resolved. The spatial distribution of the electric charge in the intrinsic layer under a reverse voltage of 200 V is illustrated in Fig. 5.58b. A layer with a thickness of approximately 33 μm was successfully obtained.

Various energies of $^{3}$H and α particles were introduced into the surface of the P-region side of the reverse-biased 4H–SiC PiN diode at varying angles. The diode output pulses resulting from the introduction of 2.05 MeV α particles and 2.73 MeV $^{3}$H particles with a − 200 V bias can be observed in Fig. 5.59a, b. Each pulse corresponds to the incidence of a single particle, and the initiation of each pulse is deliberately shifted by a distinct amount to distinguish and facilitate the observation of each pulse. According to Fig. 5.59a, the pulse height decreases as the emission angle increases. The occurrence of this phenomena is attributed to the limited distance

**Fig. 5.56** Simulation of $^6$LiF conversion layer

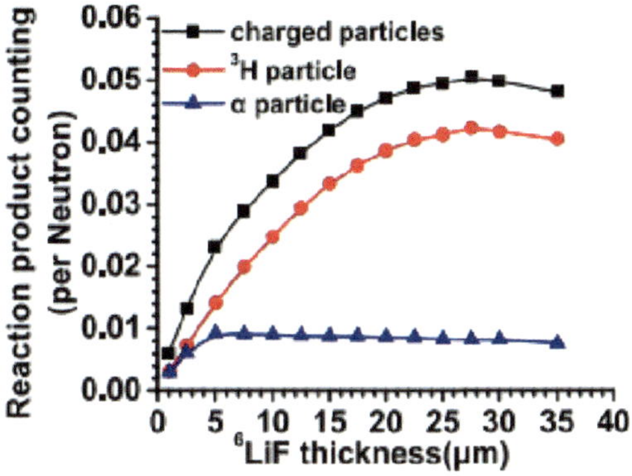


**(a)** Relative neutron sensitivity versus conversion layer thickness

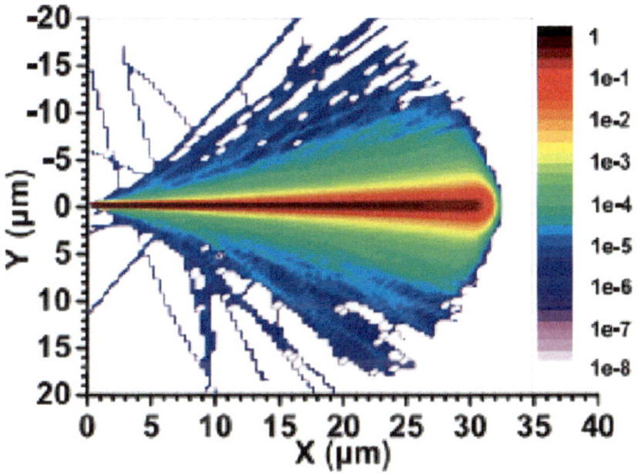


**(b)** Flux of vertically incident 2.73 MeV$^3$ H in$^6$ LiF

that α particles can travel in SiC. The existence of a P+ dead layer causes a decrease in energy deposition in the effective volume when the incidence angle is bigger, leading to a reduction in pulse height. In Fig. 5.59b, a distinct phenomena is observed. Initially, the pulse height marginally drops as the incidence angle increases. It then remains about constant between 40° and 66°, after which it rapidly falls. This could be attributed to the narrow margin between the thickness of the sensitive region at the present bias and the distance traveled by a 2.73 MeV $^3$H particle in SiC. Based on the calculation, the majority of the energy from a charged particle will be concentrated at the end of its range. Consequently, when the incidence angle is extremely tiny, the region with lower electric field strength will generate the highest number of electron–hole pairs. However, despite the fact that a significant angle of incidence leads to less efficient energy deposition, it also generates a greater number of electron–hole pairs in the area with a strong electric field, so producing the outcome seen in Fig. 5.59b.

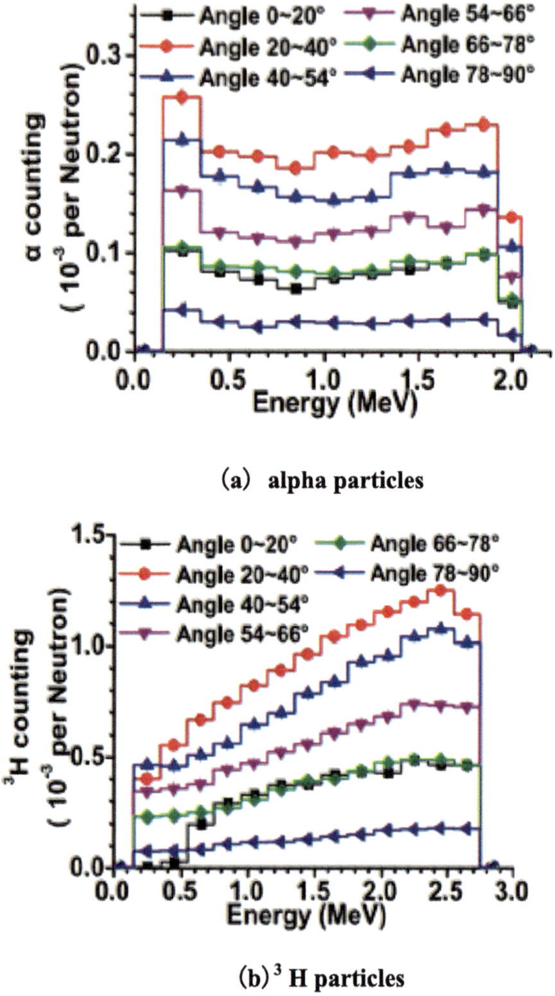


**Fig. 5.57** Charged particle energy spectrum

Applying a reverse bias of 400 V (equivalent to a sensitive region thickness of 48 μm), the pulse heights generated by 2.73 MeV $^{3}$H particles at various angles of incidence will follow the same pattern as seen in Fig. 5.59a.

After calculating the transient current response, the output charge (which is proportional to the height of the output voltage pulse under ideal conditions) is obtained by integrating the current times the time:

$$V \propto Q = \int I \mathrm{d}t \tag{5.33}$$

The integration time was adjusted to 1 μs in order to reduce the transient currents to a level similar to the leakage current, which is less than 1pA. The outcomes

**Fig. 5.58** Characteristics of 4H–SiC PiN Diode for TCAD Simulation

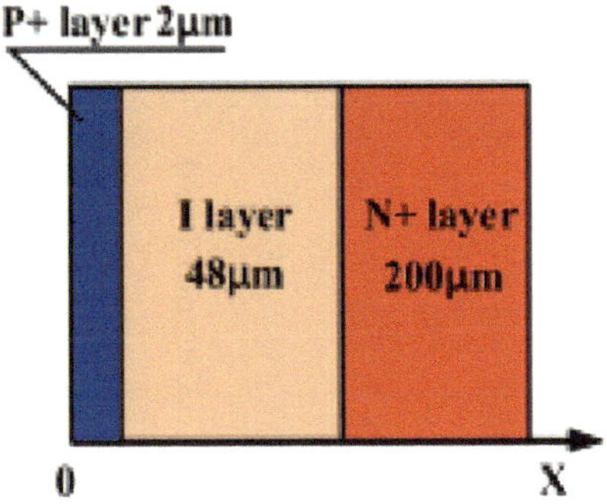


**(a)** Schematic **diagram of SiC PiN diode for pulse amplitude spectrum calculation**

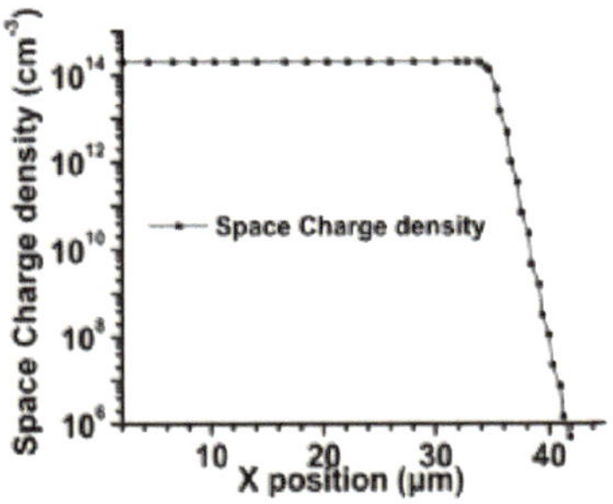


**(b)Charge density in the i-layer space along the x-axis direction**

obtained at a reverse bias voltage of 200 V are presented in Fig. 5.60. The relative amplitudes of the output pulses were directly compared to the integrating charge, without employing the circuit simulation software mentioned in Sect. 5.1.3.

3. Pulse amplitude spectrum calculation

To obtain the pulse amplitude spectrum of the device seen in Fig. 5.58a, one can combine Fig. 5.57 with Fig. 5.60 using the calculation method outlined in Sect. 5.1.3. Previously, the accuracy of the simulation was confirmed by comparing it to experimental test results obtained from the device depicted in Fig. 5.54. The aforementioned simulation procedure was replicated for the device depicted in Fig. 5.54 (with a bias of 0 V). Additionally, the FLUKA software was employed to compute the angular-energy distribution of the secondary particles, as illustrated in Fig. 5.61 (following the interpolation process). Afterwards, the aforementioned calculation procedure is repeated. This time, the focus is on the high channel counts, which are less influenced by noise. The channel address where the second-highest eigenpeak is found is used as the reference point for comparison. The simulation results are then adjusted on the $x$-axis to match the same channel address as the experimental results for the

**Fig. 5.59** Output pulse generated by a single charged particle at different incidence angles

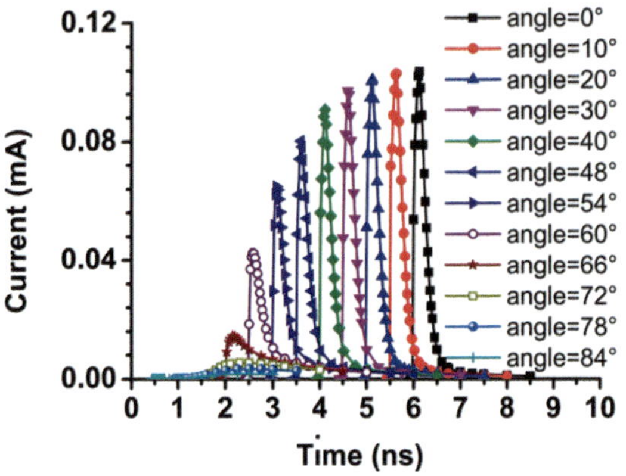


**(a) 2.05 MeV alpha particles**

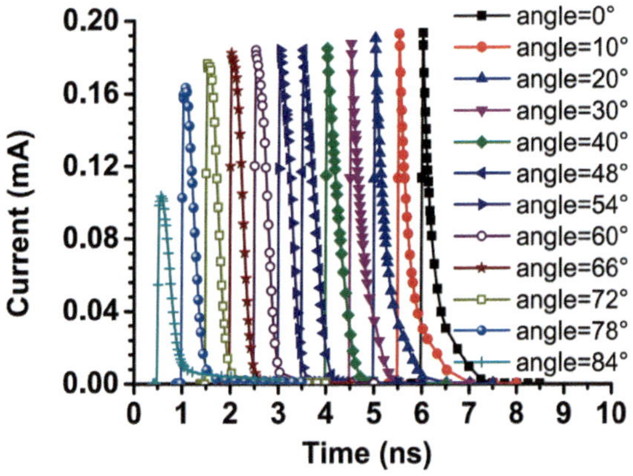


**(b) 2.73 MeV$^3$ H particles**

second-highest peaks. The main peaks' counts are taken as the benchmark for normalization. The outcomes are presented in Fig. 5.62. The simulation shows a slightly greater number of channels with the main peak compared to the experimental results. Additionally, the simulation yields slightly higher relative counts for the secondary peaks compared to the experimental values. It is worth noting that the experimental secondary peaks appear smoother, while the simulation results exhibit a small peak between the main peak and the secondary peaks. The disparity in the heights of the secondary peaks observed in the simulation and experiment can likely be attributed to noise. Additionally, the presence of a small peak may be a result of calculation errors in the interpolation method, particularly at the boundary of the region where the probability density of the distribution of the $^3$H particles is zero. It is worth noting that the simulation does not have a set threshold, causing the counts of the simulation

**Fig. 5.60** Emission angle versus output charge for different particle energies (integration time of 1 μs)

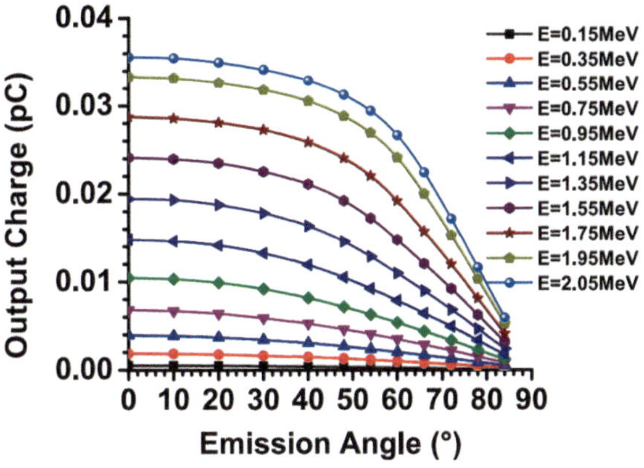


**(a) 2.05 MeV alpha particles**

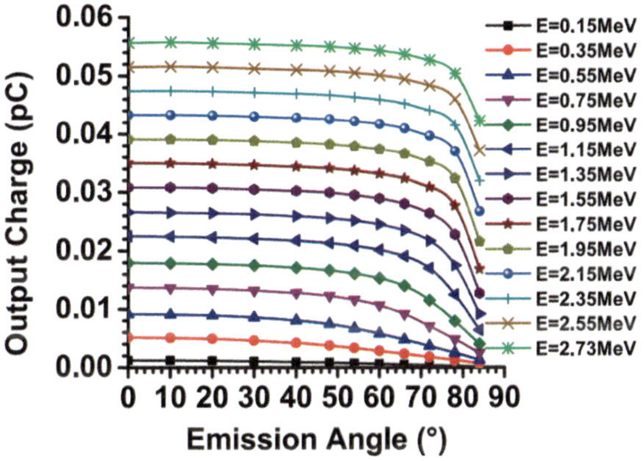


**(b) 2.73 MeV$^3$ H particles**

to rapidly increase as the number of passes decreases below 100. Apart from these differences, the other characteristics remain largely consistent.

The pulse amplitude spectrum of the structure depicted in Fig. 5.58a is computed following validation by simulation and experimental comparison. The energy spectrum for detecting thermal neutrons with an energy of 0.026 eV is exhibited at various reverse bias voltages (0, 15, 45, 200, 400 V) in Fig. 5.63a. The related energy precipitation spectrum is displayed in Fig. 5.63b. At both a reverse bias of 200 and 400 V, the thickness of the sensitive zone exceeds the distance that a 2.73 MeV 3H particle can travel in SiC. The energy spectrum for the 200 V reverse bias scenario is displayed in Fig. 5.63c, along with its α and $^3$H energy spectra. The values of the α and 3H energy spectra are obtained by adding up their respective energy spectra.

The energy spectrum exhibits a comparable form to that of a detector possessing a sufficiently extensive sensitive volume at a reverse bias voltage of no less than

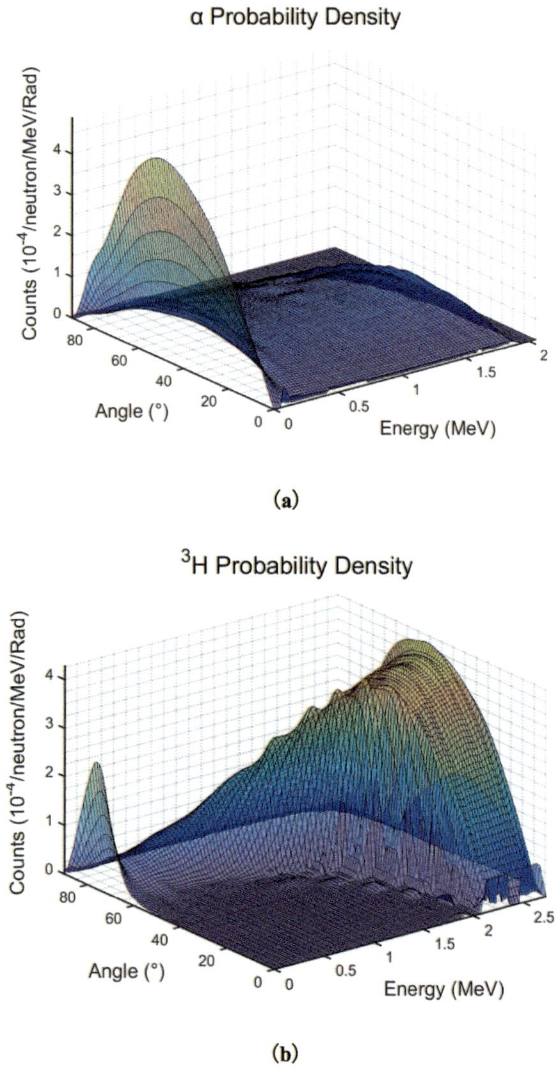


**Fig. 5.61** Angular-energy probability density distribution of the secondary particles corresponding to the 9 μm$^6$ LiF layer. **a** α-particle probability density; **b** $^3$H-particle probability density

200 V. This indicates the potential for using conventional simulation tools to calculate the energy spectrum or analyze its shape. The simulations accurately depict the disparity between the energy deposition spectrum and the actual pulse amplitude spectrum. They also forecast the alterations in the energy spectrum resulting from modifications in detector structure and operating conditions. This information is valuable for guiding detector design.

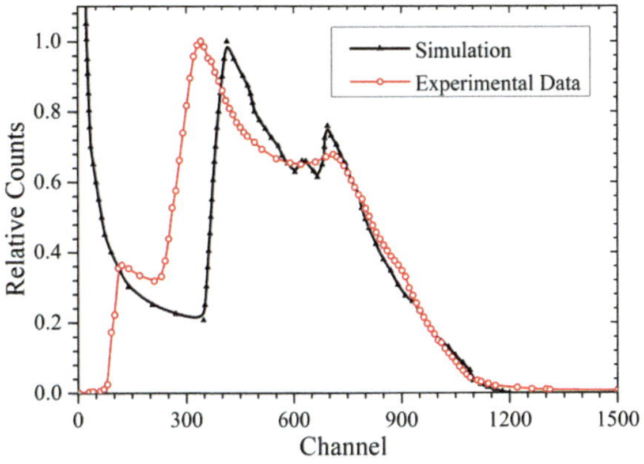


**Fig. 5.62** Comparison of the pulse amplitude spectrum between the simulation results and the experimental results reported in Ref. [2]

### 5.4.2 *14.1 MeV D-T Neutron Detection Performance Study*

Accurate detection and continuous monitoring of D-T neutrons play a crucial role in the power monitoring of upcoming ITER nuclear reactors. The measurement of D-T neutrons provides insight into the rate at which tritium particles are spreading in the reactor, hence facilitating comprehension of the reactor's functioning [40]. The detection of fast neutrons using SiC nuclear radiation detectors primarily depends on the charged particles generated by nuclear reactions between fast neutrons and C and Si atoms. While it is true that both elastic and inelastic scattering of neutrons can result in the creation of charged particles like Si+ and C+, the energies of these charged particles are not specific. Additionally, their energy deposition spectra show a wide range of counts, which makes it challenging to determine the energies of the incoming neutrons using them. Similarly, the scenario where a neutron interacts with a C atom resulting in the production of three α-particles involves a wide range of energy spectra. Therefore, to comprehend the energy of the incoming neutron and analyze its energy spectra, it is customary to identify two sets of nuclear reactions: $^{28}Si(n, p)^{25}$ Mg and $^{12}C(n, \alpha)^{9}Be$. The products resulting from the individual nuclear reactions in these two series remain consistent. However, the products of each reaction exist in distinct excited states and possess varying amounts of energy. Simultaneously, the total energy of the reaction products for each nuclear reaction is determined solely by the energy of the incoming neutron. Therefore, in theory, these nuclear reactions should appear in the pulse amplitude spectrum as a sequence of distinct peaks [7].

1. D-T neutron detection experiment

The detector utilized in the experiment was fabricated by Yuming Zhang's research team at Xidian University, and its configuration is illustrated in Fig. 5.35. The experiment utilized a neutron source that was modeled as a point source emitting neutrons in all directions. The detector was positioned at a distance of approximately 80 cm

**Fig. 5.63** Spectrum of 4H–SiC PiN thermal neutron detector obtained by calculation

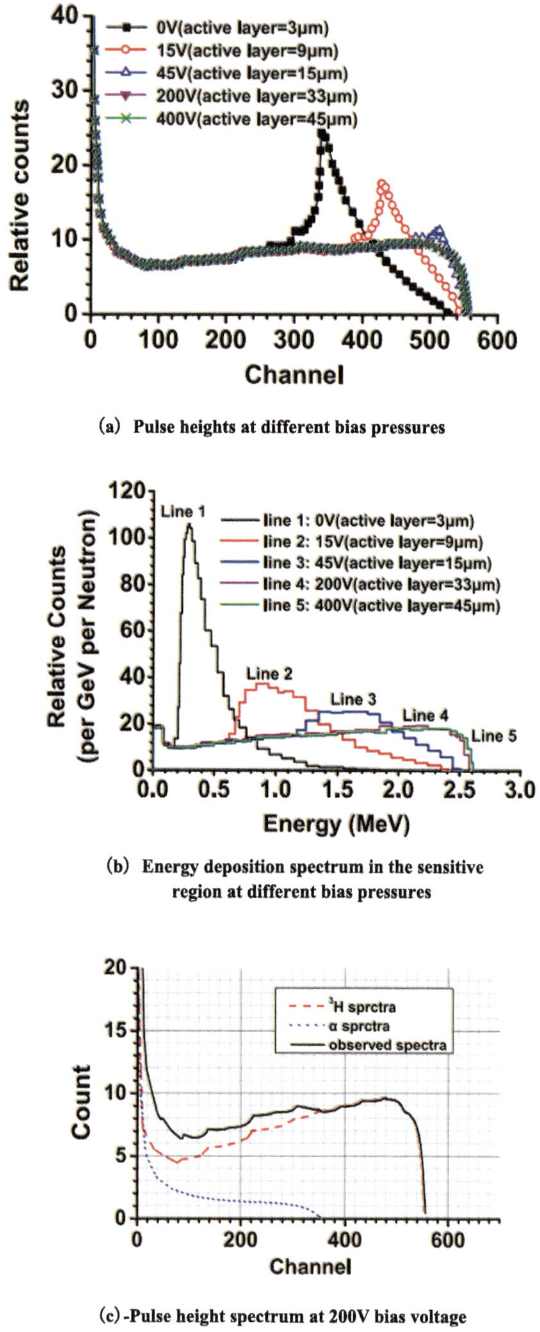

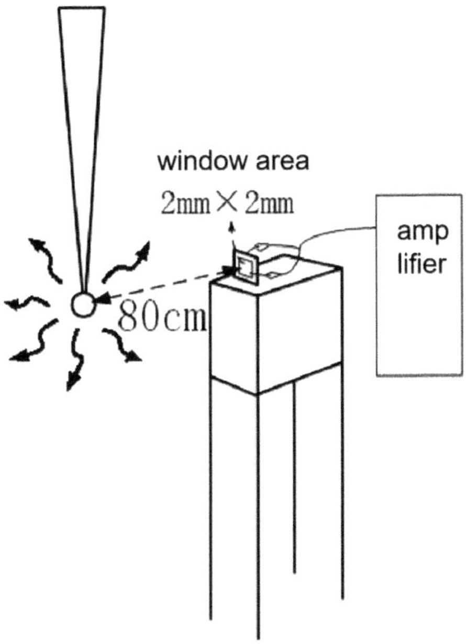


**Fig. 5.64** Schematic diagram of D-T neutron irradiation experiment

from the source. The neutron flux, or the rate at which neutrons passed through a given area, was estimated to be around $3 \times 10^{10}$/s. Figure 5.64 displays the experimental setup. The device F in Table 5.3 was chosen with a reverse bias of 600 V, resulting in a depletion zone width of approximately 45 μm. During testing, the gadget was enclosed within a compact shielding box, which was then covered with a copper wire mesh.

Figure 5.65 gives the measured energy spectra of device F at different measurement durations, including:

(1) Ambient noise background of 222 s before irradiation;
(2) Cumulative 414s/932s/1479s/1981s/2503s after the start of irradiation;
(3) 427s after the end of irradiation.

There were a total of seven spectral lines above. Initially, the detector produces signals that correspond to X-γ rays emitted by different metal supports in the laboratory. These supports are activated by neutron irradiation. The number of channels where the background signals occur before the experiment is much greater than the number of channels where the background signals occur after the experiment. In the figure, the cut-off channel number for the background signals before the experiment is higher. Additionally, the counting of signals stops at a certain time before the neutron beam current is released by the gas pedal. The initial determination is that the electrical noise crosstalk is caused by the preheating that occurs before to operating the gas pedal. Furthermore, the spectral lines acquired subsequent to the initiation of neutron irradiation, when adjusted for the duration of data collection, exhibit significant overlap, as depicted in Fig. 5.65b. This suggests that both the device and the

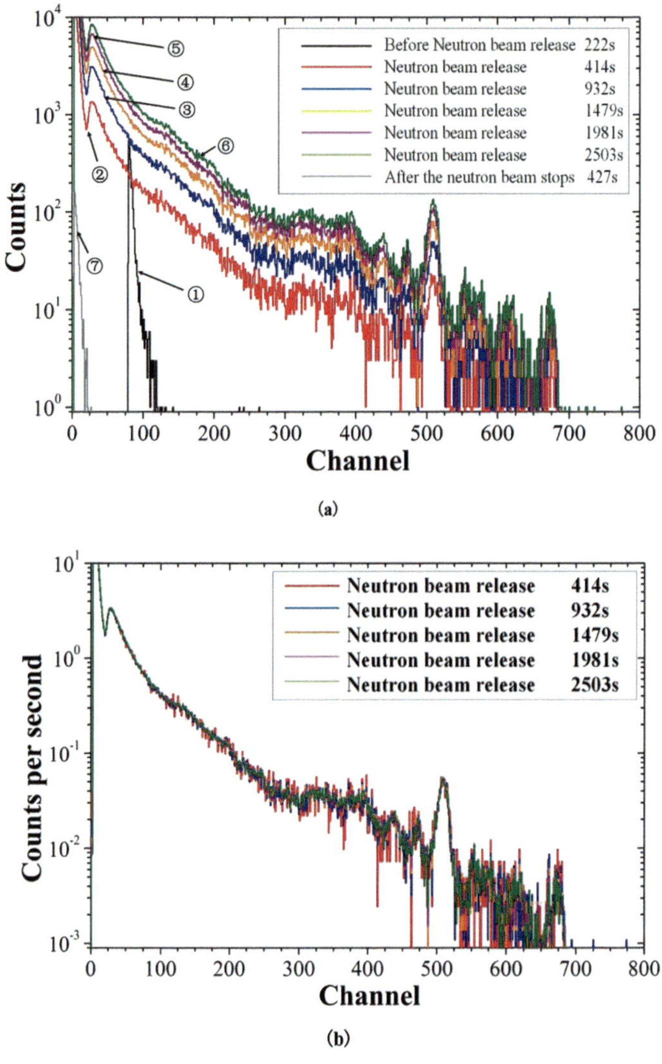


**Fig. 5.65** Neutron irradiation energy spectra obtained from different time duration statistics. **a** Detector measured energy spectrum as a function of time; **b** detector measured energy spectrum normalized by time

gas pedal are functioning in a rather steady condition, devoid of substantial variations. Furthermore, the device's susceptibility to damage is primarily attributed to the relatively low neutron influx it receives, approximately $4.244 \times 10^7$. This value remains much below the threshold of neutron irradiation necessary to cause harm to the device, typically measured at $10^{14}/cm^2$.

2. Analysis of D-T neutron detection results

The experimental findings are contrasted with the outcomes of overseas investigations conducted by Ruddy et al. [41, 42] and Issa et al. [40]. Prior to comparing the data from the literature with the current measurements, it is important to acknowledge that the article by Ruddy exhibits inconsistencies in the relative heights of certain distinct peaks found at the end of the energy spectrum. This inconsistency is observed even for the same device structure. For instance, Fig. 5.66b, c present different experimental results obtained by F. Ruddy et al. using a device structure similar to the one depicted in Fig. 5.66a.

However, when comparing the pulse amplitude spectra presented by Ruddy et al. and the energy deposition spectra derived from the energy obtained from Geant4 as reported by Issa et al., as depicted in Fig. 5.67a, b, along with the corresponding experimental results, it can be contended that the energy spectrum displayed in Fig. 5.66c is more closely aligned with the ideal scenario. Therefore, subsequent comparisons will be conducted using this particular spectrum.

Given that incomplete carrier collection can cause high-channel-address counts to appear as low-channel-address counts, and that noise does not typically cause low-channel-address counts to become high-channel-address counts, it is reasonable to assume that the characteristic peaks of the $^{28}$Si(n, α0)$^{25}$ Mg reaction near the 670 channel are "clean". Thus, the counts of the characteristic peaks of the $^{28}$Si(n, α0)$^{25}$ Mg reaction were used as the counts normalization criterion (denoted as α0-Gaussian fit), while the channel sites where the characteristic peaks of the $^{12}$C(n, α)$^{9}$Be reaction were located were used as the channel site normalization criterion (denoted as α5-Gaussian fit), and the result is shown in Fig. 5.68a. At this point, the counts of several major characteristic peaks are all about twice the reported values. If the $^{12}$C(n, α)$^{9}$Be reaction feature peak is used exclusively as the normalization criterion, the results are shown in Fig. 5.68b, and except for the counts of the $^{28}$Si(n, α0)$^{25}$ Mg reaction feature peaks, the channel sites where the other feature peaks are located and the corresponding number of counts are in almost perfect agreement with the results reported by Ruddy et al. The green line in the figure shows the pulse amplitude spectrum obtained from real measurements, the red line the Gaussian fit of the characteristic peak of the $^{12}$C(n, α)$^{9}$Be reaction, the orange line the Gaussian fit of the characteristic peak of the $^{28}$Si(n, α0)$^{25}$ Mg reaction, and the black line the results reported in the literature. It should be noted here that a Gaussian fit to the above characteristic peaks needs to be performed first before the channel sites and corresponding counts given by the fit results can be used as a normalization criterion.

The correlation between the outcomes of this experiment and the documented values is significantly strong. The measured peak for the $^{12}$C(n, α) $^{9}$Be reaction has a FWHM of approximately 2.4%, which is slightly larger than the reported value of 2.2% [41]. All other distinct peaks can be identified, but due to the low number of counts in the detector for that specific channel, the statistical properties are not sufficiently clear for accurate analysis. The first factor to be considered is the significant distance between the device and the radioactive source, which leads to a minimal count of neutrons received in total. Given that the highest number of channels in the background energy spectrum did not surpass 20 by the conclusion of the experiment, it is widely accepted that the dip seen corresponds precisely

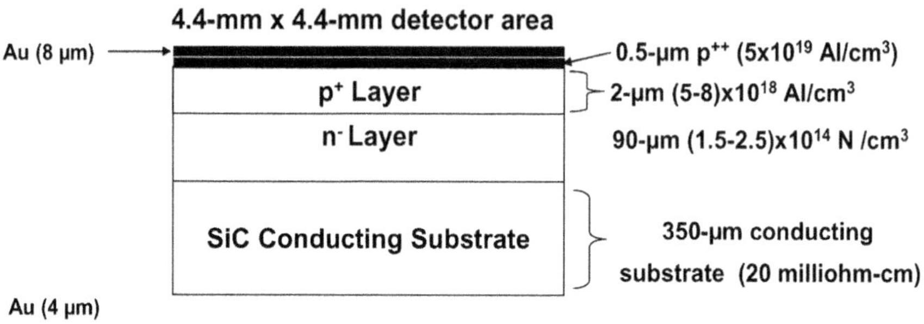


(a) Schematic diagram of SiC PiN diode structure[43]

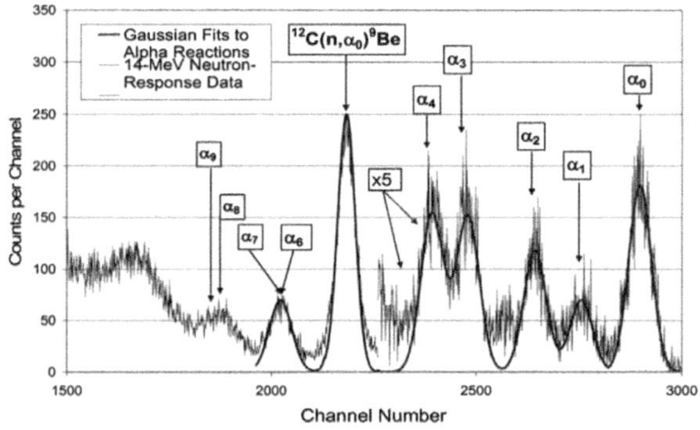


(b) Neutron energy spectrum reported in the literature [42]

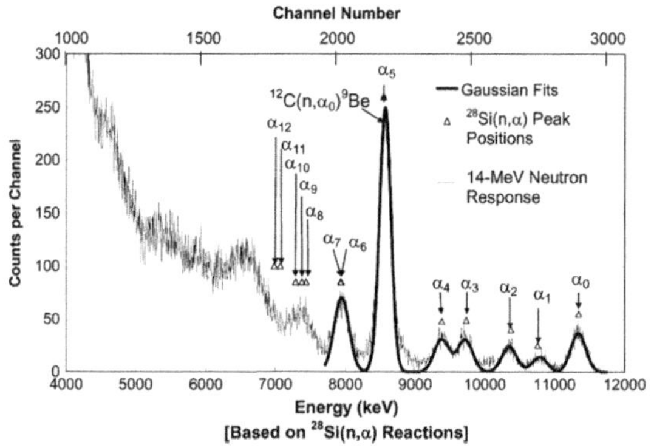


(c) Neutron energy spectrum reported in the literature [43]

**Fig. 5.66** Experimental results given by Ruddy et al. with different relative heights of the characteristic peaks, despite the similar device and experimental conditions

**Fig. 5.67** Energy spectrum obtained by Issa et al. [40]

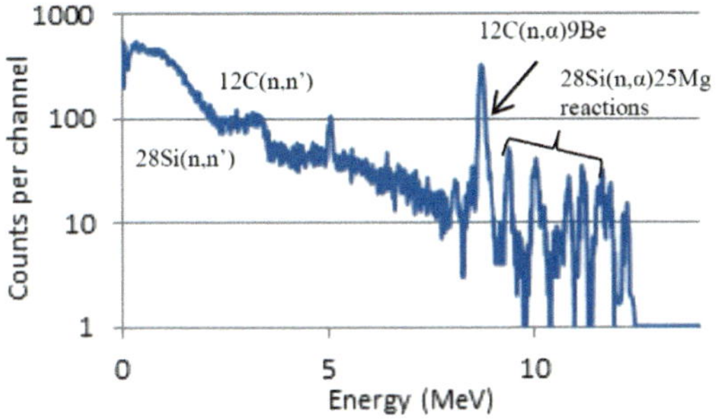


**(a) Geant4 simulated energy spectrum**

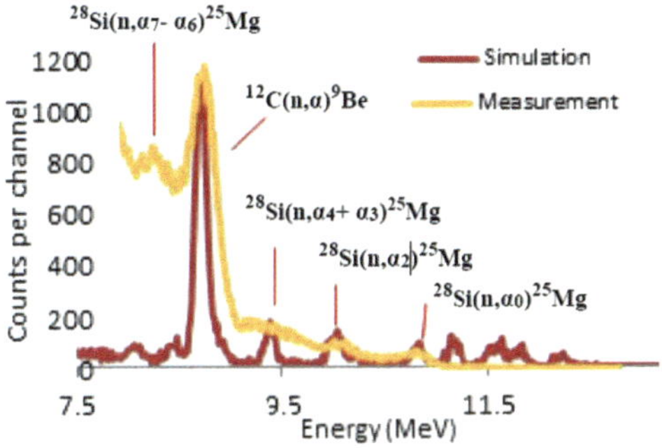


**(b) Comparison of simulated energy spectrum with experiment**

to the energy threshold for $^{12}$C(n, n')$^{12}$C production. By considering the undercove around 20 channels as a benchmark and assuming the accuracy of counts for channels above 20, the detector's efficiency is approximately 0.095%. This figure represents roughly 75% of the theoretical maximum absorption efficiency [43]. Given the bias voltage of the device is currently 600 V, it is feasible for the detector to achieve 100% of the theoretical value. Additionally, by optimizing the structural parameters of the detector, improving the quality of the epitaxial material, and enhancing the encapsulation process, the characteristic peak resolution is expected to be further enhanced. These improvements aim to enable the device to operate at a higher voltage.

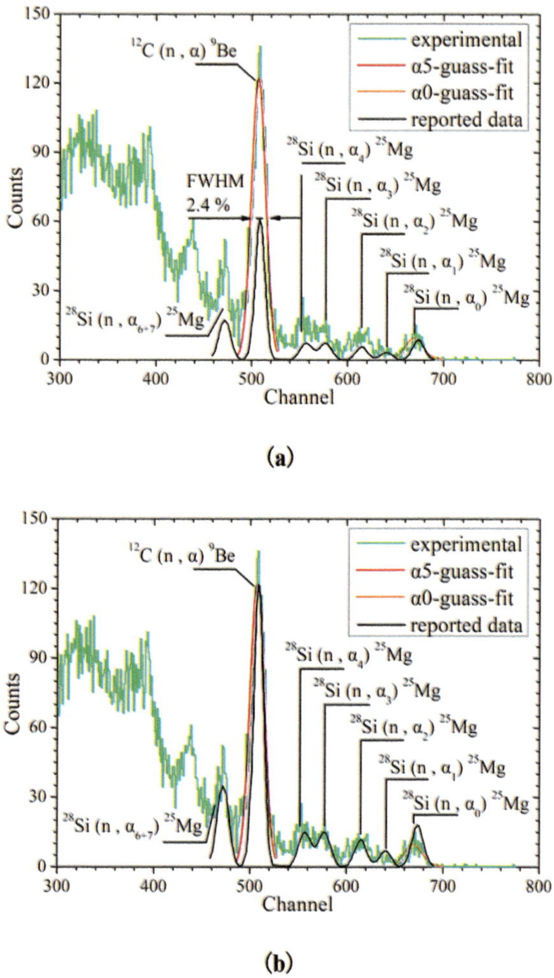


**Fig. 5.68** Comparison of the experimentally obtained D-T neutron energy spectrum with the results reported in the literature

# References


1. Huang H (2019) Research on the performance of SiC particle irradiation detectors and their performance degradation. Xidian University
2. Kim HS, Ha JH, Park SH et al (2011) Characteristics of fabricated neutron detectors based on a SiC semiconductor. J Nucl Sci Technol 48(10):1343–1347
3. Giudice AL, Fasolo F, Durisi E et al (2007) Performances of 4H-SiC Schottky diodes as neutron detectors. Nucl Instrum Methods Phys Res 583(1):177–180
4. X-5 Monte Carlo Team (2000) MCNP-A general Monte Carlo N-particle transport code, Version 5. Los Alamos National Laboratory
5. Yang S (2014) Research on 4H-SiC PiN diode alpha detectors. Xidian University
6. TCAD Sentaurus (2013) Version H-2013.03. Synopsys Inc
7. Franceschini F, Ruddy FH, Petrović B (2008) Simulation of the response of silicon carbide fast neutron detectors [C]// Reactor dosimetry state of the art. World Scientific 2009:128–135

8. Fageeha O, Howard J, Block RC (1994) Distribution of radial energy deposition around the track of energetic charged particles in silicon. J Appl Phys 75(5):2317
9. Quinn T, Bates R, Bruzzi M et al (2004) Comparison of bulk and epitaxial 4H-SiC detectors for radiation hard particle tracking. Nuclear science symposium conference record, vol 2. IEEE, pp 1028–1033
10. A275 Pulse Amplifier [EB/OL]. [2021-4-6]. http://amptek.com/products/a275-pulse-amplifier
11. A250 Charge Sensitive Preamplifier [EB/OL]. [2021-4-6]. http://amptek.com/products/a250-charge-sensitive-preamplifier
12. Wang S (2018) Research on 4H-SiC PiN diode neutron detectors. Xidian University
13. Wang W (2017) Design and simulation of wide bandgap semiconductor 4H-SiC nuclear radiation detectors. Dalian University of Technology
14. Ye X (2018) Fabrication and performance study of 4H-SiC Schottky barrier diode (SBD) alpha particle detectors. Dalian University of Technology
15. Wang W, Xia X, Liang H et al (2017) Simulation and optimization of electrical performance of 4H-SiC SBD devices for radiation detection applications
16. Ding H (2010) Nuclear radiation detectors. Harbin Engineering University Press
17. Han C (2016) Development of 4H-SiC PiN power diodes and research on key technologies. Xidian University
18. Grove AS, Leistiko O, Hooper WW (1967) Effect of surface fields on the breakdown voltage of planar silicon p-n junctions. IEEE Trans Electron Devices 14(3):157–162
19. Kao YC, Wolley ED (1967) High-voltage planar p-n junctions. Proc IEEE 55(8):1409–1414
20. Temple VAK (1977) Junction termination extension (JTE), A new technique for increasing avalanche breakdown voltage and controlling surface electric fields in P-N junctions. International Electron Devices Meeting. IEEE, pp 423–426
21. Temple VAK (1983) Increased avalanche breakdown voltage and controlled surface electric fields using a junction termination extension (JTE) technique. IEEE Trans Electron Dev 30(8):954–957
22. Jiang Q (2019) Research on the impact of Z_(1/2) defects on the performance of SiC nuclear radiation detectors. Xidian University
23. Chen Y, Jiang Y, Wu J et al (2013) Response of SiC-based neutron detectors to thermal neutrons. High Power Laser and Particle Beams 2013(10): 2711–2716
24. 142A/B/C Preamplifiers [EB/OL]. [2021-4-6]. https://www.ortec-online.com/products/electronics/preamplifiers/142a-b-c
25. DualAmplifier [EB/OL]. [2021-4-6]. https://www.ortec-online.com/products/electronics/amplifiers/855
26. Chen Y, Xiaoqiang F, Yong J et al (2013) Research on 4H-SiC Schottky diode alpha detectors. Nucl Electron Detect Technol 01:57–61
27. Jian W, Yong J, Lei G et al (2015) High energy resolution alpha particle detector based on 4H-SiC. High Power Laser Particle Beams 1:151–154
28. Ruddy FH et al (2006) High-resolution alpha-particle spectrometry using 4H silicon carbide semiconductor detectors. IEEE Trans Nuclear Sci 53(3):1713–1718
29. Knoll GF (2010) Radiation detection and measurement. Wiley
30. Chaudhuri SK, Zavalla KJ, Mandal KC (2013) Experimental determination of electron-hole pair creation energy in 4H-SiC epitaxial layer: an absolute calibration approach. Appl Phys Lett 102(3):1713
31. Zat'ko B, Dubecký F, Šagátová A et al (2015) High resolution alpha particle detectors based on 4H-SiC epitaxial layer. J Instrum 10(04):C04009–C04009
32. Xiaoping O (2008) Pulsed radiation detection technology. Chin J Eng Sci 10(4):44–55
33. Wang L (2008) Development of current-type CVD diamond detectors. Tsinghua University
34. Seshadri S, Dulloo AR (1999) Demonstration of an SiC neutron detector for high-radiation environments. IEEE Trans Electron Dev Ed 46(3):567–571
35. Das A, Duttagupta SP (2015) TCAD simulation for alpha-particle spectroscopy using SiC Schottky diode. Radiat Prot Dosimetry 167(4):443

36. Dulloo A, Ruddy F, Seidel J et al (1999) The neutron response of miniature silicon carbide semiconductor detectors—ScienceDirect. Nucl Instrum Methods Phys Res Sect A 422(1–3):47–48
37. Dulloo AR, Ruddy FH, Seidel JG et al (2003) The thermal neutron response of miniature silicon carbide semiconductor detectors. Nucl Instrum Methods Phys Res Sect A 498(1–3):415–423
38. Ha JH, Kang SM, Park SH et al (2009) A self-biased neutron detector based on an SiC semiconductor for a harsh environment. Appl Radiat Isot 67(7–8):1204–1207
39. Manfredotti C, Giudice AL, Fasolo F et al (2005) SiC detectors for neutron monitoring. Nucl Instrum Methods Phys Res Sect A 552(1–2):131–137
40. Szalkai D, Issa F, Klix A et al (2013) First tests of Silicon-Carbide semiconductors as candidate neutron detector for the ITER Test Blanket Modules. In: International conference on advancements in nuclear instrumentation measurement methods and their applications. IEEE, pp 1–4
41. Ruddy FH, Seidel JG, Dulloo AR (2006) Fast neutron dosimetry and spectrometry using silicon carbide semiconductor detectors. J ASTM Int 3(3):8
42. Ruddy FH, Dulloo AR, Seidel JG et al (2006) The fast neutron response of 4H silicon carbide semiconductor radiation detectors. IEEE Trans Nucl Sci 53(3):1666–1670
43. Seidel K, Angelone M, Batistoni P et al (2003) Measurement and analysis of neutron and gamma-ray flux spectra in SiC. Fusion Eng Des 69(1–4):379–383

GPSR Compliance

*The European Union's (EU) General Product Safety Regulation (GPSR) is a set of rules that requires consumer products to be safe and our obligations to ensure this.*

*If you have any concerns about our products, you can contact us on ProductSafety@springernature.com*

In case Publisher is established outside the EU, the EU authorized representative is:

Springer Nature Customer Service Center GmbH
Europaplatz 3
69115 Heidelberg, Germany

**Batch number: 10704307**

Printed by Printforce, the Netherlands